Advanced Studies in Theoretical and Applied Econometrics

Volume 53

This book series aims at addressing the most important and relevant current issues in theoretical and applied econometrics. It focuses on how the current data revolution has affected econometric modeling, analysis and forecasting, and how applied work has benefitted from this newly emerging data-rich environment. The series deals with all aspects of macro-, micro-, financial-, and econometric methods and related disciplines, like, for example, program evaluation or spatial analysis.

The volumes in the series are either monographs or edited volumes, mainly targeting researchers, policymakers, and graduate students.

This book series is listed in Scopus.

Felix Chan • László Mátyás
Editors

Econometrics with Machine Learning

Springer

Editors
Felix Chan
School of Accounting, Economics
& Finance
Curtin University
Bentley, Perth, WA, Australia

László Mátyás
Department of Economics
Central European University
Budapest, Hungary and Vienna, Austria

ISSN 1570-5811 ISSN 2214-7977 (electronic)
Advanced Studies in Theoretical and Applied Econometrics
ISBN 978-3-031-15151-4 ISBN 978-3-031-15149-1 (eBook)
https://doi.org/10.1007/978-3-031-15149-1

This Springer imprint is published by the registered company Springer Nature Switzerland AG
The registered company address is: Gewerbestrasse 11, 6330 Cham, Switzerland

Foreword

Felix Chan and László Mátyás have provided great service to the profession by writing and editing this volume of contributions to econometrics and machine learning. This is a remarkably fast moving area of research with long-term consequences for the analysis of high dimensional, 'big data', and its prospects for model and policy evaluation.

The book reflects well the current direction of research focus that is relevant to professionals who are more concerned with 'partial effects' in the statistical analysis of data, so-called 'causal analysis'. Partial effects are of central interest to policy and treatment effect evaluation, as well as optimal decision making in all applied fields, such as market research, evaluation of treatment outcomes in economics and health, finance, in counterfactual analysis, and model building in all areas of science. The holy grail of quantitative economics/econometrics research in the last 100 years has been the identification and development of 'causal' models, with a primary focus on conditional expectation of one or more target variables/outcomes, conditioned on several 'explanatory' variables, 'features' in the ML jargon. This edifice depends crucially on two decisions: correct functional form and a correct set of both target explanatory variables and additional control variables, reflecting various degrees of departure from the gold standard randomized experiment and sampling.

This holy grail has been a troubled path since there is little or no guidance in substantive sciences on functional forms, and certainly little or no indication on sampling and experimental failures, such as selection. Most economists would admit, at least privately, that quantitative models fail to perform successfully in a consistent manner. This failure is especially prominent in out of sample forecasting and rare event prediction, that is, in counterfactual analysis, a central activity in policy evaluation and optimal decision making. The dominant linear, additively separable multiple regression, the workhorse of empirical research for so many decades, has likely created a massive reservoir of misleading 'stylised facts' which are the artefacts of linearity, and its most obvious flaw, constant partial effect (coefficients).

Nonlinear, nonparametric, semiparametric and quantile models and methods have developed at a rapid pace, with related approximate/asymptotic inference theory, to deal with these shortcomings. This development has been facilitated by rapidly expanding computing capacity and speed, and greater availability of rich data samples.

These welcome developments and movements are meant to provide more reliable and 'robust' empirical findings and inferences. For a long time, however, these techniques have been limited to a small number of conditioning variables, or a small set of 'moment conditions', and subject to the curse of dimensionality in nonparametric and other robust methods.

The advent of regularization and penalization methods and algorithms has opened the door to allow for model searches in which an impressive degree of allowance may be made for both possible nonlinearity of functional forms (filters, 'learners'), and potential explanatory/predictive variables, even possibly larger in number than the sample size. Excitement about these new 'machine learning (ML)' methods is understandable with generally impressive prediction performance. Stupendously fast and successful 'predictive text' search is a common point of contact with this experience for the public. Fast and cheap computing is the principal facilitator, some consider it 'the reason' for mass adoption.

It turns out that an exclusive focus on prediction criteria has some deliterious consequences for the identification and estimation of partial effects and model objects that are central to economic analysis, and other substantive areas. Highly related 'causes' and features are quite easily removed in 'sparsity' techniques, such as LASSO, producing 'biased' estimation of partial effects. In my classes, I give an example of a linear model with some exact multicollinear variables, making the identification of some partial effects impossible ('biased'?), without impacting the estimation of the conditional mean, the 'goodness of fit', or the prediction criteria. The conditional mean is an identified/estimable function irrespective of the degree of multicollinearity!

The general answer to this problem has been to, one way or other, withhold a set of target features from 'selection' and possible elimination by LASSO and other naive 'model selection' techniques. Double machine learning (DML), Random Forests, and subset selection and aggregation/averaging are examples of such methods, some being pre-selection approaches and others being various types of model averaging. There are numerous variations to these examples, but I choose the taxonomy of 'model selection' vs 'model averaging' approaches. These may be guided by somewhat traditional econometrics thinking and approaches, or algorithmic, computer science approaches that are less concerned with rigorous statistical 'inference'.

Econometricians are concerned with rigorous inference and rigorous analysis of identification. Given their history of dealing with the examination of poorly collected 'observational' data, on the one hand, and the immediacy of costly failures of models in practical applications and governance, econometricians are well placed to lead in the development of 'big data' techniques that accommodate the dual goals of accurate prediction and identification (unbiased?) of partial effects.

This volume provides a very timely set of 10 chapters that help readers to appreciate the nature of the challenges and promise of big data methods (data science!?), with a timely and welcome emphasis on 'debiased' and robust estimation of partial and treatment effects. The first three chapters are essential reading, helped by a bridge over the ocean of the confusing new naming of old concepts and objects in statistics (see the Appendix on the terminology). Subsequent chapters contribute by delving further into some of the rapidly expanding techniques, and some key application areas, such as the health sciences and treatment effects.

Doubly robust ML methods are introduced in several places in this volume, and will surely be methods of choice for sometime to come. The book lays bare the challenges of causal model identification and analysis which reflect familiar challenges of model uncertainty and sampling variability. It makes clear that larger numbers of variables and moment conditions, as well as greater flexibility in functrional forms, ironically, produce new challenges (e,g., highly dependent features, reporting and summary of partial effects which are no longer artificial constant, increased danger of endogeneity,...). Wise and rigorous model building and expert advice will remain an art, especially as long as we deal with subfield (economics) models which cannot take all other causes into account. Including everything and the kitchen sink turns out to be a hindrance in causal identification and counterfactual analysis, but a boon to black box predictive algorithms. A profit making financial returns algorithm is 'king' until it fails. When it fails, we have tough time finding out why! Corrective action will be by trial and error. That is a hard pill when economic policy decisions take so long to take effect, if any.

A cautionary note, however, is that rigorous inference is also challenging, as in the rather negative, subtle finding of a lack of 'uniformity' in limit results for efficiency bounds. This makes 'generalisability' out of any subsample and sample problematic since the limiting results are pointwise, not uniform.

The authors are to be congratulated for this timely contribution and for the breadth of issues they have covered. Most readers will find this volume to be of great value, especially in teaching, even though they all surely wish for coverage of even more techniques and algorithms.

I enjoyed reading these contributions and have learned a great deal from them. This volume provides invaluable service to the profession and students in econometrics, statistics, computer science, data sciences, and more.

Emory University, Atlanta, USA, June 2022 *Esfandiar Maasoumi*

Preface

In his book *The Invention of Morel*, the famous Argentinian novelist, Adolfo Bioy Casares, creates what we would now call a parallel universe, which can hardly be distinguished from the real one, and in which the main character gets immersed and eventually becomes part of. Econometricians in the era of Big Data feel a bit like Bioy Casares' main character: We have a hard time making up our mind about what is real and imagined, what is a fact or an artefact, what is evidence or just perceived, what is a real signal or just noise, whether our data is or represents the reality we are interested in, or whether they are just some misleading, meaningless numbers. In this book, we aim to provide some assistance to our fellow economists and econometricians in this respect with the help of machine learning. What we hope to add is, as the German poet and scientist, Johann Wolfgang von Goethe said (or not) in his last words: Mehr Licht (More light).

In the above spirit, the volume aims to bridge the gap between econometrics and machine learning and promotes the use of machine learning methods in economic and econometric modelling. Big Data not only provide a plethora of information, but also often the kind of information that is quite different from what traditional statistical and econometric methods are grown to rely upon. Methods able to uncover deep and complex structures in (very) large data sets, let us call them machine learning, are ripe to be incorporated into the econometric toolbox. However, this is not painless as machine learning methods are rooted in a different type of enquiry than econometrics. Unlike econometrics, they are not focused on causality, model specification, or hypothesis testing and the like, but rather on the underlying properties of the data. They often rely on algorithms to build models geared towards prediction. They represent two cultures: one motivated by prediction, the other by explanation. Mutual understanding is not promoted by their use of different terminology. What in econometrics is called sample (or just data) used to estimate the unknown parameters of a model, in machine learning is often referred to as a training sample. The unknown parameters themselves may be known as weights, which are estimated through a learning or training process (the 'machine' or algorithm itself). Machine learning talks about supervised learning where both the covariates (explanatory variables, features, or predictors) and the dependent (outcome) variables are observed and unsupervised learning, where only the covariates are observed. Machine learning's focus on prediction is most often structural, does not involve time, while in econometrics this mainly means forecasting (about some future event).

The purpose of this volume is to show that despite this different perspective, machine learning methods can be quite useful in econometric analyses. We do not claim to be comprehensive by any means. We just indicate the ways this can be done, what we know and what we do not and where research should be focused.

The first three chapters of the volume lay the foundations of common machine learning techniques relevant to econometric analysis. Chapter 1 on Linear Econometrics Models presents the foundation of shrinkage estimators. This includes ridge, Least Absolute Shrinkage and Selection Operator (LASSO) and their variants as well as their applications to linear models, with a special focus on valid statistical inference for model selection and specification. Chapter 2 extends the discussion to nonlinear models and also provides a concise introduction to tree based methods including random forest. Given the importance of policy evaluation in economics, Chapter 3 presents the most recent advances in estimating treatment effects using machine learning. In addition to discussing the different machine learning techniques in estimating average treatment effects, the chapter presents recent advances in identifying the treatment effect heterogeneity via a conditional average treatment effect function.

The next parts extend and apply the foundation laid by the first three chapters to specific problems in applied economics and econometrics. Chapter 4 provides a comprehensive introduction to Artificial Neural Networks and their applications to economic forecasting with a specific focus on rigorous evaluation of forecast performances between different models. Building upon the knowledge presented in Chapter 3, Chapter 5 presents a comprehensive survey of the applications of causal treatment effects estimation in Health Economics. Apart from Health Economics, machine learning also appears in development economics, as discussed in Chapter 9. Here a comprehensive survey of the applications of machine learning techniques in development economics is presented with a special focus on data from Geographical Information Systems (GIS) as well as methods in combining observational and (quasi) experimental data to gain an insight into issues around poverty and inequality.

In the era of Big Data, applied economists and econometricians are exposed to a large number of additional data sources, such as data collected by social media platforms and transaction data captured by financial institutions. Interdependence between individuals reveals insights and behavioural patterns relevant to policy makers. However, such analyses require technology and techniques beyond the traditional econometric and statistical methods. Chapter 6 provides a comprehensive review of this subject. It introduces the foundation of graphical models to capture network behaviours and discusses the most recent procedures to utilise the large volume of data arising from such networks. The aim of such analyses is to reveal the deep patterns embedded in the network. The discussion of graphical models and their applications continues in Chapter 8, which discusses how shrinkage estimators presented in Chapters 1 and 2 can be applied to graphical models and presents its applications to portfolio selection problems via a state-space framework. Financial applications of machine learning are not limited to portfolio selection, as shown in Chapter 10, which provides a comprehensive survey of the contribution of machine learning techniques in identifying the relevant factors that drive empirical asset pricing.

Since data only capture historical information, any bias or prejudice induced by humans in their decision making is also embedded into the data. Predictions from these data therefore continue with such bias and prejudice. This issue is becoming increasingly important and Chapter 7 provides a comprehensive survey of recent techniques to enforce fairness in data-driven decision making through Structural Econometric Models.

Perth, June 2022 — *Felix Chan*

Budapest and Vienna, June 2022 — *László Mátyás*

Acknowledgements

We address our thanks to all those who have facilitated the birth of this book: the contributors who produced quality work, despite onerous requests and tight deadlines; Esfandiar Maasoumi, who supported this endeavour and encouraged the editors from the very early planning stages; and last but not least, the Central European University and Curtin University, who financially supported this project. Some chapters have been polished with the help of Eszter Timár. Her English language editing made them easier and more enjoyable to read.

The final camera–ready copy of the volume has been prepared with LaTeX and Overleaf by the authors, the editors and some help from Sylvia Soltyk and the LaTeX wizard Oliver Kiss.

Contents

List of Contributors

Maria Victoria Anauati
Universidad de San Andres and CEDH-UdeSA, Buenos Aires, Argentina, e-mail: vanauati@udesa.edu.ar

Wendy Brau
Universidad de San Andres and CEDH-UdeSA, Buenos Aires, Argentina, e-mail: wbrau@udesa.edu.ar

Samuele Centorrino
Stony Brook University, Stony Brook, New York, USA, e-mail: samuele.centorrino@stonybrook.edu

Felix Chan
Curtin University, Perth, Australia, e-mail: felix.chan@cbs.curtin.edu.au

William Crown
Brandeis University, Waltham, Massachusetts, USA, e-mail: wcrown@brandeis.edu

Marcelo Cunha Medeiros
Pontifical Catholic University of Rio de Janeiro, Rio de Janeiro, Brasil, e-mail: mcm@econ.puc-rio.br

Jean-Pierre Florens
Toulouse School of Economics, Toulouse, France, e-mail: jean-pierre.florens@tse-fr.eu

Mark N, Harris
Curtin University, Perth, Australia, e-mail: mark.harris@curtin.edu.au

Yu-Chin Hsu
Academia Sinica, National Central University and National Chengchi University, Taiwan, e-mail: ychsu@econ.sinica.edu.tw

Oliver Kiss
Central European University, Budapest, Hungary and Vienna, Austria, e-mail:

Kiss_Oliver@phd.ceu.edu

Robert Lieli
Central European University, Budapest, Hungary and Vienna, Austria, e-mail: LieliR@ceu.edu

Jean-Michel Loubes
Institut de Mathematiques de Toulouse, Toulouse, France, e-mail: loubes@math.univ-toulouse.fr

László Mátyás
Central European University, Budapest, Hungary and Vienna, Austria, e-mail: matyas@ceu.edu

Ágoston Reguly
Central European University, Budapest, Hungary and Vienna, Austria, e-mail: regulyagoston@gmail.com

Gyorgy Ruzicska
Central European University, Budapest, Hungary and Vienna, Austria e-mail: Ruzicska_Gyorgy@phd.ceu.edu

Ekaterina Seregina
Colby College, Waterville, ME, USA, e-mail: eseregin@colby.edu

Ranjodh Singh
Curtin University, Perth, Australia, e-mail: Ranjodh.Singh@curtin.edu.au

Walter Sosa-Escudero
Universidad de San Andres, CONICET and Centro de Estudios para el Desarrollo Humano (CEDH-UdeSA), Buenos Aires, Argentina, e-mail: wsosa@udesa.edu.ar

Jantje Sönksen
Eberhard Karls University, Tübingen, Germany, e-mail: jantje.soenksen@uni-tuebingen.de

Ben Weiern Yeo
Curtin University, Perth, Australia, e-mail: weiern.yeo@student.curtin.edu.au

Chapter 1
Linear Econometric Models with Machine Learning

Felix Chan and László Mátyás

Abstract This chapter discusses some of the more popular shrinkage estimators in the machine learning literature with a focus on their potential use in econometric analysis. Specifically, it examines their applicability in the context of linear regression models. The asymptotic properties of these estimators are discussed and the implications on statistical inference are explored. Given the existing knowledge of these estimators, the chapter advocates the use of partially penalized methods for statistical inference. Monte Carlo simulations suggest that these methods perform reasonably well. Extensions of these estimators to a panel data setting are also discussed, especially in relation to fixed effects models.

1.1 Introduction

This chapter has two main objectives. First, it aims to provide an overview of the most popular and frequently used shrinkage estimators in the machine learning literature, including the Least Absolute Shrinkage and Selection Operator (LASSO), Ridge, Elastic Net, Adaptive LASSO and Smoothly Clipped Absolute Deviation (SCAD). The chapter covers their definitions and theoretical properties. Then, the usefulness of these estimators is explored in the context of linear regression models from the perspective of econometric analysis. While some of these shrinkage estimators, such as, the Ridge estimator as proposed by Hoerl and Kennard (1970b), have a long history in Econometrics, the evolution of the shrinkage estimators has become one of the main focuses in the development of machine learning techniques. This is partly due to their excellent results in obtaining superior predictive models when the number of covariates (explanatory variables) is large. They are also particularly

Felix Chan ✉
Curtin University, Perth, Australia, e-mail: felix.chan@cbs.curtin.edu.au

László Mátyás
Central European University, Budapest, Hungary and Vienna, Austria, e-mail: matyas@ceu.edu

F. Chan, L. Mátyás (eds.), *Econometrics with Machine Learning*, Advanced Studies in Theoretical and Applied Econometrics 53, https://doi.org/10.1007/978-3-031-15149-1_1

useful when traditional estimators, such as the Ordinary Least Squares (OLS), are no longer feasible e.g., when the number of covariates is larger than the number of observations. In these cases and, in the absence of any information on the relevance of each covariate, shrinkage estimators provide a feasible approach to potentially identify relevant variables from a large pool of covariates. This feature highlights the fundamental problem in sparse regression, i.e., a linear regression model with a large parameter vector that potentially contains many zeros. The fundamental assumption here is that, while the number of covariates is large, perhaps much larger than the number of observations, the number of associated non-zero coefficients is relatively small. Thus, the fundamental problem is to identify the non-zero coefficients.

While this seems to be an ideal approach to identify economic relations, it is important to bear in the mind that the fundamental focus of shrinkage estimators is to construct the best approximation of the response (dependent) variable. The interpretation and statistical significance of the coefficients do not necessarily play an important role from the perspective of machine learning. In practice, a zero coefficient may manifest itself from two different scenarios in the context of shrinkage estimators: (i) its true value is 0 in the data generating process (DGP) or (ii) the true value of the coefficient is close enough to 0 that shrinkage estimators cannot identify its importance, e.g., because of the noise in the data. The latter is related to the concept of uniform signal strength, which is discussed in Section 1.4. Compared to conventional linear regression analysis in econometrics, zero coefficients are often inferred from statistical inference procedures, such as the F-t or t-tests. An important question is whether shrinkage estimators can provide further information to improve conventional statistical inference typically used in econometric analysis. Putting this into a more general framework, machine learning often views data as 'pure information', while in econometrics the signal to noise ratio plays an important role. When using machine learning techniques in econometrics this 'gap' has to be bridged in some way.

In order to address this issue, this chapter explores the possibility of conducting valid statistical inference for shrinkage estimators. While this question has received increasing attention in recent times, it has not been the focus of the literature. This, again, highlights the fundamental difference between machine learning and econometrics in the context of linear models. Specifically, machine learning tends to focus on producing the best approximation of the response variable, while econometrics often focuses on the interpretations and the statistical significance of the coefficient estimates. This clearly highlights the above gap in making shrinkage estimators useful in econometric analysis. This chapter explores this gap and provides an evaluation of the scenarios in which shrinkage estimators can be useful for econometric analysis. This includes the overview of the most popular shrinkage estimators and their asymptotic behaviour, which is typically analyzed in the form of the so-called Oracle Properties. However, the practical usefulness of these properties seems somewhat limited as argued by Leeb and Pötscher (2005) and Leeb and Pötscher (2008).

Recent studies show that valid inference may still be possible on the subset of the parameter vector that is not part of the shrinkage. This means shrinkage estimators are particular useful in identifying variables that are not relevant from an economics/econometrics interpretation point of view. Let us call them control variables, when the number of such potential variables is large, especially when larger than the number of observations. In this case, one can still conduct valid inference on the variables of interests by applying shrinkage (regularization) on the list of potential control variables only. This chapter justifies this approach by obtaining the asymptotic distribution of this partially 'shrunk' estimator under the Bridge regularizer, which has LASSO and Ridge as special cases. Monte Carlo experiments show that the result may also be true for other shrinkage estimators such as the adaptive LASSO and SCAD.

The chapter also discusses the use of shrinkage estimators in the context of fixed effects panel data models and some recent applications for deriving an 'optimal' set of instruments from a large list of potentially weak instrumental variables (see Belloni, Chen, Chernozhukov & Hansen, 2012). Following similar arguments, this chapter also proposes a novel procedure to test for structural breaks with unknown breakpoints.[1]

The overall assessment is that shrinkage estimators are useful when the number of potential covariates is large and conventional estimators, such as OLS, are not feasible. However, because statistical inference based on shrinkage estimators is a delicate and technically demanding problem that requires careful analysis, users should proceed with caution.

The chapter is organized as follows. Section 1.2 introduces some of the more popular regularizers in the machine learning literature and discusses their properties. Section 1.3 provides a summary of the algorithms used to obtain the shrinkage estimates and the associated asymptotic properties are discussed in Section 1.4. Section 1.5 provides some Monte Carlo simulation results examining the finite sample performance of the partially penalized (regularized) estimators. Section 1.6 discusses three econometric applications using shrinkage estimators including fixed effects estimators with shrinkage and testing for structural breaks with unknown breakpoints. Some concluding remarks are made in Section 1.7.

1.2 Shrinkage Estimators and Regularizers

This section introduces some of the more popular shrinkage estimators in the machine learning literature. Interestingly, some of them have a long history in econometrics. The goal here is to provide a general framework for the analysis of these estimators and to highlight their connections. Note that the discussion focuses solely on linear models which can be written as

$$y_i = \mathbf{x}_i'\boldsymbol{\beta}_0 + u_i, \qquad u_i \sim D(0, \sigma_u^2), \quad i = 1, \ldots, N, \tag{1.1}$$

[1] The theoretical properties and the finite sample performance of the proposed procedure are left for further researchers. The main objective here is to provide an example of other potential useful applications of the shrinkage estimators in econometrics.

where $\mathbf{x}_i = (x_{1i}, \ldots, x_{pi})'$ is a $p \times 1$ vector containing p explanatory variables, which in the machine learning literature, are often referred to as *covariates*, *features* or *predictors* with the parameter vector $\boldsymbol{\beta}_0 = (\beta_{10}, \ldots, \beta_{p0})'$. The response variable is denoted by y_i, which is often called the endogenous, or dependent, variable in econometrics and u_i denotes the random disturbance term with finite variance, i.e., $\sigma_u^2 < \infty$. Equation (1.1) can also be expressed in matrix form

$$y = \mathbf{X}\boldsymbol{\beta}_0 + u, \tag{1.2}$$

where $\mathbf{y} = (y_1, \ldots, y_N)$, $\mathbf{X} = (\mathbf{x}_1, \ldots, \mathbf{x}_p)$, and $\mathbf{u} = (u_1, \ldots, u_N)$.

An important deviation from the typical econometric textbook setting is that some of the elements in $\boldsymbol{\beta}_0$ can be 0 with the possibility that $p \geq N$. Obviously, the familiar Ordinary Least Square (OLS) estimator

$$\hat{\boldsymbol{\beta}}_{OLS} = (\mathbf{X}'\mathbf{X})^{-1}\mathbf{X}'\mathbf{y} \tag{1.3}$$

cannot be computed when $p > N$ since the *Gram Matrix*, $\mathbf{X}'\mathbf{X}$, does not have full rank in this case. Note that in the shrinkage estimator literature, it is often, if not always, assumed that $p_1 << N < p$ or $p_1 << p < N$ where p_1 denotes the number of non-zero coefficients. In other words, it is assumed that the OLS estimator could be computed if the p_1 covariates with non-zero coefficients could be detected a priori.

At this point, it may be useful to define some terminology and notations that would aid subsequent discussions. The size of a parameter vector $\boldsymbol{\beta}$ is the number of elements in the vector and the length of $\boldsymbol{\beta}$ is the length of the vector as measured by an assigned *norm*. While the L_γ norm[2] is the most popular family of norms, there are others in the literature, such as the *maximum norm* and *zero norm*. The L_γ norm of a vector $\boldsymbol{\beta} = (\beta_1, \ldots, \beta_p)'$ denoted by the notation, $||\boldsymbol{\beta}||_\gamma$, is defined as

$$||\boldsymbol{\beta}||_\gamma = \left(\sum_{i=1}^{p} |\beta_i|^\gamma\right)^{1/\gamma} \qquad \gamma > 0, \tag{1.4}$$

where $|\beta|$ denotes the absolute value of β. When $\gamma = 2$, the L_γ norm is known as the *Euclidean Norm*, or *Euclidean Distance*, which is perhaps more familiar to econometricians. As a simple example, let $\boldsymbol{\beta} = (\beta_1, \beta_2)$, then Euclidean Norm, of L_2 norm of $\boldsymbol{\beta}$ is $||\boldsymbol{\beta}||_2 = \sqrt{|\beta_1|^2 + |\beta_2|^2}$.

The idea of a shrinkage estimator is to impose a restriction on the length of the estimated parameter vector $\hat{\boldsymbol{\beta}}$. Since the length of the vector is fixed, if the objective is to construct the best approximation to the response variable, then the coefficients corresponding to the 0 elements in $\boldsymbol{\beta}_0$ should be among the first to reach 0 before the other coefficient estimates, as these variables would not be useful in predicting y_i and setting their coefficients to 0 would allow more 'freedom' for the other non-zero coefficients. In other words, the idea is to 'shrink' the parameter vector in order to identify the 0 elements in $\boldsymbol{\beta}_0$. This can be framed as the following optimization

---

[2] The literature tends to refer to this as the L_p norm, γ is used instead of p in this chapter as p is used to denote the number of covariates, i.e., the size of the vector.

problem

$$\hat{\boldsymbol{\beta}} = \underset{\boldsymbol{\beta}}{\arg\min}\; g\left(\boldsymbol{\beta};\mathbf{y},\mathbf{X}\right) \tag{1.5}$$

$$\text{s.t. } p\left(\boldsymbol{\beta};\boldsymbol{\alpha}\right) \leq c, \tag{1.6}$$

where $g\left(\boldsymbol{\beta};\mathbf{y},\mathbf{X}\right)$ denotes the objective (loss) function, $p\left(\boldsymbol{\beta};\boldsymbol{\alpha}\right)$ denotes a function that can be used to regulate the total length of the estimated parameter vector and is often called the *regularizer* in the machine learning literature. However, it can also be interpreted as the *penalty* function, which may be more familiar to econometricians. As indicated in the constraint, the total length of the estimated parameter vector is bounded by the constant $c > 0$, which has to be determined (or selected) by the researchers *a priori*. Unsurprisingly, this choice is of fundamental importance as it affects the ability of the shrinkage estimator to correctly identify those coefficients whose value is indeed 0. If c is too small, then it is possible that coefficients with a sufficiently small magnitude are incorrectly identified as coefficients with zero value. Such mis-identification is also possible when the associated variables are noisy, such as those suffering from measurement errors. In addition to setting zero values incorrectly, small c may also induce substantial bias to the estimates of non-zero coefficients.

In the case of a least squares type estimator, $g\left(\boldsymbol{\beta};\mathbf{y},\mathbf{X}\right) = \left(\mathbf{y}-\mathbf{X}\boldsymbol{\beta}\right)'\left(\mathbf{y}-\mathbf{X}\boldsymbol{\beta}\right)$ but there can also be other objective functions, such as a log-likelihood function in the case of non-linear models or a quadratic form typically seen in the Generalized Method of Moments estimation. Unless otherwise stated, in this chapter the focus is solely on the least squares loss. Different regularizers, i.e., different definitions of $p(\boldsymbol{\beta};\boldsymbol{\alpha})$, lead to different shrinkage estimators, including the *Least Absolute Shrinkage and Selection Operator* (LASSO) i.e., $p(\boldsymbol{\beta}) = \sum_{j=1}^{p} |\beta_j|$, the *Ridge Estimator* i.e., $p(\boldsymbol{\beta}) = \sum_{j=1}^{p} |\beta_j|^2$, *Elastic Net* i.e., $p(\boldsymbol{\beta}) = \sum_{j=1}^{p} \alpha|\beta_j| + (1-\alpha)|\beta_j|^2$, Smoothly Clipped Absolute Deviation (SCAD), and other regularizers. The theoretical properties of some of these estimators are discussed later in the chapter, including their *Oracle Properties* and the connection of these properties to the more familiar concepts such as consistency and asymptotic distributions.

The optimization as defined in Equations (1.5) – (1.6) can be written in its Lagrangian form as

$$\hat{\boldsymbol{\beta}} = \underset{\boldsymbol{\beta}}{\arg\min}\; g\left(\boldsymbol{\beta};\mathbf{y},\mathbf{X}\right) + \lambda p\left(\boldsymbol{\beta};\boldsymbol{\alpha}\right). \tag{1.7}$$

A somewhat subtle difference between the Lagrangian as defined Equation (1.7) is that the *Lagrange* multiplier, λ, is fixed by the researcher, rather than being a choice variable along with $\boldsymbol{\beta}$. This reflects the fact that c, the length of the parameter vector, is fixed by the researcher a priori before the estimation procedure. It can be shown that there is a one-to-one correspondence between λ and c with λ being a decreasing function of c. This should not be surprising if one interprets λ as the penalty induced by the constraints. In the extreme case that $\lambda \to 0$ as $c \to \infty$, Equation

(1.7) approaches the familiar OLS estimator, under the assumption that $p < N$.[3] Since λ is pre-determined, it is often called the *tuning parameter*. In practice, λ is often selected by *cross validation* which is discussed further in Section 1.3.2.

1.2.1 L_γ norm, Bridge, LASSO and Ridge

A particularly interesting class of regularizers is called the *Bridge* estimator as defined by Frank and Friedman (1993), which proposed the following regularizer in Equation (1.7)

$$p(\boldsymbol{\beta};\gamma) = ||\boldsymbol{\beta}||_\gamma^\gamma = \sum_{j=1}^{p} |\beta_j|^\gamma, \qquad \gamma \in \mathbb{R}^+. \tag{1.8}$$

The Bridge estimator encompasses at least two popular shrinkage estimators as special cases. When $\gamma = 1$, the Bridge estimator becomes the *Least Absolute Shrinkage and Selection Operator* (LASSO) as proposed by Tibshirani (1996) and when $\gamma = 2$, tbe Bridge estimator becomes the *Ridge* estimator as defined by Hoerl and Kennard (1970b, 1970a). Perhaps more importantly, the asymptotic properties of the Bridge estimator were examined by Knight and Fu (2000) and subsequently, the results shed light on the asymptotic properties of both the LASSO and Ridge estimators. This is discussed in more details in Section 1.4.

As indicated in Equation (1.8), the Bridge uses the L_γ norm as the regularizer. This leads to the interpretation that LASSO ($\gamma = 1$) regulates the length of the coefficient vector using the L_1 (absolute) norm while Ridge ($\gamma = 2$) regulates the length of the coefficient vector using the L_2 (Euclidean) norm. An advantage of the L_1 norm i.e., LASSO, is that it can produce estimates with exactly zero values, i.e., elements in $\hat{\boldsymbol{\beta}}$ can be exactly 0, while the L_2 norm, i.e., Ridge, does not usually produce estimates with values that equal exactly 0. Figure 1.1 illustrates the difference between the three regularizers for $p = 2$. Figure 1.1a gives the plot of LASSO when $|\beta_1| + |\beta_2| = 1$ and as indicated in the figure, if one of the coefficients is in fact zero, then it is highly likely that the contour of the least squares will intersect with one of the corners first and thus identifies the appropriate coefficient as 0.

In contrast, the Ridge contour does not have the 'sharp' corner as indicated in Figure 1.1b and hence the likelihood of reaching exactly 0 in one of the coefficients is low even if the true value is 0. However, the Ridge does have a computational advantage over other variations of the Bridge estimator. When $\gamma = 2$, there is a close form solution, namely

$$\hat{\boldsymbol{\beta}}_{Ridge} = (\mathbf{X}'\mathbf{X} + \lambda\mathbf{I})^{-1}\mathbf{X}'\mathbf{y}. \tag{1.9}$$

[3] While the optimization problem for shrinkage estimator approaches the optimization problem for OLS regardless on the relative size between p and N, the solution of the latter does not exist when $p > N$.

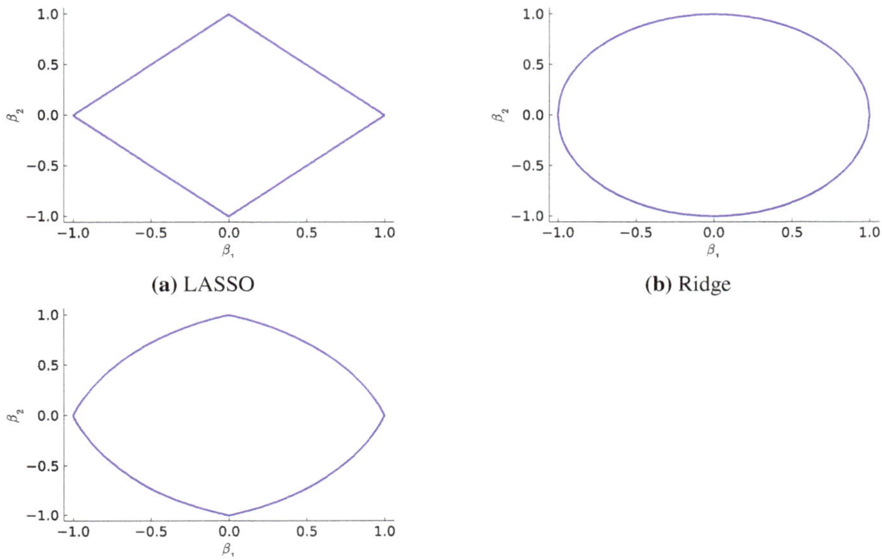

(a) LASSO

(b) Ridge

(c) Elastic Net

Fig. 1.1: Contour plots of LASSO, Ridge and Elastic Net

When $\gamma \neq 2$, there is no closed form solution to the associated constrained optimization problems so it must be solved numerically. The added complexity is that when $\gamma \geq 1$, the regularizer is a convex function, which means a whole suite of algorithms is available for solving the optimization as defined in Equations (1.5) and (1.6), at least in the least squares case. When $\gamma < 1$, the regularizer is no longer convex and algorithms for solving this problem are less straightforward. Interestingly, this also affects the asymptotic properties of the estimators (see Knight & Fu, 2000). Specifically, the asymptotic distributions are different when $\gamma < 1$ and $\gamma \geq 1$. This is discussed briefly in Section 1.4.

1.2.2 Elastic Net and SCAD

The specification of the regularizer can be more general than a norm. One example is *Elastic Net* as proposed by Zou and Hastie (2005), which is a linear combination between L_1 and L_2 norms. Specifically,

$$p(\boldsymbol{\beta};\alpha) = \alpha_1||\boldsymbol{\beta}||_1 + \alpha_2||\boldsymbol{\beta}||_2^2, \qquad \alpha \in [0,1]. \tag{1.10}$$

Clearly, Elastic Net has both LASSO and Ridge as special cases. It reduces to the former when $(\alpha_1, \alpha_2) = (1, 0)$ and the latter when $(\alpha_1, \alpha_2) = (0, 1)$. The exact value of (α_1, α_2) is to be determined by the researchers, along with λ. Thus, Elastic Net requires more than one tuning parameter. While these can be selected via cross validation (see, for example, Zou & Hastie, 2005), a frequent choice is $\alpha_2 = 1 - \alpha_1$ with $\alpha_1 \in [0, 1]$. In this case, the Elastic Net is an *affine combination* of the L_1 and L_2 regularizers which reduces the number of tuning parameters. The motivation of Elastic Net is to overcome certain limitations of the LASSO by striking a balance between LASSO and the Ridge. Figure 1.1c contains the contour of Elastic Net. Note that the contour is generally smooth but there is a distinct corner in each of the four cases when one of the coefficients is 0. As such, Elastic Net also has the ability to identify coefficients with 0 values. However, unlike the LASSO, Elastic Net can select more than one variable from a group of highly correlated covariates.

An alternative to LASSO is *Smoothly Clipped Absolute Deviation* regularizer as proposed by Fan and Li (2001). The main motivation of SCAD is to develop a regularizer that satisfies the following three conditions:

1. Unbiasedness. The resulting estimates should be unbiased, or at the very least, nearly unbiased. This is particularly important when the true unknown parameter is large with a relatively small c.
2. Sparsity. The resulting estimator should be a thresholding rule. That is, it satisfies the role of a selector by setting the coefficient estimates of all 'unnecessary' variables to 0.
3. Continuity. The estimator is continuous in data.

Condition 1 is to address a well-known property of LASSO, namely that it often produces biased estimates. Under the assumption that $\mathbf{X}'\mathbf{X} = \mathbf{I}$, Tibshirani (1996) showed the following relation between LASSO and OLS

$$\hat{\beta}_{LASSO,i} = \mathrm{sgn}\,\hat{\beta}_{OLS,i}\left(|\hat{\beta}_{OLS,i}| - \lambda\right), \tag{1.11}$$

where $\mathrm{sgn}\,x$ denotes the sign of x. The equation above suggests that the greater λ is (or the smaller c is) the larger is the bias in LASSO, under the assumption that the OLS is unbiased or consistent. It is worth noting that the distinction between unbiased and consistent is not always obvious in the machine learning literature. For the purposes of the discussion in this chapter, Condition 1 above is treated as *consistency* as typically defined in the econometric literature. Condition 2 in this context refers to the ability of a shrinkage estimator to produce estimates that are exactly 0. While LASSO satisfies this condition, Ridge in generally does not produce estimates that are exactly 0. Condition 3 is a technical condition that is typically assumed in econometrics to ensure the continuity of the loss (objective) function and is often required to prove consistency.

Conditions 1 and 2 are telling. In the language of conventional econometrics and statistics, if an estimator is consistent, then it should automatically satisfy these two conditions, at least asymptotically. The introduction of these conditions suggests that shrinkage estimators are not generally consistent, at least not in the traditional sense. In fact, while LASSO satisfies *sparsity*, it is not unbiased. Equation 1.11 shows that LASSO is a *shifted* OLS estimator when $\mathbf{X}'\mathbf{X} = \mathbf{I}$. Thus, if OLS is unbiased or consistent, then LASSO will be biased (or inconsistent) with the magnitude of the bias determined by λ, or equivalently, c. This should not be a surprise, since if $c < ||\boldsymbol{\beta}||_1$ then it is obviously not possible for $\hat{\boldsymbol{\beta}}_{LASSO}$ to be unbiased (or consistent) as the total length of the estimated parameter vector is less than the total length of the true parameter vector. Even if $c > ||\boldsymbol{\beta}||_1$, the unbiasedness of LASSO is not guaranteed as shown by Fan and Li (2001). The formal discussion of these properties leads to the development of the *Oracle Properties*, which are discussed in Section 1.4. For now, it is sufficient to point out that a motivation for some of the alternative shrinkage estimators is to obtain, in a certain sense, unbiased or consistent parameter estimates, while having the ability to identify the unnecessary explanatory variable by assigning 0 to their coefficients i.e., *sparsity*.

The SCAD regularizer can be written as

$$p(\beta_j; a, \lambda) = \begin{cases} |\beta| & \text{if} \quad |\beta| \le \lambda, \\ \dfrac{2a\lambda|\beta| - \beta^2 - \lambda^2}{2(a-1)\lambda} & \text{if} \quad \lambda < |\beta| \le a\lambda, \\ \dfrac{\lambda(a+1)}{2} & \text{if} \quad |\beta| > a\lambda, \end{cases} \tag{1.12}$$

where $a > 2$. The SCAD has two interesting features. First, the regularizer is itself a function of the tuning parameter λ. Second, SCAD divides the coefficient into three different regions namely, $|\beta_j| \le \lambda$, $\lambda < |\beta_j| < a\lambda$ and $|\beta_j| \ge a\lambda$. When $|\beta_j|$ is less than the tuning parameter, λ, the penalty is equivalent to the LASSO. This helps to ensure the *sparsity* feature of LASSO i.e., it can assign zero coefficients. However, unlike the LASSO, the penalty does not increase when the magnitude of the coefficient is large. In fact, when $|\boldsymbol{\beta}_j| > a\lambda$ for some $a > 2$, the penalty is constant. This can be better illustrated by examining the derivative of the SCAD regularizer,

$$p'(\beta_j; a, \lambda) = I(|\beta_j| \le \lambda) + \frac{(a\lambda - |\beta_j|)_+}{(a-1)\lambda} I(|\beta_j| > \lambda), \tag{1.13}$$

where $I(A)$ is an indicator function that equals 1 if A is true and 0 otherwise and $(x)_+ = x$ if $x > 0$ and 0 otherwise. As shown in the expression above, when $|\beta_j| \le \lambda$, the rate of change of the penalty is constant, when $|\beta_j| \in (\lambda, a\lambda]$, the penalty increases linearly and becomes 0 when $|\beta_j| > a\lambda$. Thus, there is no additional penalty when $|\beta_j|$ exceeds a certain magnitude. This helps to ease the problem of *biased* estimates related to the standard LASSO. Note that the derivative as shown in Equation (1.13) exists for all $|\beta_j| > 0$ including the two boundary points. Thus SCAD can be interpreted as a quadratic spline with knots at λ and $a\lambda$.

Figure 1.2 provides some insight on the regularizers through their plots. As shown in Figure 1.1a, the penalty increases as the coefficient increases. This means more penalty is applied to coefficients with a large magnitude. Moreover, the rate of change of the penalty equals the tuning parameter, λ. Both Ridge and Elastic Net exhibit similar behaviours for large coefficients as shown in Figures 1.1b and 1.1c but they behave differently for coefficients close to 0. In case of the Elastic Net, the rate of change of the penalty for coefficients close to 0 is larger than in the case of the Ridge, which makes it more likely to push small coefficients to 0. In contrast, the SCAD is quite different from the other penalty functions. While it behaves exactly like the LASSO when the coefficients are small, the penalty is a constant for coefficients with large magnitude. This means once coefficients exceed a certain limit, there is no additional penalty imposed, regardless how much larger the coefficients are. This helps to alleviate the bias imposed on large coefficients, as in the case of LASSO.

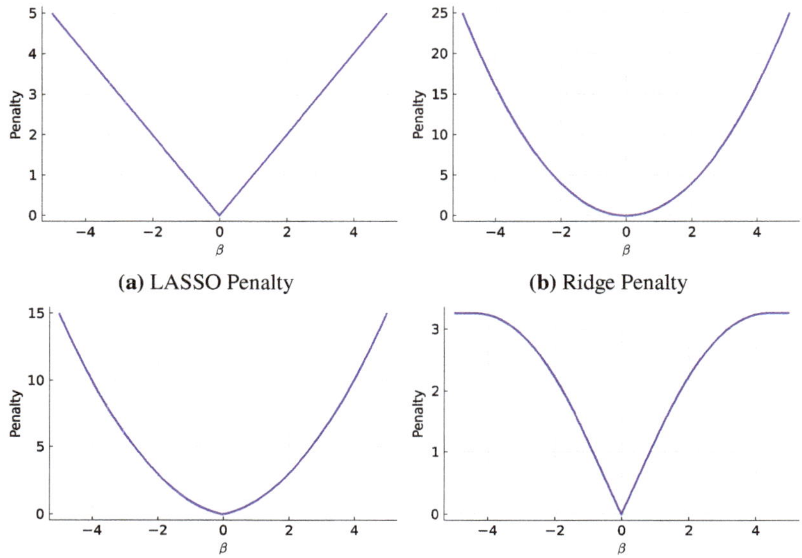

(a) LASSO Penalty

(b) Ridge Penalty

(c) Elastic Net Penalty

(d) SCAD Penalty

Fig. 1.2: Penalty Plots for LASSO, Ridge, Elastic Net and SCAD

1.2.3 Adaptive LASSO

While LASSO does not possess the Oracle Properties in general, a minor modification of it can lead to a shrinkage estimator with Oracle Properties, while also satisfying sparsity and continuity. The Adaptive LASSO (adaLASSO) as proposed in Zou (2006) can be defined as

$$\hat{\boldsymbol{\beta}}_{ada} = \underset{\boldsymbol{\beta}}{\arg\min}\; g(\boldsymbol{\beta}; \mathbf{X}, \mathbf{Y}) + \lambda \sum_{j=1}^{p} w_j |\beta_j|, \tag{1.14}$$

where $w_j > 0$ for all $j = 1, \ldots, p$ are weights to be pre-determined by the researchers. As shown by Zou (2006), an appropriate data-driven determination of w_j would lead to adaLASSO with Oracle Properties. The term adaptive reflects the fact the weight vector $\mathbf{w} = (w_1, \ldots, w_p)'$ is based on any consistent estimator of $\boldsymbol{\beta}$. In other words, the adaLASSO takes the information provided by the consistent estimator and allocates significantly more penalty to the coefficients that are close to 0. This is often achieved by assigning $\hat{\mathbf{w}} = 1./|\hat{\boldsymbol{\beta}}|^{\eta}$, where $\eta > 0$, ./ indicates element-wise division and $|\boldsymbol{\beta}|$ denotes the element-by-element absolute value operation. $\hat{\boldsymbol{\beta}}$ can be chosen based on any consistent estimator of $\boldsymbol{\beta}$. OLS is a natural choice under standard assumptions but it is only valid when $p < N$. Note that using a consistent estimator to construct the weight is a limitation particularly for the case $p > N$, where consistent estimator is not always possible to obtain. In this case, LASSO can actually be used to construct the weight but suitable adjustments must be made for the case when $\hat{\beta}_{LASSO,j} = 0$ (see Zou, 2006 for one possible adjustment).

Both LASSO and adaLASSO have been widely used and extended in various settings in recent times, especially for time series applications in terms of lag order selection. For examples, see, Wang, Li and Tsai (2007), Hsu, Hung and Chang (2008) and Huang, Ma and Zhang (2008). Two particularly interesting studies are by Medeiros and Mendes (2016), and Kock (2016). The former extended the adaLASSO for time series models with non-Guassian and conditional heteroskedastic errors, while the latter established the validity of using adaLASSO with non-stationary time series data.

A particularly convenient feature of the adaLASSO in the linear regression setting is that under $\hat{\mathbf{w}} = 1./|\boldsymbol{\beta}|^{\eta}$, the estimation can be transformed into a standard LASSO problem. Thus, adaLASSO imposes no additional computational cost other than obtaining initial consistent estimates. To see this, rewrite Equation (1.14) using the least squares objective

$$\hat{\boldsymbol{\beta}}_{ada} = \underset{\boldsymbol{\beta}}{\arg\min}\ (\mathbf{y} - \mathbf{X}\boldsymbol{\beta})'(\mathbf{y} - \mathbf{X}\boldsymbol{\beta}) + \lambda \mathbf{w}'|\boldsymbol{\beta}|, \tag{1.15}$$

since $w_j > 0$ for all $j = 1, \ldots, p$, $\mathbf{w}'|\boldsymbol{\beta}| = |\mathbf{w}'\boldsymbol{\beta}|$. Define $\boldsymbol{\theta} = (\theta_1, \ldots, \theta_p)'$ with $\theta_j = w_j \beta_j$ for all j and $\mathbf{Z} = (\mathbf{x}_1/w_1, \ldots, \mathbf{x}_p/w_p)$ is a $N \times p$ matrix transforming each column in $\mathbf{X}$ by dividing the appropriate element in $\mathbf{w}$. Thus, adaLASSO can now be written as

$$\hat{\boldsymbol{\theta}}_{ada} = \underset{\boldsymbol{\theta}}{\arg\min}\ (\mathbf{y} - \mathbf{Z}\boldsymbol{\theta})'(\mathbf{y} - \mathbf{Z}\boldsymbol{\theta}) + \lambda \sum_{j=1}^{p} |\theta_j|, \tag{1.16}$$

which is a standard LASSO problem with $\hat{\beta}_j = \hat{\theta}_j / w_j$.

1.2.4 Group LASSO

There are frequent situations when the interpretation of the coefficients make sense only if all of them in a subset of variables are non-zero. For example, if an explanatory variable is a categorical variable (or factor) with M options, then a typical approach is to create M dummy variables,[4] each representing a single category. This leads to M columns in $\mathbf{X}$ and M coefficients in $\boldsymbol{\beta}$. The interpretation of the coefficients can be problematic if some of these coefficients are zeros, which often happens in the case of LASSO, as highlighted by Yuan and Lin (2006). Thus, it would be more appropriate to 'group' these coefficients together to ensure that the sparsity happens at the categorical variable (or factor) level, rather than at the individual dummy variables level. One way to capture this is to rewrite Equation (1.2) as

$$\mathbf{y} = \sum_{j=1}^{J} \mathbf{X}_j \boldsymbol{\beta}_{0j} + \mathbf{u}, \tag{1.17}$$

where $\mathbf{X}_j = \left(\mathbf{x}_{j1}, \ldots, \mathbf{x}_{jM_j}\right)$ and $\boldsymbol{\beta}_{0j} = \left(\beta_{0j1}, \ldots, \beta_{0jM_j}\right)$. Let $\mathbf{X} = (\mathbf{X}_1, \ldots, \mathbf{X}_J)$ and $\boldsymbol{\beta} = \left(\boldsymbol{\beta}'_1, \ldots, \boldsymbol{\beta}'_J\right)'$, the *group LASSO* as proposed by Yuan and Lin (2006) is defined as

$$\hat{\boldsymbol{\beta}}_{group} = \underset{\boldsymbol{\beta}}{\arg\min}\ (\mathbf{y} - \mathbf{X}\boldsymbol{\beta})'(\mathbf{y} - \boldsymbol{\beta}\mathbf{X}) + \lambda \sum_{j=1}^{J} ||\boldsymbol{\beta}_j||_{K_j}, \tag{1.18}$$

where the penalty function is the root of the quadratic form

$$||\boldsymbol{\beta}_j||_{K_j} = \left(\boldsymbol{\beta}'_j K_j \boldsymbol{\beta}_j\right)^{1/2} \qquad j = 1, \ldots, J \tag{1.19}$$

for some positive semi-definite matrix K_j to be chosen by the researchers. Note that when $M_j = 1$ with $K_j = \mathbf{I}$ for all j, then group LASSO is reduced into a standard LASSO. Clearly, it is possible for $M_j = 1$ and $K_j = \mathbf{I}$ for some j only. Thus, group LASSO allows the possibility of mixing categorical variables (or factors) with continuous variables. Intuitively, the construction of the group LASSO imposes a L_2 norm, like the Ridge regularizer, to the coefficients that are being 'grouped' together, while imposing an L_1 norm, like the LASSO, to each of the coefficients of the continuous variables and the *collective coefficients* of the categorical variables. This helps to ensure that if all categorical variables are relevant, all associated coefficients are likely to have non-zero estimates.

While SCAD and adaLASSO may have more appealing theoretical properties, LASSO, Ridge and Elastic Net remain popular due to their computational convenience. Moreover, LASSO, Ridge and Elastic Net have been implemented in several software packages and programming languages, such as R, Python and Julia, which also explains their popularity. Perhaps more importantly, these routines are typically capable of identifying appropriate tuning parameters, i.e., λ, which makes them more appealing to researchers. In general, the choice of regularizers is still an open question as the finite sample performance of the associated estimator can vary with problems. For further discussion and comparison see Hastie, Tibshirani and Friedman (2009).

---

[4] Assuming there is no intercept.

1.3 Estimation

This section provides a brief overview of the computation of various shrinkage estimators discussed earlier, including the determination of the tuning parameters.

1.3.1 Computation and Least Angular Regression

It should be clear that each shrinkage estimator introduced above is a solution to a specific (constrained) optimization problem. With the exception of the Ridge estimator, which has a closed form solution, some of these optimization problems are difficult to solve in practice. The popularity of the LASSO is partly due to its computation convenience via the *Least Angular Regression* (LARS, proposed by Efron, Hastie, Johnstone & Tibshirani, 2004). In fact, LARS turns out to be so flexible and powerful that the solutions to most of the regularization problems above can be solved using a variation of LARS. This applies also to regularizers such as SCAD, where it is a nonlinear function. As shown by Zou and Li (2008), it is possible to obtain SCAD estimates by using LARS with *local linear approximation*. The basic idea is to approximate Equations (1.7) and (1.12) using Taylor approximation, which gives a LASSO type problem. Then, iteratively solve the associated LASSO problem until convergence. Interestingly, in the context of linear models, the number of iterations required until convergence is often a single step! This greatly facilitates the estimation process using SCAD.

There are some developments of LARS focusing on improving the objective function by including an additional variable to better the fitted value of $\mathbf{y}$, $\hat{\mathbf{y}}$. In other words, the algorithm focuses on constructing the best approximation of the response variable, $\hat{\mathbf{y}}$, the accuracy of which is measured by the objective function, rather than the coefficient estimates, $\hat{\boldsymbol{\beta}}$. This, once again, highlights the difference between machine learning and econometrics. The former focuses on the *predictions* of $\mathbf{y}$ while the latter also focuses on the coefficient estimate $\hat{\boldsymbol{\beta}}$. This can also be seen via the determination of the tuning parameter λ which is discussed in Section 1.3.2.

The outline of the LARS algorithm can be found below:[5]

Step 1. Standardize the predictors to have mean zero and unit norm. Start with the residual $\hat{\mathbf{u}} = \mathbf{y} - \bar{\mathbf{y}}$, where $\bar{\mathbf{y}}$ denotes the sample mean of $\mathbf{y}$ with $\boldsymbol{\beta} = \mathbf{0}$.
Step 2. Find the predictor x_j most correlated with $\hat{\mathbf{u}}$.
Step 3. Move β_j from 0 towards its least-squares coefficient until some other predictors, x_k has as much correlation with the current residuals as does x_j.
Step 4. Move β_j and β_k in the direction defined by their joint least squares coefficient of the current residual on (x_j, x_k) until some other predictor, x_l, has as much correlation with the current residuals.
Step 5. Continue this process until all p predictors have been entered. After $\min(N-1, p)$ steps, this arrives at the full least-squares solution.

[5] This has been extracted from Hastie et al. (2009)

The computation of LASSO requires a small modification to Step 4 above, namely:
4a. If a non-zero coefficient hits zero, drop its variable from the active set of variables and recompute the current joint least squares direction.

This step allows variables to join and leave the selection as the algorithm progresses and, thus, allows a form of 'learning'. In other words, every step revises the current variable selection set, and if certain variables are no longer required, the algorithm removes them from the selection. However, such variables can 're-enter' the selection set at later iterations.

Step 5 above also implies that LARS will produce at most N non-zero coefficients. This means if the intercept is non-zero, it will identify at most $N-1$ covariates with non-zero coefficients. This is particularly important in the case when $p_1 > N$ and LARS cannot identify more than N relevant covariates. The same limitation is likely to be true for any algorithms but a formal proof of this claim is still lacking and could be an interesting direction for future research.

As mentioned above, LARS has been implemented in most of the popular open source languages, such as R, Python and Julia. This implies LASSO and any related shrinkage estimators that can be computed in the form of a LASSO problem can be readily calculated in these packages.

LARS is particularly useful when the regularizer is convex. When the regularizer is non-convex, such as the case of SCAD, it turns out that it is possible to approximate the regularizer via *local linear approximation* as shown by Zou and Li (2008). The idea is to transform the estimation problem into a sequence of LARS, which then can be conducted iteratively.

1.3.2 Cross Validation and Tuning Parameters

The discussion so far has assumed that the tuning parameter, λ, is given. In practice, λ is often obtained via *K-folds* cross validation. This approach yet again highlights the difference between machine learning and econometrics, where the former focuses pre-dominantly on the prediction performance of $\mathbf{y}$. The basic idea of cross validation is to divide the sample randomly into K partitions and randomly select $K-1$ partitions to estimate the parameters. The estimated parameters can then be used to construct predictions for the remaining (unused) partition, called the *left-out* partition, and the average prediction errors are computed based on a given loss function (prediction criterion) over the left-out partition. The process is then repeated K times each with a different left-out partition. The tuning parameter, λ, is chosen by minimizing the average prediction errors over the K folds. This can be summarized as follows:

Step 1. Divide the dataset into K partitions randomly such that $\mathcal{D} = \uplus_{k=1}^{K} \mathcal{D}_k$.
Step 2. Let $\hat{\mathbf{y}}_k$ be the prediction of $\mathbf{y}$ in $\mathcal{D}_k$ based on the parameter estimates from the other $K-1$ partitions.
Step 3. The total prediction error for a given λ is

$$e_k(\lambda) = \sum_{i \in \mathcal{D}_k} (y_i - \hat{y}_{ki})^2 .$$

Step 4. For a given λ, the average prediction errors over the K-folds is

$$CV(\lambda) = K^{-1} \sum_{k=1}^{K} e_k(\lambda).$$

Step 5. The tuning parameter can then be chosen based on

$$\hat{\lambda} = \arg\min_{\lambda} CV(\lambda).$$

The process discussed here is known to be unstable for moderate sample sizes. In order to ensure robustness, the K-fold process can be repeated $N - 1$ times and the tuning parameter, λ, can be obtained as the average of these repeated K-fold cross validations.

It is important to note that the discussion in Section 1.2 explicitly assumed c is fixed and by implication, this means λ is also fixed. In practice, however, λ is obtained via statistical procedures such as cross validation introduced above. The implication is that λ or c should be viewed as a random variable in practice, rather than being fixed. This impacts on the properties of the shrinkage estimators but to the best of the authors' knowledge, this issue has yet to be examined properly in the literature. Thus, this would be another interesting avenue for future research.

The methodology above explicitly assumes that the data are independently distributed. While this may be reasonable in a cross section setting, it is not always valid for time series data, especially in terms of autoregressive models. It may also be problematic in a panel data setting with time effects. In those cases, the determination of λ is much more complicated. It often reverts to evaluating some forms of goodness-of-fit via information criteria for different values of λ. For examples of such approaches, see Wang et al. (2007), Zou, Hastie and Tibshirani (2007) and Y. Zhang, Li and Tsai (2010). In general, if prediction is not the main objective of a study, then these approaches can also be used to determine λ. See also, Hastie et al. (2009) and Fan, Li, Zhang and Zou (2020) for more comprehensive treatments of cross validation.

1.4 Asymptotic Properties of Shrinkage Estimators

Valid statistical inference often relies on the asymptotic properties of estimators. This section provides a brief overview of the asymptotic properties of the shrinkage estimators presented in Section 1.2 and discusses their implications for statistical inference. The literature in this area is highly technical, but rather than focusing on these aspects, the focus here is on the extent to which these can facilitate valid statistical inference typically employed in econometrics, with an emphasis on the qualitative aspects of the results (see the references for technical details).

The asymptotic properties in the shrinkage estimators literature for linear models can broadly be classified into three focus areas, namely:

1. *Oracle Properties*,
2. Asymptotic distribution of shrinkage estimators, and
3. Asymptotic properties of estimators for parameters that are not part of the shrinkage.

1.4.1 Oracle Properties

The asymptotic properties of the shrinkage estimators presented above are often investigated through the so-called *Oracle Properties*. Although the origin of the term *Oracle Properties* can be traced back to Donoho and Johnstone (1994), Fan and Li (2001) are often credited as the first to formalize its definition mathematically. Subsequent presentation can be found in Zou (2006) and Fan, Xue and Zou (2014), with the latter having possibly the most concise definition to-date. The term *Oracle* is used to highlight the feature that Oracle estimator shares the same properties as estimators with the correct set of covariates. In other words, the Oracle estimator can 'foresee' the correct set of covariates. While presentations can be slightly different, the fundamental idea is very similar. To aid the presentation, let us rearrange the true parameter vector, $\boldsymbol{\beta}_0$, so that all parameters with non-zero values can be grouped into a sub-vector and all the parameters with zero values can be grouped into another sub-vector. It is also helpful to create a set that contains the indexes of all non-zero coefficients as well as another one that contains the indexes of all zero-coefficients. Formally, let $\mathcal{A} = \{j : \beta_{0j} \neq 0\}$ and $\hat{\mathcal{A}} = \{j : \hat{\beta}_j \neq 0\}$, and without loss of generality, partition $\boldsymbol{\beta}_0 = \left(\boldsymbol{\beta}'_{0\mathcal{A}}, \boldsymbol{\beta}'_{0\mathcal{A}^c}\right)'$ and $\hat{\boldsymbol{\beta}} = \left(\hat{\boldsymbol{\beta}}'_{\mathcal{A}}, \hat{\boldsymbol{\beta}}'_{\mathcal{A}^c}\right)'$, where $\boldsymbol{\beta}_{0\mathcal{A}}$ denotes the sub-vector of $\boldsymbol{\beta}_0$ containing all the non-zero elements of $\boldsymbol{\beta}_0$ i.e., those with indexes that belong to $\mathcal{A}$, while $\boldsymbol{\beta}_{\mathcal{A}^c}$ is the sub-vector of $\boldsymbol{\beta}_0$ containing all the zero elements i.e., those with indexes that do not belong to $\boldsymbol{\beta}_{\mathcal{A}}$. Similar definitions apply to $\hat{\boldsymbol{\beta}}_{\mathcal{A}}$ and $\hat{\boldsymbol{\beta}}_{\mathcal{A}^c}$. Then the estimator $\hat{\boldsymbol{\beta}}$ is said to have the Oracle Properties if it has

1. Selection Consistency: $\lim_{N\to\infty} \Pr\left(\hat{\mathcal{A}} = \mathcal{A}\right) = 1$, and
2. Asymptotic normality: $\sqrt{N}\left(\hat{\boldsymbol{\beta}}_{\mathcal{A}} - \boldsymbol{\beta}_{\mathcal{A}}\right) \xrightarrow{d} N(0, \boldsymbol{\Sigma})$,

where $\boldsymbol{\Sigma}$ is the variance-covariance matrix of the following estimator

$$\hat{\boldsymbol{\beta}}_{oracle} = \underset{\boldsymbol{\beta}:\boldsymbol{\beta}_{\mathcal{A}^c}=\mathbf{0}}{\arg\min}\, g(\boldsymbol{\beta}). \tag{1.20}$$

Equation (1.20) is called the *Oracle estimator* by Fan et al. (2014) because it shares the same properties as the estimator that contains only the variables with non-zero coefficients. A shrinkage estimator is said to have the Oracle Properties if it is selection consistent, i.e., is able to identify a variable with zero and non-zero coefficients, and has the same asymptotic distribution as a consistent estimator that contains only the correct set of variables. Note that selection consistency is a weaker condition than consistency in the traditional sense. The requirement of selection consistency is to discriminate coefficients with zero and non-zero values but it does not require the estimates to be consistent if they have non-zero values.

It should be clear from the previous discussions that neither LASSO nor Ridge have Oracle Properties in general since LASSO is typically inconsistent and Ridge does not usually have selection consistency. However, adaLASSO, SCAD and group LASSO have been shown to possess Oracle Properties (for technical details, see Zou, 2006, Fan & Li, 2001 and Yuan & Lin, 2006, respectively). While these shrinkage estimators possess Oracle Properties, their proofs usually rely on the following three assumptions:

Assumption 1. $\mathbf{u}$ is a vector of independent, identically distributed random variables with finite variance.

Assumption 2. There exist a matrix $\mathbf{C}$ with finite elements such that $N^{-1}\sum_{i=1}^{N}\mathbf{x}_i\mathbf{x}_i' - \mathbf{C} = o_p(1)$.

Assumption 3. $\lambda \equiv \lambda_N = O(N^q)$ for some $q \in (0, 1]$.

While Assumption 2 is fairly standard in the econometric literature, Assumption 1 appears to be restrictive as it does not include common issues in econometrics such as serial correlation or heteroskedasticity. However, several recent studies, such as Wang et al. (2007) and Medeiros and Mendes (2016) have attempted to relax this assumption to non-Gaussian, serial correlated or conditional heteroskedastic errors.

Assumption 3 is perhaps the most interesting. First, it once again highlights the importance of the tuning parameter, not just for the performance of the estimators, but also for their asymptotic properties. Then, the assumption requires that $N^{-q}\lambda_N - \lambda_0 = o_p(1)$ for some $\lambda_0 \geq 0$. This condition trivially holds when $\lambda_N = \lambda$ stays constant regardless of the sample size. However, this is unlikely to be the case in practice when λ_N is typically chosen based on cross validation as described in Section 1.3.2. Indeed, when λ_N remains constant, $\lambda_0 = 0$, which implies that the constraint imposes no penalty on the loss asymptotically and in the least squares case, the estimator collapses to the familiar OLS estimator asymptotically. In the case when λ_N is a non-decreasing function of N, Assumption 3 assets an upper bound of the growth rate. This should not be surprising since λ_N is an inverse function of c. If λ_N increases, the total length of $\boldsymbol{\beta}$ decreases and that may increase the amount of bias in the estimator. Another perspective of this assumption is its relation to the *uniform signal strength condition* as discussed by C.-H. Zhang and Zhang (2014). The condition asserts that all non-zero coefficients must be greater in magnitude than an inflated level of noise which can be expressed by

$$C\sigma\sqrt{\frac{2\log p}{N}}, \tag{1.21}$$

where C is the inflation factor (see C.-H. Zhang and Zhang (2014)). Essentially, the coefficients are required to be 'large' enough (relative to the noise) to ensure selection consistency and this assumption has also been widely used in some of the literature (for examples, see the references within C.-H. Zhang & Zhang, 2014). While there seems to be an intuitive link between Assumption 3 and the uniform signal strength condition, there does not appear to be any formal discussion about their connection and this could be an interesting area for future research. Perhaps more importantly, the relation between choosing λ via cross validation, Assumption 3, and the uniform signal strength condition is still somewhat unclear. This also means that if the true value of the coefficient is sufficiently small, there is a good chance that shrinkage estimators will be unable to identify them and it is unclear how to statistically verify this situation in finite samples. Theoretically, however, if the shrinkage estimator is selection consistent, then it should be able to discriminate coefficients with a small magnitude and coefficients with zero value.

The Oracle Properties are appealing because they seem to suggest that one can select the right set of covariates via selection consistency, and simultaneously estimate the non-zero parameters consistently, which are also asymptotically normal. It is therefore tempting to

1. conduct statistical inference directly on the shrinkage estimates in the usual way or
2. use a shrinkage estimator as a variable selector, then estimate the coefficients using OLS on model with only the selected covariates and conduct statistical inference on the OLS estimates in the usual way.

The last approach is particularly tempting especially for the shrinkage estimators that satisfy selection consistency but not necessarily asymptotic normality, such as the original LASSO. This approach is often called *Post-Selection OLS* or in the case of LASSO, *Post-LASSO OLS*. These two approaches turn out to be overly optimistic in practice as argued by Leeb and Pötscher (2005) and Leeb and Pötscher (2008). The main issue has to do with the mode of convergence in proving the Oracle Properties. In most cases, the convergence is *pointwise* rather than *uniform*. An implication of the pointwise convergence is that a different $\boldsymbol{\beta}_0$ may require different sample sizes before the asymptotic distribution becomes a reasonable approximation. Since the sample size is typically fixed in practice with $\boldsymbol{\beta}_0$ unknown, blindly applying the asymptotic result for inference may lead to misleading conclusion, especially if the sample size is not large enough for the asymptotic to 'kick in'. Worse still, it is typically not possible to examine if the sample size is large enough since it depends on the unknown $\boldsymbol{\beta}_0$. Thus, the main message is that while Oracle Properties are interesting theoretical properties and provide an excellent framework to aid the understanding of different shrinkage estimators, it does not necessary provide the assurance one wishes for in practice for statistical inference.

1.4.2 Asymptotic Distributions

Next, let us examine the distributional properties of shrinkage estimators directly. The seminal work of Knight and Fu (2000) derived the asymptotic distribution of the Bridge estimator as defined in Equations (1.7) and (1.8). Under Assumptions 1 – 3 with $g(\boldsymbol{\beta}) = (\mathbf{y}-\mathbf{X}\boldsymbol{\beta})'(\mathbf{y}-\mathbf{X}\boldsymbol{\beta})$, $q = 0.5$ and $\gamma \geq 1$, Knight and Fu (2000) showed that

$$\sqrt{N}\left(\hat{\boldsymbol{\beta}}_{Bridge} - \boldsymbol{\beta}_0\right) \xrightarrow{d} \arg\min V(\boldsymbol{\omega}), \tag{1.22}$$

where

$$V(\boldsymbol{\omega}) = -2\boldsymbol{\omega}'\mathbf{W}\boldsymbol{\omega} + \boldsymbol{\omega}'\mathbf{C}\boldsymbol{\omega} + \lambda_0 \sum_{j=1}^{p} u_j \operatorname{sgn} \beta_{0j} |\beta_{0j}|^{\gamma-1}, \tag{1.23}$$

with $\mathbf{W}$ having a $N(\mathbf{0}, \sigma^2\mathbf{C})$ distribution. For $\gamma < 1$, Assumption 3 needs to be adjusted such that $q = \gamma/2$ with $V(\boldsymbol{\omega})$ changes to

$$V(\boldsymbol{\omega}) = -2\boldsymbol{\omega}'\mathbf{W}\boldsymbol{\omega} + \boldsymbol{\omega}'\mathbf{C}\boldsymbol{\omega} + \lambda_0 \sum_{j=1}^{p} |u_j|^{\gamma} I(\beta_{0j} = 0). \tag{1.24}$$

Recall that the Bridge regularizer is convex for $\gamma \geq 1$, but not for $\gamma < 1$. This is reflected by the difference in the asymptotic distributions implied by Equations (1.23) and (1.24). Interestingly, but perhaps not surprisingly, the main difference between the two expressions is the term related to the regularizer. Moreover, the growth rate of λ_N is also required to be much slower for the $\gamma < 1$ case than for the $\gamma \geq 1$ one.

Let $\boldsymbol{\omega}^* = \arg\min V(\boldsymbol{\omega})$, then for $\gamma \geq 1$ it is straightforward to show that

$$\boldsymbol{\omega}^* = \mathbf{C}^{-1}\left[\mathbf{W} - \lambda_0 \operatorname{sgn} \boldsymbol{\beta}_0 |\boldsymbol{\beta}_0|^{\gamma-1}\right], \tag{1.25}$$

where $\operatorname{sgn} \boldsymbol{\beta}$, $|\boldsymbol{\beta}|^{\gamma-1}$ and the product of the two terms are understood to be taken element wise. This means

$$\sqrt{N}\left(\hat{\boldsymbol{\beta}}_{Bridge} - \boldsymbol{\beta}_0\right) \xrightarrow{d} \mathbf{C}^{-1}\left[\mathbf{W} - \lambda_0 \operatorname{sgn} \boldsymbol{\beta}_0 |\boldsymbol{\beta}_0|^{\gamma-1}\right]. \tag{1.26}$$

Note that setting $\gamma = 1$ and $\gamma = 2$ yield the asymptotic distribution of LASSO and Ridge, respectively. However, as indicated in Equation (1.25), the distribution depends on $\boldsymbol{\beta}_0$. the true parameter vector. The result has two important implications. First, Bridge estimator is generally inconsistent and second, the asymptotic distribution depends on the true parameter vector, which means it is subject to the criticism of Leeb and Pötscher (2005). The latter means that the sample size required for the asymptotic distribution to be a 'reasonable' approximation to the finite sample distribution depends on the true parameter vector. Since the true parameter vector is not typically known, this means the asymptotic distribution may not be particularly helpful in practice, at least not in the context of hypothesis testing.

While directly statistical inference on the shrinkage estimates appears to be difficult, there are other studies that take slightly different approaches. The two studies that are worth mentioning are Lockhart, Taylor, Tibshirani and Tibshirani (2014) and Lee, Sun, Sun and Taylor (2016). The former developed a test statistics for coefficients as they enter the model during the LARS process, while the latter derived the asymptotic distribution of the least squares estimator conditional on model selection from shrinkage estimator, such as the LASSO.

In addition to the two studies above, it is also worth noting that in some cases, such as LASSO, where the estimates can be bias-corrected and through such correction, the asymptotic distribution of the bias-corrected estimator can be derived (for examples, see C.-H. Zhang & Zhang, 2014 and Fan et al., 2020). Despite the challenges as shown by Leeb and Pötscher (2005) and Leeb and Pötscher (2008), the asymptotic properties of the shrinkage estimators, particularly for purposes of valid statistical inference, remains an active area of research.

Overall, the current knowledge regarding the asymptotic properties of various shrinkage estimators can be summarized as follows:

1. Some shrinkage estimators have shown to possess Oracle Properties, which means asymptotically they can select the right covariates i.e., correctly assign 0 to the coefficients with true 0 value. It also means that the estimators have an asymptotic normal distribution.
2. Despite the Oracle Properties and other asymptotic results, such as those of Knight and Fu (2000), the practical usefulness of these results are still somewhat limited. The sample size required for the asymptotic results to 'kick in' depends on the true parameter vector, which is unknown in practice. Thus, one can never be sure about the validity of using the asymptotic distribution for a given sample size.

1.4.3 Partially Penalized (Regularized) Estimator

While valid statistical inference on shrinkage estimators appears to be challenging, there are situations where the parameter of interests may not be part of the shrinkage. This means the regularizer does not have to be applied to the entire parameter vector. Specifically, let us rewrite Equation (1.2) as

$$\mathbf{y} = \mathbf{X}_1\boldsymbol{\beta}_1 + \mathbf{X}_2\boldsymbol{\beta}_2 + \mathbf{u}, \tag{1.27}$$

where $\mathbf{X}_1$ and $\mathbf{X}_2$ are $N \times p_1$ and $N \times p_2$ matrices such that $p = p_1 + p_2$ with $\mathbf{X} = [\mathbf{X}_1, \mathbf{X}_2]$, and $\boldsymbol{\beta} = \left(\boldsymbol{\beta}_1', \boldsymbol{\beta}_2'\right)'$ such that $\boldsymbol{\beta}_1$ and $\boldsymbol{\beta}_2$ are $p_1 \times 1$ and $p_2 \times 1$ parameter vectors. Assume that only $\boldsymbol{\beta}_2$ is sparse i.e., contains elements with zero value, and consider the following shrinkage estimator

$$\left(\hat{\boldsymbol{\beta}}_1', \hat{\boldsymbol{\beta}}_2'\right)' = \underset{\boldsymbol{\beta}_1, \boldsymbol{\beta}_2}{\arg\min}\ (\mathbf{y} - \mathbf{X}_1\boldsymbol{\beta}_1 - \mathbf{X}_2\boldsymbol{\beta}_2)'(\mathbf{y} - \mathbf{X}_1\boldsymbol{\beta}_1 - \mathbf{X}_2\boldsymbol{\beta}_2) + \lambda p(\boldsymbol{\beta}_2). \tag{1.28}$$

Note that the penalty function (regularizer) applies only to $\boldsymbol{\beta}_2$ but not $\boldsymbol{\beta}_1$. A natural question in this case is whether the asymptotic properties of $\hat{\boldsymbol{\beta}}_1$ could facilitate valid statistical inference in the usual manner?

In the case of the Bridge estimator, it is possible to show that $\hat{\boldsymbol{\beta}}_1$ has an asymptotic normal distribution similar to the OLS estimator. This is formalized in Proposition 1.1.

Proposition 1.1 *Consider the linear model as defined in Equation (1.27) and the estimator as defined in Equation (1.28) with* $p(\boldsymbol{\beta}_2) = \sum_{j=p_1+1}^{p} |\beta_j|^\gamma$ *for some* $\gamma > 0$. *Under Assumptions 1 and 2 along with* $\lambda_N/\sqrt{N} \to \lambda_0 \geq 0$ *for* $\gamma \geq 1$ *and* $\lambda_N/\sqrt{N^\gamma} \to \lambda_0 \geq 0$, *then*

$$\sqrt{N}\left(\hat{\boldsymbol{\beta}}_1 - \boldsymbol{\beta}_{01}\right) \xrightarrow{d} \omega_1^*$$

where

$$\omega_1^* = \mathbf{C}_1^{-1}\mathbf{W}_1,$$

with $\mathbf{W}_1 \sim N(0, \sigma_u^2\mathbf{I})$ *denoting a* $p_1 \times 1$ *random vector and* C_1^{-1} *the* $p_1 \times p_1$ *matrix consisting of the first* p_1 *rows and columns of* $\mathbf{C}$.

Proof See the Appendix. □

The implication of Proposition 1.1 is that valid inference in the usual manner should be possible for parameters that are not part of the shrinkage process, at least for the Bridge estimator. This is verified by some Monte Carlo experiment in Section 1.5.1. This leads to the question on whether this idea can be generalized further. For example, consider

$$\mathbf{y} = \mathbf{X}_1\boldsymbol{\beta}_1 + \mathbf{u}, \tag{1.29}$$

where $\mathbf{X}_1$ is a $N \times p_1$ matrix containing (some) endogenous variables. Now assume there are p_2 potential instrumental variables where p_2 can be very large. In a Two Stage Least Squares setting, one would typically construct instrumental variables by first estimating

$$\mathbf{X}_1 = \mathbf{X}_2\boldsymbol{\Pi} + \mathbf{v}$$

and set the instrumental variables $\mathbf{Z} = \mathbf{X}_2\hat{\boldsymbol{\Pi}}$. Given the estimation of $\boldsymbol{\Pi}$ is separate from $\boldsymbol{\beta}_1$ and perhaps more importantly, the main target is $\mathbf{Z}$, which in a sense, should be the best approximation of $\mathbf{X}_1$ given $\mathbf{X}_2$. As shown by Belloni et al. (2012), when p_2 is large, it is possible to

1. leverage shrinkage estimators to produce the best instruments i.e., the best approximation of $\mathbf{X}_1$ given $\mathbf{X}_2$ and
2. reduce the number of instruments given the sparse nature of shrinkage estimators and thus alleviate the issue of well too many instrumental variables.

The main message is that it is possible to obtain a consistent and asymptotically normal estimator for $\boldsymbol{\beta}_1$ by constructing optimal instruments from shrinkage estimators using a large number of potential instrumental variables. There are two main ingredients which make this approach feasible. The first is that it is possible to obtain 'optimal' instruments $\mathbf{Z}$ based on *Post-Selection OLS*, that is, OLS after a shrinkage procedure as shown by Belloni and Chernozhukov (2013). Given $\mathbf{Z}$, Belloni et al. (2012) show that the usual IV type estimators, such as $\hat{\boldsymbol{\beta}}_{IV} = (\mathbf{Z}'\mathbf{X}_1)^{-1}\mathbf{Z}'\mathbf{y}$, follow standard asymptotic results. This makes intuitive sense, as the main target in this case is the best approximation of $\mathbf{X}_1$ rather than the quality of the estimator for $\boldsymbol{\Pi}$. Thus, in a sense, this approach leverages the intended usage of shrinkage estimators of producing optimal approximation and use this approximation as instruments to resolve endogeneity.

This idea can be expanded further to analyze a much wider class of econometric problems. Chernozhukov, Hansen and Spindler (2015) generalized this approach by considering the estimation problem where the parameters would satisfy the following system of equations

$$M(\boldsymbol{\beta}_1,\boldsymbol{\beta}_2) = \mathbf{0}. \tag{1.30}$$

Note that least squares, maximum likelihood and Generalized Method of Moments can be captured in this framework. Specifically, the system of equation as denoted by M can be viewed as the first order derivative of the objective function and thus Equation (1.30) represents the First Order Necessary Condition for a wide class of M-estimators. Along with some relatively mild assumptions, Chernozhukov et al. (2015) show that the following:

$$\frac{\partial}{\partial \boldsymbol{\beta}_2'} M(\boldsymbol{\beta}_1,\boldsymbol{\beta}_2)\big|_{\boldsymbol{\beta}_2=\hat{\boldsymbol{\beta}}_2} = 0, \tag{1.31}$$

where $\hat{\boldsymbol{\beta}}_2$ denotes a good quality shrinkage estimator of $\boldsymbol{\beta}_2$, are sufficient to ensure valid statistical inference on $\hat{\boldsymbol{\beta}}_1$. Equation (1.31) is often called the *immunization condition*. Roughly speaking, it asserts that if the system of equations as defined in Equation (1.30) is not sensitive, and therefore immune, to small changes of $\boldsymbol{\beta}_2$ for a given estimator of $\boldsymbol{\beta}_2$, then statistical inference based on $\hat{\boldsymbol{\beta}}_1$ is possible.[6]

Using the IV example above, $\boldsymbol{\Pi}$, the coefficient matrix for constructing optimal instruments, can be interpreted as $\boldsymbol{\beta}_2$ in Equation (1.30). Equation (1.31) therefore requires the estimation of $\boldsymbol{\beta}_1$ not to be sensitive to small changes in the shrinkage estimators used to estimate $\boldsymbol{\Pi}$. In other words, the condition requires that a small change in $\hat{\boldsymbol{\Pi}}$ does not affect the estimation of $\boldsymbol{\beta}_1$. This is indeed the case as long as the small changes in $\hat{\boldsymbol{\Pi}}$ do not affect the estimation of the instruments $\mathbf{Z}$ significantly.

1.5 Monte Carlo Experiments

This section provides a brief overview of some of the possible applications of shrinkage estimators in econometrics. A handful of Monte Carlo Simulation results are also presented to shed more light on a few issues discussed in the Chapter.

---

[6] Readers are referred to Chernozhukov et al. (2015) and Belloni and Chernozhukov (2013) for technical details and discussions.

1.5.1 Inference on Unpenalized Parameters

As seen in Section 1.4.3, while valid inference remains challenging for Shrinkage estimators in general, it is possible to obtain valid inference on parameters that are not part of the shrinkage. In other words, if the parameters do not appear in the regularizer, then inference of these parameter estimates in the case of the Bridge estimator is asymptotically normal, as shown in Proposition 1.1. Here some Monte Carlo evidence is provided about the finite sample performance of this estimator. In addition to the Bridge estimator, this section also examines the finite sample properties of the *unpenalized* (unregularized) parameter with other regularizers, including Elastic Net, adaLASSO and SCAD. This may be helpful in identifying if these regularizers share the same properties as the Bridge for the unpenalized parameters.

Unless otherwise stated, all Monte Carlo simulations in this section involve simulating six covariates, x_i, $i = 1,\ldots,6$. The six covariates are generated through a multivariate lognormal distribution, $MVLN(\mathbf{0},\mathbf{\Gamma})$. The semi-positive definite matrix $\mathbf{\Gamma}$ controls the degree of correlations between the six covariates. In this section, $\mathbf{\Gamma} = \{\rho_{ij}\}$ with

$$\rho_{ij} = \begin{cases} \sigma_i & \text{if} \quad i = j, \\ \dfrac{\rho^{i+j-2}}{\sigma_i\sigma_j} & \text{if} \quad i \neq j, \end{cases}$$

where σ_i^2 denotes the diagonal of $\mathbf{\Gamma}$, which is assigned to be $\{2, 1, 1.5, 1.25, 1.75, 3\}$. Three different values of ρ are considered, namely $\rho = 0, 0.45, 0.9$, which covers cases of non-correlated, moderately correlated and highly correlated covariates. The rationale in choosing the multivariate lognormal distribution is to allow further investigations on the impact of variable transformations, such as the logarithmic transformation frequently used in economic and econometric analyses, on the performance of shrinkage estimators. This is the focus in Section 1.5.2.

In each case, the Monte Carlo experiment has 5000 replication over five different sample sizes: 30, 50, 100, 500 and 1000. The collection of estimators considered includes OLS, LASSO, Ridge, Elastic Net, adaLASSO with $\eta = 1$ and $\eta = 2$, and SCAD. The tuning parameters are selected based on five-fold cross validation as discussed in Section 1.3.2.

The first set of experiments examines the size and power of a simple t test. The data generating process (DGP) in this case is

$$y_i = \beta_1 \log x_{1i} + 2\log x_{2i} - 2\log x_{3i} + u_i, \qquad u_i \sim N(0,1).$$

Four different values of β_1 are considered namely, 0, 0.1, 1 and 2. In each case, the following model is estimated

$$y_i = \beta_0 + \mathbf{x}_i'\boldsymbol{\beta} + u_i, \tag{1.32}$$

where $\mathbf{x}_i = (\log x_{1i}, \ldots, \log x_{6i})'$ with $\boldsymbol{\beta} = (\beta_1, \ldots, \beta_6)'$. In this experiment, the coefficient of $\log x_1$, β_1, is not part of the shrinkage. In other words, the regularizers are applied only to $\beta_2, \ldots, \beta_6$. After the estimation, the t-test statistics for testing $H_0 : \beta_1 = 0$ are computed in the usual way. This means for the case $\beta_1 = 0$, the experiment examines the size of the simple t-test on β_1. In all other cases, the experiments examine the finite sample power of the simple t-test on β_1 with varying signal-to-noise ratio.

The results of each case are summarized in Tables 1.1 – 1.4. Given that the significance level of the test is set at 0.05, the t-test has the expected size as shown in Table 1.1 for all estimators when the covariates are uncorrelated. However, when the correlation between the covariates increases, the test size for the Ridge estimator seems to be higher than expected, albeit not drastically. This is related to Theorem 4 by Knight and Fu (2000) where the authors considered the case when **C** (the variance-covariance matrix of the covariates) is nearly singular. The near singularity of **C** is reflected through the increasing value of ρ in the experiment. As a result, a different, more restrictive bound, on λ_n is required for the asymptotic result to remain valid in the case when $\gamma > 1$. Interestingly, the test size under all other estimators remained reasonably close to expectation as N increased for the values of ρ considered in the study.

Table 1.1: Size of a Simple t-test for Unpenalized Parameter

ρ	N	OLS	LASSO	adaLASSO	adaLASSO2	Ridge	Elastic Net	SCAD
	30	6.02	5.4	5.44	5.4	5.62	5.52	5.62
	50	5.74	5.46	5.54	5.46	5.62	5.6	5.52
0	100	5.14	5.34	5.28	5.26	5.26	5.26	5.16
	500	5.42	5.34	5.38	5.34	5.32	5.46	5.24
	1000	5.2	5.18	5.14	5.08	5.08	5.16	5.12
	30	6.04	6.02	6.12	6.22	5.8	5.94	6.06
	50	5.52	5.46	5.42	5.54	5.94	5.62	5.5
0.45	100	5.8	6.16	6.0	6.16	6.84	6.14	5.98
	500	5.48	5.96	5.86	5.98	6.44	5.8	7.4
	1000	5.06	5.0	4.96	5.04	6.24	4.9	8.98
	30	6.44	5.86	5.84	5.72	3.8	5.6	5.92
	50	5.78	5.18	5.26	5.16	3.66	4.88	4.94
0.9	100	5.22	5.22	5.26	5.28	4.3	5.26	4.68
	500	5.14	4.68	4.84	4.9	6.74	4.92	4.62
	1000	4.76	4.26	4.22	4.2	6.08	4.4	4.92

Table 1.2: Power of a Simple t-test for Unpenalized Parameter: $\beta_1 = 2$

ρ	N	OLS	LASSO	adaLASSO	adaLASSO2	Ridge	Elastic Net	SCAD
	30	100.0	100.0	100.0	100.0	100.0	100.0	100.0
	50	100.0	100.0	100.0	100.0	100.0	100.0	100.0
0	100	100.0	100.0	100.0	100.0	100.0	100.0	100.0
	500	100.0	100.0	100.0	100.0	100.0	100.0	100.0
	1000	100.0	100.0	100.0	100.0	100.0	100.0	100.0
	30	100.0	100.0	100.0	100.0	100.0	100.0	100.0
	50	100.0	100.0	100.0	100.0	100.0	100.0	100.0
0.45	100	100.0	100.0	100.0	100.0	100.0	100.0	100.0
	500	100.0	100.0	100.0	100.0	100.0	100.0	100.0
	1000	100.0	100.0	100.0	100.0	100.0	100.0	100.0
	30	98.38	99.08	98.92	98.94	99.04	98.92	98.78
	50	99.98	100.0	100.0	100.0	100.0	100.0	100.0
0.9	100	100.0	100.0	100.0	100.0	100.0	100.0	100.0
	500	100.0	100.0	100.0	100.0	100.0	100.0	100.0
	1000	100.0	100.0	100.0	100.0	100.0	100.0	100.0

When $\beta_1 = 2$, the power of the test is nearly 1 for all cases as shown in Table 1.2. As the true value decreases, the performance of the t-test from all shrinkage estimators remained comparable to OLS as shown in Tables 1.3 - 1.4. This provides some evidence to support Proposition 1.1 as well as an indication that its theoretical results may also be applicable to Adaptive LASSO, Elastic Net, and SCAD.

1.5.2 Variable Transformations and Selection Consistency

Another important issue examined in this section is the finite sample performance of *selection consistency*. Specifically, this section examines the selection consistency in the presence of correlated covariates and covariates with a different signal to noise ratio in the form of different variable transformations as well as different variances. The variable transformations considered include logarithmic and quadratic transforms.

The DGP considered is

$$y_i = 2\log x_{2i} - 0.1\log x_{3i} + 1.2x_{2i}^2 - x_{3i}^3 + u_i, \qquad u_i \sim N(0,1).$$

Table 1.3: Power of a Simple t-test for Unpenalized Parameter: $\beta_1 = 1$

ρ	N	OLS	LASSO	adaLASSO	adaLASSO2	Ridge	Elastic Net	SCAD
	30	90.94	90.2	90.24	90.1	88.86	90.2	91.02
	50	99.24	99.26	99.3	99.26	98.92	99.24	99.32
0	100	100.0	100.0	100.0	100.0	100.0	100.0	100.0
	500	100.0	100.0	100.0	100.0	100.0	100.0	100.0
	1000	100.0	100.0	100.0	100.0	100.0	100.0	100.0
	30	84.12	85.86	85.94	85.88	86.04	85.98	85.32
	50	97.8	98.4	98.5	98.46	98.66	98.34	98.46
0	100	100.0	100.0	100.0	100.0	100.0	100.0	100.0
	500	100.0	100.0	100.0	100.0	100.0	100.0	100.0
	1000	100.0	100.0	100.0	100.0	100.0	100.0	100.0
	30	21.94	25.18	24.9	25.06	23.08	23.84	23.0
	50	34.28	40.16	40.0	40.0	40.22	38.08	37.34
0	100	61.3	68.32	68.32	68.06	72.98	65.98	66.48
	500	99.88	99.94	99.94	99.94	100.0	99.94	99.98
	1000	100.0	100.0	100.0	100.0	100.0	100.0	100.0

The estimated model is similar to Equation (1.32) but the specification of $\mathbf{x}_i$ changes to $\mathbf{x}_i = \left(x_{1i}, \ldots, x_{6i}, \log x_{1i}, \ldots, \log x_{6i}, x_{1i}^2, \ldots, x_{2i}^2\right)'$ and $\boldsymbol{\beta} = (\beta_1, \ldots, \beta_{18})'$.

Tables 1.5 and 1.6 contains the percentage of the replications where the estimators identified the coefficients of $\log x_2$ and x_2^2 as zero, respectively. Note that the DGP suggests that neither coefficients should be zero. Thus, these results highlight the finite sample performance of the selection consistency. Given the size of each coefficient and the variance of x_2, it is perhaps not surprising that shrinkage estimators generally assigned non-zero coefficient to x_2^2 but they tend to penalize the coefficient of $\log x_2$ to 0 for large portion of the replications. In fact, as sample size increases, the number of replications where zero coefficients had been assigned to $\log x_2$ increases, but the opposite is true for x_2^2. This can be explained by two factors. First, $\log x_2$ and x_2^2 are highly correlated and it is well known that LASSO type estimators will generally select only one variable from a set of highly correlated covariates, see for example Tibshirani (1996). Second, the variance of $\log x_2$ is smaller than that of x_2^2, which implies a lower signal to noise ratio. This also generally affects the ability of shrinkage estimators to select such variables, as discussed earlier.

These results seem to suggest that one should not take selection consistency for granted. While the theoretical results are clearly correct, the assumptions underlying the result, specifically the assumption on the tuning parameter, λ, and its relation to

Table 1.4: Power of a Simple t-test for Unpenalized Parameter: $\beta_1 = 0.1$

ρ	N	OLS	LASSO	adaLASSO	adaLASSO2	Ridge	Elastic Net	SCAD
	30	12.54	11.6	11.68	11.6	11.56	12.22	12.16
	50	16.22	15.56	15.86	15.88	15.44	15.9	16.18
0	100	28.48	27.88	28.1	27.92	27.24	28.06	28.08
	500	87.64	87.4	87.4	87.38	86.3	87.42	87.18
	1000	99.42	99.44	99.46	99.46	99.16	99.46	99.46
	30	11.18	12.54	12.26	12.4	13.54	12.48	12.0
	50	14.9	17.58	17.78	17.42	19.62	17.4	16.7
0.45	100	23.9	28.6	28.54	28.42	33.94	27.86	28.58
	500	77.3	82.26	81.94	81.96	93.02	81.2	87.6
	1000	97.12	98.2	98.14	98.08	99.88	98.0	99.38
	30	7.8	7.74	7.94	7.78	5.58	7.32	7.68
	50	6.78	7.42	7.44	7.24	6.46	6.96	6.8
0.9	100	7.84	9.1	9.24	9.28	10.14	8.7	8.54
	500	17.56	22.62	22.36	22.38	42.82	20.8	25.5
	1000	31.12	38.32	38.1	38.12	74.44	35.64	48.08

the signal to noise ratio in the form of uniform signal strength condition clearly plays an extremely important role in variable selection using shrinkage estimators. Thus, caution and further research in this area seems warranted.

1.6 Econometrics Applications

Next, some recent examples of econometric application using shrinkage estimators are provided. The first application examines shrinkage estimators in a distributed lag model setting. The second example discusses the use of shrinkage estimators in panel data with fixed effects, and the third one proposes a new test for structural breaks with unknown break points.

Table 1.5: Percentage of Replications Excluding $\log x_2$

ρ	N	OLS	LASSO	adaLASSO	adaLASSO2	Ridge	Elastic Net	SCAD
	30	0.0	52.24	52.22	52.26	0.0	46.82	79.58
	50	0.0	56.44	56.44	56.44	0.0	50.56	86.48
0	100	0.0	64.3	64.3	64.3	0.0	62.98	94.16
	500	0.0	88.08	88.08	88.08	0.0	89.16	99.98
	1000	0.0	95.02	95.02	95.02	0.0	96.26	100.0
	30	0.0	50.08	50.18	50.22	0.0	47.9	79.74
	50	0.0	56.0	55.96	56.0	0.0	53.7	85.8
0.45	100	0.0	64.08	64.08	64.08	0.0	63.72	94.34
	500	0.0	89.88	89.88	89.88	0.0	91.28	99.96
	1000	0.0	96.64	96.64	96.64	0.0	96.7	100.0
	30	0.0	47.46	47.68	47.66	0.0	47.06	73.68
	50	0.0	53.06	53.34	53.12	0.0	54.16	79.42
0.9	100	0.0	65.44	65.5	65.42	0.0	64.94	89.76
	500	0.0	93.98	93.98	93.98	0.0	92.38	99.62
	1000	0.0	98.82	98.82	98.82	0.0	97.62	99.94

1.6.1 Distributed Lag Models

The origin of the distributed lag model can be traced back to Tinbergen (1939). While there have been studies focusing on lag selection in an Autoregressive-Moving Average (ARMA) setting (for examples, see Wang et al., 2007; Hsu et al., 2008 and Huang et al., 2008), the application of the partially penalized estimator as discussed in Section 1.4.3 has not been discussed in this context.

Consider the following DGP

$$y_i = \mathbf{x}_i'\boldsymbol{\beta} + \sum_{j=1}^{L} \mathbf{x}_{i-j}'\boldsymbol{\alpha}_j + u_i, \tag{1.33}$$

where $L < N$. Choosing the appropriate lag order for a particular variable is a challenging task. Perhaps more importantly, as the number of observations increases, the number of potential (lag) variables increases and this creates additional difficulties in identifying a satisfactorily model. If one is only interested in statistical inference on the estimates of $\boldsymbol{\beta}$, then the results from Section 1.4.3 may seem to be useful. In this case, one can apply a Partially Penalized Estimator and obtain the parameter

Table 1.6: Percentage of Replications Excluding x_2^2

ρ	N	OLS	LASSO	adaLASSO	adaLASSO2	Ridge	Elastic Net	SCAD
	30	0.0	3.32	3.28	3.3	0.0	2.08	7.72
	50	0.0	2.08	2.06	2.06	0.0	1.18	5.16
0	100	0.0	1.16	1.16	1.16	0.0	0.68	3.58
	500	0.0	0.64	0.62	0.62	0.0	0.18	1.88
	1000	0.0	0.5	0.5	0.5	0.0	0.1	1.88
	30	0.0	4.22	4.22	4.24	0.0	3.4	7.96
	50	0.0	2.4	2.42	2.4	0.0	1.52	5.08
0.45	100	0.0	1.22	1.2	1.2	0.0	0.5	4.02
	500	0.0	0.7	0.7	0.7	0.0	0.12	1.94
	1000	0.0	0.34	0.34	0.34	0.0	0.1	1.28
	30	0.0	14.28	14.3	14.32	0.0	13.6	19.36
	50	0.0	9.6	9.52	9.5	0.0	8.46	14.32
0.9	100	0.0	5.62	5.6	5.62	0.0	5.3	9.52
	500	0.0	2.54	2.54	2.58	0.0	2.0	5.24
	1000	0.0	1.9	1.84	1.88	0.0	1.3	4.24

estimates as follows

$$\left(\hat{\boldsymbol{\beta}},\hat{\boldsymbol{\alpha}}\right) = \underset{\boldsymbol{\beta},\boldsymbol{\alpha}}{\arg\min} \sum_{i=1}^{N}\left(y_i - \mathbf{x}_i'\boldsymbol{\beta} - \sum_{j=1}^{L}\mathbf{x}_{i-j}'\boldsymbol{\alpha}_j\right)^2 + \lambda p\left(\boldsymbol{\alpha}\right), \tag{1.34}$$

where $\boldsymbol{\alpha} = \left(\boldsymbol{\alpha}_1',\ldots,\boldsymbol{\alpha}_L'\right)'$ and the $p(\boldsymbol{\alpha})$ is a regularizer applied only to $\boldsymbol{\alpha}$. Since $\hat{\boldsymbol{\beta}}$ is not part of the shrinkage, under the Bridge regularizer and the assumptions in Proposition 1.1, $\hat{\boldsymbol{\beta}}$ has an asymptotically normal distribution which facilitates valid inferences on $\boldsymbol{\beta}$. Obviously, $\boldsymbol{\beta}$ does not have to be the coefficients associated with the covariates at the same time period as the response variable. The argument above should apply to any coefficients of interests. The main idea is that if a researcher is interested in conducting statistical inference on a particular set of coefficients with a potentially high number of possible control variables, as long as the coefficients of interests are not part of the shrinkage, valid inference on these coefficients may still be possible.

1.6.2 Panel Data Models

Following the idea above, another potentially useful application is the panel data model with fixed effect. Consider

$$y_{it} = \mathbf{x}_{it}'\boldsymbol{\beta} + \alpha_i + u_{it}, \qquad i = 1,\ldots,N, \quad t = 1,\ldots,T. \tag{1.35}$$

The parameter vector $\boldsymbol{\beta}$ is typically estimated by the *fixed effect* estimator

$$\hat{\boldsymbol{\beta}}_{FE} = \operatorname*{arg\,min}_{\boldsymbol{\beta}} \sum_{i=1}^{N}\sum_{t=1}^{T} \left(\dot{y}_{it} - \dot{\mathbf{x}}_{it}'\boldsymbol{\beta}\right)^2, \tag{1.36}$$

where $\dot{y}_{it} = y_{it} - \bar{y}_i$ with $\bar{y}_i = T^{-1}\sum_{t=1}^{T} y_{it}$ and $\dot{x}_{it} = \mathbf{x}_{it} - \bar{\mathbf{x}}_i$ with $\bar{\mathbf{x}}_i = T^{-1}\sum_{t=1}^{T}\mathbf{x}_{it}$. In practice, the estimator is typically computed using the dummy variable approach. However, when N is large, the number of α_i is also large. Since it is common for $N >> T$ in panel data, the number of dummy variables required for the fixed effect estimator can also be unacceptably large. Since the main focus is $\boldsymbol{\beta}$, under the assumption that α_i are constants for $i = 1,\ldots,N$, it seems also possible to apply the methodology from the previous example and consider the following estimator

$$\left(\hat{\boldsymbol{\beta}},\hat{\boldsymbol{\alpha}}\right) = \operatorname*{arg\,min}_{\boldsymbol{\beta},\boldsymbol{\alpha}} \sum_{i=1}^{N}\sum_{t=1}^{T} \left(y_{it} - \mathbf{x}_{it}'\boldsymbol{\beta} - \alpha_i\right)^2 + \lambda p\left(\boldsymbol{\alpha}\right). \tag{1.37}$$

where $\boldsymbol{\alpha} = (\alpha_1,\ldots,\alpha_N)'$ and $p(\boldsymbol{\alpha})$ denotes a regularizer applied only to the coefficients of the fixed effect dummies. Proposition 1.1 should apply in this case without any modification.

This can be extended to a higher dimensional panel with more than two indexes. In such cases, the number of dummies required grows exponentially. While it is possible to obtain a fixed effect estimators in a higher dimensional panel through various transformations proposed by Balazsi, Mátyás and Wansbeek (2018), these transformations are not always straightforward to derive and the dummy variable approach could be more practically convenient. The dummy variable approach, however, suffers from the curse of dimensionality and the proposed method here seems to be a feasible way to resolve this issue. Another potential application is to incorporate interacting fixed effects of the form α_{it} into model (1.35). This is, of course, not possible in a usual two-dimensional panel data setting, but feasible with this approach.

Another possible application, which has been proposed in the literature, is to incorporate a regularizer in Equation (1.36) and therefore define a shrinkage estimator in a standard fixed effects framework. Specifically, *fixed effects with shrinkage* can be defined as

$$\hat{\boldsymbol{\beta}}_{FE} = \operatorname*{arg\,min}_{\boldsymbol{\beta}} \sum_{i=1}^{N}\sum_{t=1}^{T} \left(\dot{y}_{it} - \dot{\mathbf{x}}_{it}'\boldsymbol{\beta}\right)^2 + \lambda p(\boldsymbol{\beta};\boldsymbol{\alpha}), \tag{1.38}$$

where $p(\boldsymbol{\beta})$ denotes the regularizer, which in principle, can be any regularizers such as those introduced in Section 1.2. Given the similarity between Equations (1.38) and all the other shrinkage estimators considered so far, it seems reasonable to assume that the results in Knight and Fu (2000) and Proposition 1 would apply possibly with only some minor modifications. This also means that fixed effect models with a shrinkage estimator are not immune to the shortfall of shrinkage estimators in general. The observations and issues highlighted in this chapter would apply equally in this case.

1.6.3 Structural Breaks

Another econometric example where shrinkage type estimators could be helpful is the testing for structural breaks with unknown breakpoints. Consider the following DGP

$$y_i = \mathbf{x}_i'\boldsymbol{\beta} + \mathbf{x}_i'\boldsymbol{\delta}_0 I(i > t_1) + u_i, \qquad u_i \sim D(0, \sigma_u^2), \tag{1.39}$$

where the break point, t_1 is unknown. Equation (1.39) implies that the parameter vector when $i \leq t_1$ is $\boldsymbol{\beta}_0$ and when $i > t_1$, it is $\boldsymbol{\beta}_0 + \boldsymbol{\delta}_0$. In other words, a structural break occurs at $i = t_1$ and $\boldsymbol{\delta}$ denotes the shift in the parameter vector before and after the break point.

Such models have a long history in econometrics, for example, see Andrews (1993) and Andrews (2003) as well as the references within. However, the existing tests are bounded by the $p < N$ restriction. That is, the number of variables must be less than the number of observations. Given that these tests are mostly residuals based tests, this means that it is possible to obtain post-shrinkage (or post selection) residuals and use these residuals in the existing tests. To illustrate the idea, consider the simple case when t_1 is known. In this case, a typical approach is to consider the following F-test statistics as proposed by Chow (1960)

$$F = \frac{RSS_R - RSS_{UR1} - RSS_{UR2}}{RSS_{UR1} + RSS_{UR2}} \frac{N - 2p}{p}, \tag{1.40}$$

where RSS_R denotes the residual sum-of-squares from the restricted model ($\boldsymbol{\delta} = \mathbf{0}$), while RSS_{UR1} and RSS_{UR2} denote the unrestricted sum-of-squares before and after the break, respectively. Specifically, RSS_{UR1} denotes the residual sum-of-squares from the residuals $\hat{u}_t = y_t - \mathbf{x}_t'\hat{\boldsymbol{\beta}}$ for $t \leq t_1$ and RSS_{UR2} denotes the residuals sum-of-squares from the residuals $\hat{u}_i = y_t - \mathbf{x}_t'\left(\hat{\boldsymbol{\beta}} + \hat{\boldsymbol{\delta}}\right)$ for $t > t_1$.

It is well known that under the null hypothesis $H_0 : \boldsymbol{\delta} = \mathbf{0}$, the F-test statistics in Equation (1.40) follows an F distribution under the usual regularity conditions. When t_1 is not known, Andrews (1993) derived the asymptotic distribution for

$$F = \sup_s F(s), \tag{1.41}$$

where $F(s)$ denotes the F-statistics as defined in Equation (1.40) assuming s as the breakpoint for $s = p+1, \ldots, N-p-1$. The idea is to select a breakpoint s, such that the test has the highest chance to reject the null of $H_0 : \boldsymbol{\delta} = \mathbf{0}$. The distribution based on this approach is non-standard as shown by Andrews (1993) and must therefore be tabulated or simulated.

Note that the statistics in Equation (1.40) is based on the residuals rather than the individual coefficient estimates, so it is possible to use the arguments by Belloni et al. (2012) and construct the statistics as follows:

Step 1. Estimate the parameter vector $y_i = \mathbf{x}_i'\boldsymbol{\beta} + u_i$ using a LASSO type estimator, called it $\hat{\boldsymbol{\beta}}_{LASSO}$.
Step 2. Obtain a Post-Selection OLS. That is, estimate the linear regression model using OLS with the covariates selected in the previous step.
Step 3. Construct the residuals using the estimates from the previous step $\hat{u}_{R,i} = y_i - \hat{y}_i$ where $\hat{y}_i = \mathbf{x}_i'\hat{\boldsymbol{\beta}}_{OLS}$.
Step 4. Compute $RSS_{UR} = \sum_{i=1}^{N} \hat{u}_{R,i}^2$.
Step 5. Estimate the following model using a LASSO-type estimator

$$y_i = \mathbf{x}_i' i\boldsymbol{\beta} + \sum_{j=2}^{N-1} \delta_j \mathbf{x}_i \boldsymbol{\beta} I(i \leq j) + u_i \tag{1.42}$$

and denotes the estimates for $\boldsymbol{\beta}$, $\boldsymbol{\delta}$ and j as $\hat{\boldsymbol{\beta}}_{LASSO-UR}$, $\hat{\delta}_{LASSO}$ and $\hat{j}$, respectively. Under the assumption there is only one break $\delta_j = 0$ for all j except when $j = t_1$.
Step 6. Obtain the Post-Selection OLS for the pre-break unrestricted model, $\hat{\boldsymbol{\beta}}_{UR1-OLS}$. That is, estimate the linear regression model using OLS with the covariates selected in Step 5 for $i \leq \hat{j}$.
Step 7. Obtain the Post-Selection OLS for the post-break unrestricted model, $\hat{\boldsymbol{\beta}}_{UR2-OLS}$. That is, estimate the linear regression model using OLS with the covariates selected in Step 5 for $i > \hat{j}$.
Step 8. Construct the pre-break residuals using $\hat{\boldsymbol{\beta}}_{UR1-OLS}$. That is, $\hat{u}_{UR1,i} = y_i - \hat{y}_i$, where $\hat{y}_i = \mathbf{x}_i'\hat{\boldsymbol{\beta}}_{UR1-OLS}$ for $i \leq \hat{j}$.
Step 9. Construct the post-break residuals using $\hat{\boldsymbol{\beta}}_{UR2-OLS}$. That is, $\hat{u}_{UR2,i} = y_i - \hat{y}_i$, where $\hat{y}_i = \mathbf{x}_i'\hat{\boldsymbol{\beta}}_{UR2-OLS}$ for $i > \hat{j}$.
Step 10. Compute $RSS_{UR1} = \sum_{i=1}^{\hat{j}} \hat{u}_{UR1,i}^2$ and $RSS_{UR2} = \sum_{i=\hat{j}+1}^{N} \hat{u}_{UR2,i}^2$.
Step 11. Compute the test statistics as defined in Equation (1.40).

Essentially, the proposal above uses LASSO as a variable selector as well as a break point identifier. It then generates the residuals sum-of-squares using OLS based on the selection given by LASSO. This approach can potentially be justified by the results by Belloni et al. (2012) and Belloni and Chernozhukov (2013). Unlike the conventional approach when the breakpoint is unknown, such as those studied by Andrews (1993) whose test statistics have non-standard distributions, the test statistics proposed here is likely to follow the F distribution similar to the original test statistics as proposed by Chow (1960) and can accommodate the case when $p > N$. To the best of the authors' knowledge, this approach is novel with both the theoretical properties and finite sample performance to be further evaluated. However, given the results of Belloni et al. (2012), this seems like a plausible approach to tackle of the problem of detecting structural breaks with unknown breakpoints.

1.7 Concluding Remarks

This chapter has provided a brief overview of the most popular shrinkage estimators in the machine learning literature and discussed their potential applications in econometrics. While valid statistical inference may be challenging to obtain directly for shrinkage estimators, it seems possible, at least in the case of the Bridge estimator, to conduct valid inference on the statistical significance of a subset of the parameter vector, for the elements which are not part of the regularization. In the case of the Bridge estimator, this chapter has provided such a result by modifying the arguments of Knight and Fu (2000). Monte Carlo evidence suggested that similar results may also be applicable to other shrinkage estimators, such as the adaptive LASSO and SCAD. However, the results also highlighted that the finite sample performance of a simple t-test for an unregularized parameter is no better than those obtained directly from OLS. Thus, if it is possible to obtain OLS estimates, shrinkage estimators do not seem to add value for inference purposes. However, when OLS is not possible, as in the case when $p > N$, shrinkage estimators provide a feasible way to conduct statistical inference on the coefficients of interests, as long as they are not part of the regularization.

Another interesting and useful result from the literature is that while the theoretical properties of shrinkage estimators may not be useful in practice, shrinkage estimators do lead to superior fitted values, especially in the case of post-shrinkage OLS. That is, fitted values obtained by using OLS with the covariates selected by a shrinkage estimator. The literature relies on this result to obtain optimal instruments in the presence of many (possibly weak) instrumental variables. Using a similar idea, this chapter has also proposed a new approach to test for a structural break when the break point is unknown.[7] Finally, the chapter has also highlighted the usefulness of these methods in a panel data framework.

Table 1.7 contains a brief summary of the various shrinkage estimators introduced in this chapter. Overall, machine learning methods in the framework of shrinkage estimators seem to be quite useful in several cases when dealing with linear econometric models. However, users have to be careful, mostly with issues related to estimation and variable selection consistency.

---

[7] The theoretical properties and the finite sample performance of this test may be an interesting area for future research.

Appendix

Proof of Proposition 1.1

For $\gamma \geq 1$, using the same argument as Theorem 2 in Knight and Fu (2000), it is straightforward to show that

$$\sqrt{N}\left(\hat{\boldsymbol{\beta}} - \boldsymbol{\beta}_0\right) \xrightarrow{d} \arg\min V(\boldsymbol{\omega}),$$

where

$$V(\boldsymbol{\omega}) = -2\boldsymbol{\omega}'\mathbf{W} + \boldsymbol{\omega}'\mathbf{C}\boldsymbol{\omega} + \lambda_0 \sum_{j=p_1+1}^{p} \omega_j \operatorname{sgn} \beta_{0j} |\beta_{0j}|^{\gamma-1}.$$

Let $\boldsymbol{\omega}^* = \arg\min V(\boldsymbol{\omega})$, which can be obtained by solving the First Order Necessary Condition and this gives

$$\omega_j^* = \mathbf{c}_j \left(w_j - \frac{\lambda_0}{2} \operatorname{sgn} \beta_{0j} |\beta_{0j}|^{\gamma-1} I(j > p_2) \right), \qquad j = 1, \ldots, p,$$

where $\mathbf{c}_j$ denotes the jth row of $\mathbf{C}^{-1}$ and w_j denotes the jth element in $\mathbf{W}$. Note that the last term in the expression above is 0 for $j \leq p_1$. Thus, collect the first p_1 elements in $\boldsymbol{\omega}^*$ gives the result. The argument for $\gamma < 1$ is essentially the same with the definition of $V(\boldsymbol{\omega})$ being replaced by Theorem 3 in Knight and Fu (2000). This completes the proof.

Table 1.7: Summary of Shrinkage Estimators

Estimator	Advantages	Disadvantages	Software
LASSO	• Can assign 0 to coefficients • Computationally convenient	• Lack Oracle Properties in general. • Asymptotic distribution is not practically useful • Sensitive to the choice of tuning parameter • Estimator has no closed form solution and must rely on numerical methods.	• Available in R, Python and Julia
Ridge	• Closed form solution exists • Lots of theoretical results	• Lack Oracle Properties. • Cannnot assign 0 to coefficients, • Sensitive to the choice of tuning parameter	• Available in almost all software packages. • Also easy to implement given the closed form solution
Elastic Net	• Aims to strike a balance between LASSO and Ridge • Can assign 0 to coefficients • Researchers can adjust the balance between LASSO and Ridge	• Lack Oracle properties. • Need to choose two tuning parameters. The weight between LASSO and Ridge and the penalty factor • No closed form solution and must rely on numerical methods	• Available in R, Python and Julia. • Need to adjust the LARS algorithms
Adaptive LASSO	• Possesses Oracle Properties • Can assign 0 to coefficients • Less biased than LASSO • It is convenient to compute for linear model	• Require initial estimates. • Practical usefulness of Oracle Properties is limited	• Not widely available in standard software packages • For linear model, it is straightforward to compute based on variable transformations

Table 1.7 Cont.:Summary of Shrinkage Estimators

Estimator	Advantages	Disadvantages	Software
SCAD	• Possess Oracle properties • Can assign 0 to coefficients • Estimates are generally less biased than LASSO	• Practical usefulness of Oracle properties is limited • No closed form solution and must rely on numerical methods • Two tuning parameters and unclear how to determine their values in practice • Regularizer is complicated and a function of the tuning parameter	• Not widely available in standard software packages.
Group LASSO	• Useful when covariates involve categorical variables • Allow coefficients of a group of variables to be either all zeros or all non-zeros • Possess Oracle properties	• Practical usefulness of Oracle properties is limited • No closed form solution and must rely on numerical methods • An additional kernel matrix is required for each group	• Not widely available in standard software packages • Computationally intensive

References

Andrews, D. W. K. (1993). Tests for Parameter Instability and Structural Change with Unknown Change Point. *Econometrica: Journal of the Econometric Society*, 821–856.

Andrews, D. W. K. (2003). End-of-Sample Instability Tests. *Econometrica*, *71*(6), 1661–1694.

Balazsi, L., Mátyás, L. & Wansbeek, T. (2018). The Estimation of Multidimensional Fixed Effects Panel Data Models. *Econometric Reviews*, *37*, 212-227.

Belloni, A., Chen, D., Chernozhukov, V. & Hansen. (2012). Sparse Models and Methods for Optimal Instruments With an Application to Eminent Domain. *Econometrica*, *80*(6), 2369–2429. doi: 10.3982/ECTA9626

Belloni, A. & Chernozhukov, V. (2013, May). Least Squares after Model Selection in High-dimensional Sparse Models. *Bernoulli*, *19*(2), 521–547. doi: 10.3150/11-BEJ410

Chernozhukov, V., Hansen, C. & Spindler, M. (2015). Valid Post-Selection and Post-Regularization Inference: An Elementary, General Approach. *Annual Review of Economics*, *7*, 649–688.

Chow, G. C. (1960). Tests of Equality between Sets of Coefficients in Two Linear Regressions. *Econometrica: Journal of the Econometric Society*, *28*(3), 591–605.

Donoho, D. L. & Johnstone, I. M. (1994). Ideal Spatial Adaptation by Wavelet Shrinkage. *Biometrika*, *81*, 425-455.

Efron, B., Hastie, T., Johnstone, I. & Tibshirani, R. (2004). Least Angle Regression. *Annals of Statistics*, *32*(2), 407–451.

Fan, J. & Li, R. (2001). Variable Selection via Nonconcave Penalized Likelihood and its Oracle Properties. *Journal of the American Statistical Association*, *96*(456), 1348–1360. doi: 10.1198/016214501753382273

Fan, J., Li, R., Zhang, C.-H. & Zou, H. (2020). *Statistical Foundations of Data Science*. CRC Press, Chapman and Hall.

Fan, J., Xue, L. & Zou, H. (2014). Strong Oracle Optimality of Folded Concave Penalized Estimation. *Annals of Statistics*, *42*(3), 819–849.

Frank, I. & Friedman, J. (1993). A Statistical View of Some Chemometrics Regression Tools. *Technometrics*, *35*, 109-148.

Hastie, T., Tibshirani, R. & Friedman, J. (2009). *The elements of statistical learning: Data mining, inference and prediction*. Springer.

Hoerl, A. & Kennard, R. (1970a). Ridge Regression: Applications to Nonorthogonal Problems. *Technometrics*, *12*, 69-82.

Hoerl, A. & Kennard, R. (1970b). Ridge Regression: Biased Estimation for Nonorthogonal Problems. *Technometrics*, *12*, 55-67.

Hsu, N.-J., Hung, H.-L. & Chang, Y.-M. (2008, March). Subset Selection for Vector Autoregressive Processes using LASSO. *Computational Statistics & Data Analysis*, *52*(7), 3645–3657. doi: 10.1016/j.csda.2007.12.004

Huang, J., Ma, S. & Zhang, C.-H. (2008). Adaptive LASSO for Sparse High-dimensional Regression Models. *Statistica Sinica*, *18*, 1603-1618.

Knight, K. & Fu, W. (2000). Asymptotics for LASSO-Type Estimators. *The Annals of Statistics*, *28*(5), 1356–1378.

Kock, A. B. (2016, February). Consistent and Conservative Model Selection with the Adaptive LASSO in Stationary and Nonstationary Autoregressions. *Econometric Theory*, *32*(1), 243–259. doi: 10.1017/S0266466615000304

Lee, J. D., Sun, D. L., Sun, Y. & Taylor, J. E. (2016, June). Exact Post-selection Inference, with Application to the LASSO. *The Annals of Statistics*, *44*(3). doi: 10.1214/15-AOS1371

Leeb, H. & Pötscher, B. M. (2005). Model Selection and Inference: Facts and Fiction. *Econometric Theory*, *21*(1), 21–59. doi: 10.1017/S0266466605050036

Leeb, H. & Pötscher, B. M. (2008). Sparse Estimators and the Oracle Property, or the Return of Hodges' Estimator. *Journal of Econometrics*, *142*(1), 201–211. doi: 10.1016/j.jeconom.2007.05.017

Lockhart, R., Taylor, J., Tibshirani, R. J. & Tibshirani, R. (2014). A Significance Test for the LASSO. *The Annals of Statistics*, *42*(2), 413–468.

Medeiros, M. C. & Mendes, E. F. (2016). L_1-Regularization of High-dimensional Time-Series Models with non-Gaussian and Heteroskedastic Errors. *Journal of Econometrics*, *191*(1), 255–271. doi: 10.1016/j.jeconom.2015.10.011

Tibshirani, R. (1996, January). Regression Shrinkage and Selection Via the Lasso. *Journal of the Royal Statistical Society: Series B (Methodological)*, *58*(1), 267–288. doi: 10.1111/j.2517-6161.1996.tb02080.x

Tinbergen, J. (1939). *Statistical Testing of Business-Cycle Theories*. League of Nations, Economic Intelligence Service.

Wang, H., Li, G. & Tsai, C.-L. (2007, February). Regression Coefficient and Autoregressive Order Shrinkage and Selection via the LASSO. *Journal of the Royal Statistical Society: Series B (Statistical Methodology)*, *69*(1). doi: 10.1111/j.1467-9868.2007.00577.x

Yuan, M. & Lin, Y. (2006, February). Model Selection and Estimation in Regression with Grouped Variables. *Journal of the Royal Statistical Society: Series B (Statistical Methodology)*, *68*(1), 49–67. doi: 10.1111/j.1467-9868.2005.00532.x

Zhang, C.-H. & Zhang, S. S. (2014, January). Confidence Intervals for Low Dimensional Parameters in High Dimensional Linear Models. *Journal of the Royal Statistical Society: Series B (Statistical Methodology)*, *76*(1), 217–242. doi: 10.1111/rssb.12026

Zhang, Y., Li, R. & Tsai, C. (2010). Regularization Parameter Selections via Generalized Information Criterion. *Journal of the American Statistical Association*, *105*, 312-323.

Zou, H. (2006, December). The Adaptive Lasso and Its Oracle Properties. *Journal of the American Statistical Association*, *101*(476), 1418–1429. doi: 10.1198/016214506000000735

Zou, H. & Hastie, T. (2005, April). Regularization and Variable Selection via the Elastic Net. *Journal of the Royal Statistical Society: Series B (Statistical Methodology)*, *67*(2), 301–320. doi: 10.1111/j.1467-9868.2005.00503.x

Zou, H., Hastie, T. & Tibshirani, R. (2007). On the Degrees of Freedom of the LASSO. *Annals of Statistics*, *35*, 2173-2192.

Zou, H. & Li, R. (2008, August). One-step Sparse Estimates in Nonconcave Penalized Likelihood Models. *The Annals of Statistics*, *36*(4). doi: 10.1214/009053607000000802

Chapter 2
Nonlinear Econometric Models with Machine Learning

Felix Chan, Mark N. Harris, Ranjodh B. Singh and Wei (Ben) Ern Yeo

Abstract This chapter introduces machine learning (ML) approaches to estimate nonlinear econometric models, such as discrete choice models, typically estimated by maximum likelihood techniques. Two families of ML methods are considered in this chapter. The first, shrinkage estimators and related derivatives, such as the Partially Penalised Estimator, introduced in Chapter 1. A formal framework of these concepts is presented as well as a brief literature review. Additionally, some Monte Carlo results are provided to examine the finite sample properties of selected shrinkage estimators for nonlinear models. While shrinkage estimators are typically associated with parametric models, tree based methods can be viewed as their non-parametric counterparts. Thus, the second ML approach considered here is the application of tree-based methods in model estimation with a focus on solving classification, or discrete outcome, problems. Overall, the chapter attempts to identify the nexus between these ML methods and conventional techniques ubiquitously used in applied econometrics. This includes a discussion of the advantages and disadvantages of each approach. Several benefits, as well as strong connections to mainstream econometric methods are uncovered, which may help in the adoption of ML techniques by mainstream econometrics in the discrete and limited dependent variable spheres.

Felix Chan ✉
Curtin University, Perth, Australia, e-mail: felix.chan@cbs.curtin.edu.au

Mark N. Harris
Curtin University, Perth, Australia, e-mail: mark.harris@curtin.edu.au

Ranjodh B. Singh
Curtin University, Perth, Australia, e-mail: ranjodh.singh@curtin.edu.au

Wei (Ben) Ern Yeo
Curtin University, Perth, Australia, e-mail: weiern.yeo@postgrad.curtin.edu.au

F. Chan, L. Mátyás (eds.), *Econometrics with Machine Learning*, Advanced Studies in Theoretical and Applied Econometrics 53, https://doi.org/10.1007/978-3-031-15149-1_2

2.1 Introduction

This chapter aims to examine the potential applications of machine learning techniques in specifying and estimating nonlinear econometric models. The two main objectives of the chapter are to:

1. provide an overview of a suite of machine learning (ML) techniques that are relevant to the specification, estimation and testing of nonlinear econometric models; and
2. identify existing gaps that prevent these techniques from being readily applicable to econometric analyses.

Here we take 'nonlinear econometric models' to mean any of a wide range of different models used in various areas of applied economics and econometric analyses. They can generally be classified into the following categories based on the characteristics of the dependent (response) variable.

1. The dependent (response) variable is discrete.
2. The dependent variable is partly continuous, but is limited in some way; such as it can only be positive.
3. The response variable is continuous but has a nonlinear relationship, such as piece-wise step function, with one or more observed covariates.

The first case can be further divided into three different subcases namely, *cardinal*, *nominal* and *ordinal* responses. Cardinal responses typically manifest as *Count Data* in econometrics. In this case, the set of all possible outcomes is often infinite, with the Poisson model being the fundamental building block of the analysis.

Nominal and ordinal responses, however, often appear in different context in econometrics to cardinal responses, as they are typically a set of finite *discrete choices*, which are usually modelled by a suite of well-known discrete choice models, such as Logit, Probit and Multinomial models, to name just a few. In the machine learning (ML) literature, the modelling of nominal and ordinal responses is often called a *classification problem*. The aim is to model a discrete choice outcome (among a set of alternatives) of an individual/entity. Popular examples of discrete choice analyses include modelling labour force participation (Yes/No), Political beliefs (Strongly Agree, Agree, Neutral, Disagree and Strongly Disagree) or the choice of among different products or occupations. The objective of these analyses is to understand the relation between the covariates (also known as predictors, features and confounding variables) and the stated/revealed choice.

The second type of nonlinear model typically appears where a continuous variable has been truncated, or censored, in some manner. Examples abound in modelling labour supply, charitable donations and financial ratios, among others, where the response variable is always ≥ 0. This chapter is less focused on this particular type of model, as it is less common in the ML literature.

The third type of nonlinear models is more popular in the time series econometrics literature, examples include *threshold regression* and the *threshold autoregressive* (TAR) model; see Tong (2003) for a comprehensive review of threshold models in time series analysis. These models have been used extensively in modelling regime switching behaviours. Econometric applications of these models include, for example, threshold effects between inflation and growth during different phases of a business cycle (see, e.g., Teräsvirta & Anderson, 1992 and Jansen & Oh, 1999).

Given the extensive nature of nonlinear models, it is beyond the scope of this chapter to cover the details of ML techniques applicable to all such models. Instead, the chapter focuses on two ML methods which are applicable to a selected set of models popular among applied economists and econometricians. These are shrinkage estimators and tree based techniques in the form of Classification and Regression Tree (CART). The intention is to provide a flavour on the potential linkage between ML procedures and econometric models.

The choice of the two techniques is not arbitrary. Shrinkage estimators can be seen as an extension to the estimators used for *parametric* models, whereas CART can be viewed as the ML analogue to *nonparametric* estimators often used in econometrics. The connection between CART and nonparametric estimation is also discussed in this chapter.

The chapter is organised as follows. In Section 2.2 shrinkage estimators and Partially Penalised Estimators for nonlinear models are introduced. The primary focus is on variable selection and/or model specification. A brief literature review is presented in order to outline the applications of this method to a variety of disciplines. This includes recent developments on valid statistical inferences with shrinkage and Partially Penalised Estimators for nonlinear models. Asymptotic distributions of the Partial Penalised Estimators for Logit, Probit and Poisson models are derived. The results should facilitate valid inferences. Some Monte Carlo results are also provided to assess the finite sample performance of shrinkage estimators when applied to a simple nonlinear model.

Section 2.3 provides an overview of ML tree-based methods, including an outline of how trees are constructed as well as additional methods to improve their predictive performance. This section also includes an example for demonstration purposes. Section 2.3.3 highlights the potential connections between tree-based and mainstream econometric methods. This may help an applied econometrician to make an informed decision about which method/s are more appropriate for any given specific problem. Finally, Section 2.4 summarises the main lessons of the chapter.

2.2 Regularization for Nonlinear Econometric Models

This section explores shrinkage estimators for nonlinear econometric models, including the case when the response variable is discrete. In the model building phase, one challenge is the selection of explanatory variables. Especially in the era of big data, potentially the number of variables can exceed the number of observations. Thus, a natural question is how can one systematically select relevant variables? As already seen in Chapter 1, obtaining model estimates using traditional methods is not feasible when the number of covariates exceeds the number of observations.

As in the case of linear models, one possible solution is to apply shrinkage estimators with regularization. This method attempts to select relevant variables from the entire set of variables, where the number of such variables may be greater than the number of observations. In addition to variable selection, regularization may also be used to address model specification. Specifically, here we are referring to different transformations of variables entering the model: squaring and/or taking logarithms, for example.

While Chapter 1 explores shrinkage estimators for linear models, their application to nonlinear ones requires suitable modifications on the objective function. Recall, a shrinkage estimator can be expressed as solution to an optimisation as presented in Equations (1.5) and (1.6). namely,

$$\hat{\boldsymbol{\beta}} = \arg\min_{\boldsymbol{\beta}} g\left(\boldsymbol{\beta};\mathbf{y},\mathbf{X}\right)$$
$$\text{s.t. } p\left(\boldsymbol{\beta};\boldsymbol{\alpha}\right) \leq c,$$

where $\boldsymbol{\beta}$ denotes the parameter vector, $\mathbf{y}$ and $\mathbf{X}$ are the vectors containing the observations for the response variable and the covariates, respectively, $p(\boldsymbol{\beta};\boldsymbol{\alpha})$ is the regularizer where $\boldsymbol{\alpha}$ denotes additional tuning parameters, and c is a positive constant.

In the case of linear models, the least squares is often used as the objective function, that is, $g(\boldsymbol{\beta};\mathbf{y},\mathbf{X}) = (\mathbf{y}-\mathbf{X}\boldsymbol{\beta})'(\mathbf{y}-\mathbf{X}\boldsymbol{\beta})$ coupled with different regularizers, $p(\boldsymbol{\beta};\boldsymbol{\alpha})$. Popular choices include LASSO, Ridge, Adaptive LASSO and SCAD. See Chapter 1 for more information on the different regularizers under the least square objective. In the econometric literature, nonlinear least squares and maximum likelihood are the two most popular estimators for nonlinear models. A natural extension of shrinkage estimators then is to define the objective function based on nonlinear least squares or the likelihood function.

2.2.1 Regularization with Nonlinear Least Squares

Consider the following model

$$y_i = h(\mathbf{x}_i;\boldsymbol{\beta}_0) + u_i, \qquad u_i \sim D(0,\sigma_u^2), \tag{2.1}$$

where u_i is a continuous random variable, which implies that the response variable, y_i, is also a continuous random variable. The expectation of y_i conditional on the covariates, $\mathbf{x}_i$, is a twice differentiable function $h : \mathbb{R}^p \to \mathbb{R}$, which depends on the covariates $\mathbf{x}_i$ as well as the parameter vector $\boldsymbol{\beta}_0$. Similarly to the linear case in Chapter 1, while the true parameter vector $\boldsymbol{\beta}_0$ is a $p \times 1$ vector, many of the elements in $\boldsymbol{\beta}$ can be zeros. Since $\boldsymbol{\beta}_0$ is not known in practice, the objective is to estimate $\boldsymbol{\beta}_0$ using shrinkage estimators, or at the very least, using shrinkage estimator to identify which elements in $\boldsymbol{\beta}_0$ are zeros.

Under the assumption that the functional form of h is known, shrinkage estimators in this case can be defined as the solution to the following optimisation problem

$$\hat{\boldsymbol{\beta}} = \arg\min_{\boldsymbol{\beta}} \sum_{i=1}^{n} [y_i - h(\mathbf{x}_i;\boldsymbol{\beta})]^2$$
$$\text{s.t. } p(\boldsymbol{\beta};\boldsymbol{\alpha}) \leq c,$$

and similarly to the linear case, the above can be expressed in its Lagrangian form

$$\hat{\boldsymbol{\beta}} = \arg\min_{\boldsymbol{\beta}} \sum_{i=1}^{n} [y_i - h(\mathbf{x}_i;\boldsymbol{\beta})]^2 + \lambda p(\boldsymbol{\beta};\boldsymbol{\alpha}), \tag{2.2}$$

where λ denotes the tuning parameter. As mentioned in Chapter 1, there are many choices for regularizers, p. Popular choices include the Bridge, which has Ridge and LASSO as special cases, Elastic Net, Adaptive LASSO and SCAD. Each of these choices applies a different penalty to the elements of $\boldsymbol{\beta}$ (please see Chapter 1 for further discussions on the properties of these regularizers).

While shrinkage estimators, such as LASSO, are often used as a variable *selector* for linear models by identifying which elements in $\boldsymbol{\beta}$ are zeros, such interpretation does not carry through to nonlinear models in general. This is because the specification of the conditional mean, $h(\mathbf{x}_i;\boldsymbol{\beta})$, is too general to ensure that each element in $\boldsymbol{\beta}$ is associated with a particular covariate. That is, $\beta_j = 0$, does not necessary mean that a covariate x_j is to be excluded. Therefore, shrinkage estimators in this case are not necessarily a *selector*. At the very least, not a variable selector. One example is

$$y_i = \beta_1 \cos(\beta_2 x_{1i} + \beta_3 x_{2i}) + \beta_4 x_{2i} + u_i$$

and clearly $\beta_3 = 0$ does not mean x_{2i} has not been 'selected'. One exception is when the conditional mean can be expressed as a *single index* function, that is, $h(\mathbf{x}_i;\boldsymbol{\beta}) = h(\mathbf{x}_i'\boldsymbol{\beta})$. In this case, the conditional mean is a function of the linear combination of the covariates. For this reason, the interpretation of *selection consistency* and *Oracle properties* as introduced in Chapter 1 require extra care. For a nonlinear model, selection consistency should be interpreted as the ability of a shrinkage estimator to identify elements in $\boldsymbol{\beta}$ with 0 value. The ability of a shrinkage estimator as a variable selector is only applicable when the conditional mean is a single index function.

When the functional form of h is unknown, shrinkage estimators are no longer feasible. In econometrics, nonparametric estimators are typically used in this case. In the Machine Learning literature, this problem is often solved by techniques such as CART, to be discussed in Section 2.3.

2.2.2 Regularization with Likelihood Function

The maximum likelihood estimator is arguably one of the most popular estimators in econometrics, especially for nonlinear models. Regardless of the nature of the response variables, continuous or discrete, as long as the distribution of the response variable is known, or can be approximated reasonably well, then it is possible to define the corresponding shrinkage estimators by using the log-likelihood as the objective function.

However, there are two additional complications that require some minor adjustments. First, the nature of the optimisation is now a maximisation problem i.e., maximising the (log-) likelihood, rather than a minimisation problem, i.e., the least squares. Second, the distribution of the response variable often involves additional parameters. For example, under the assumption of normality, the least squares estimator generally does not estimate the variance of the conditional distribution jointly with the parameter vector. In the maximum likelihood case, the variance parameter is jointly estimated with the parameter vector. This needs to be incorporated into the formulation of shrinkage estimators. Therefore, a shrinkage estimator can be defined as the solution to the following optimisation problem

$$\left(\hat{\boldsymbol{\beta}},\hat{\boldsymbol{\gamma}}\right) = \underset{\boldsymbol{\beta},\boldsymbol{\gamma}}{\arg\max}\ \log L(\boldsymbol{\beta},\boldsymbol{\gamma};\mathbf{y},\mathbf{X}) \tag{2.3}$$

$$\text{s.t.} \quad p\left(\boldsymbol{\beta};\boldsymbol{\alpha}\right) \leq c, \tag{2.4}$$

where $\boldsymbol{\gamma}$ denotes additional parameter vector required for the distribution, $L(\boldsymbol{\beta},\boldsymbol{\gamma};\mathbf{y},\mathbf{x})$ denotes the likelihood function and, as usual, $p(\boldsymbol{\beta};\boldsymbol{\alpha})$ denotes the regularizer (penalty) function. The optimisation problem above is often presented in its Lagrangian form for a given c. That is,

$$\left(\hat{\boldsymbol{\beta}},\hat{\boldsymbol{\gamma}}\right) = \underset{\boldsymbol{\beta},\boldsymbol{\gamma}}{\arg\max}\ \left[\log L\left(\boldsymbol{\beta},\boldsymbol{\gamma};\mathbf{y},\mathbf{X}\right) - \lambda p(\boldsymbol{\beta};\alpha)\right], \tag{2.5}$$

where $\lambda \geq 0$. Note that the optimisation problem above is a maximisation problem rather than a minimisation problem, as presented in Equations (1.5) and (1.6). However, since maximising a function $f(x)$ is the same as minimising $g(x) = -f(x)$, the formulation above can be adjusted, so that it is consistent with Equations (1.5) and (1.6). namely,

$$\left(\hat{\boldsymbol{\beta}},\hat{\boldsymbol{\gamma}}\right) = \underset{\boldsymbol{\beta},\boldsymbol{\gamma}}{\arg\min}\ -\log L(\boldsymbol{\beta},\boldsymbol{\gamma};\mathbf{y},\mathbf{X}) \tag{2.6}$$

$$\text{s.t.} \quad p\left(\boldsymbol{\beta};\boldsymbol{\alpha}\right) \leq c, \tag{2.7}$$

with the corresponding Lagrangian being

$$\left(\hat{\boldsymbol{\beta}},\boldsymbol{\gamma}\right) = \underset{\boldsymbol{\beta},\boldsymbol{\gamma}}{\arg\min}\ \left[-\log L\left(\boldsymbol{\beta},\boldsymbol{\gamma};\mathbf{y},\mathbf{X}\right) + \lambda p(\boldsymbol{\beta};\alpha)\right]. \tag{2.8}$$

Continuous Response Variable

Consider the model as defined in Equation (2.1), under the assumption that $u_i \sim NID(0, \sigma_u^2)$, then a shrinkage estimator under the likelihood objective is defined as

$$\left(\hat{\boldsymbol{\beta}}, \hat{\sigma}_u^2\right) = \underset{\boldsymbol{\beta}, \sigma_u^2}{\arg\max} \; -\frac{n}{2} \log \sigma_u^2 - \sum_{i=1}^{n} \frac{[y_i - h(\mathbf{x}_i; \boldsymbol{\beta})]^2}{2\sigma_u^2} - \lambda p(\boldsymbol{\beta}; \boldsymbol{\alpha}). \tag{2.9}$$

It is well known that the least squares estimator is algebraically equivalent to the maximum likelihood estimator for $\boldsymbol{\beta}$ under normality. This relation is slightly more complicated in the context of shrinkage estimators, and it concerns mainly with the tuning parameter λ. To see this, differentiate Equation (2.2) to obtain the first order condition for the nonlinear least squares shrinkage, this gives

$$\sum_{i=1}^{n} \left[y_i - h(\mathbf{x}_i; \hat{\boldsymbol{\beta}})\right] \left.\frac{\partial h}{\partial \boldsymbol{\beta}}\right|_{\boldsymbol{\beta}=\hat{\boldsymbol{\beta}}} = \frac{\lambda_{LS}}{2} \left.\frac{\partial p}{\partial \boldsymbol{\beta}}\right|_{\boldsymbol{\beta}=\hat{\boldsymbol{\beta}}} \tag{2.10}$$

where λ_{LS} denotes the tuning parameter λ associated with the least squares objective. Repeat the process above for Equation (2.9) to obtain the first order condition for the shrinkage estimators with likelihood objective, and this gives

$$\sum_{i=1}^{n} \frac{\left[y_i - h(\mathbf{x}_i; \hat{\boldsymbol{\beta}})\right]}{\hat{\sigma}_u^2} \left.\frac{\partial h}{\partial \boldsymbol{\beta}}\right|_{\boldsymbol{\beta}=\hat{\boldsymbol{\beta}}} = \lambda_{ML} \left.\frac{\partial p}{\partial \boldsymbol{\beta}}\right|_{\boldsymbol{\beta}=\hat{\boldsymbol{\beta}}} \tag{2.11}$$

$$\hat{\sigma}_u^2 = \sum_{i=1}^{n} \frac{\left[y_i - h(\mathbf{x}_i; \hat{\boldsymbol{\beta}})\right]^2}{n}, \tag{2.12}$$

where λ_{ML} denotes the tuning parameter λ for the shrinkage estimator with likelihood objective. Compare Equation (2.10) and Equation (2.12), it is straightforward to see that the two estimators will only be algebraically equivalent if their tuning parameters satisfy

$$\lambda_{LS} = 2\hat{\sigma}_u^2 \lambda_{ML}. \tag{2.13}$$

This relation provides a link between shrinkage estimators with the least squares and likelihood objective under normality. Note that this relation holds for both linear and nonlinear least squares, since $h(\boldsymbol{\xi}; \boldsymbol{\beta})$ can theoretically be a linear function in $\mathbf{x}_i$. Given the two estimators are algebraically equivalent under the appropriate choice of tuning parameters, they are likely to share the same properties conditional on the validity of Equation (2.13). As such, this chapter focuses on shrinkage estimators with a likelihood objective.

While theoretically they may share the same properties under Equation (2.13), the choice of the tuning parameter is often based on data-driven techniques in practice, such as cross validation, see 2.2.3 below. Therefore, it is unclear if Equation (2.13) necessarily holds in practice. More importantly, since $h(\mathbf{x}_i; \boldsymbol{\beta})$ can be a linear function, this means the shrinkage estimators based on least squares may very well be different from shrinkage estimators based on maximum likelihood in practice, even under the same regularizers.

Discrete Response Variables

Another application of shrinkage estimators with a likelihood objective is the modelling of discrete random variables. There are two common cases in econometrics. The first concerns with discrete random variables with finite number of outcomes. More specifically, consider a random variable, y_i, that takes on values from a finite, countable set $\mathcal{D} = \{d_1, \ldots, d_k\}$ and the probability of $y_i = d_j$ conditional on a set of covariates, $\mathbf{x}_i$, can be written as

$$\Pr\left(y_i = d_j\right) = g_j\left(\mathbf{x}_i'\boldsymbol{\beta}_0\right), \qquad j = 1, \ldots, k, \tag{2.14}$$

where $\mathbf{x}_i = \left(x_{1i}, \ldots, x_{pi}\right)'$ is a $p \times 1$ vector of covariates, $\boldsymbol{\beta}_0 = \left(\beta_{01}, \ldots, \beta_{0p}\right)'$ is the $p \times 1$ parameter vector and $g_j(\mathbf{x})$ denotes a twice differentiable function in $\mathbf{x}$.

Two simple examples of the above are the binary choice Logit and Probit models where $\mathcal{D} = \{0, 1\}$. In the case of Logit model,

$$\begin{aligned}\Pr(y_i = 1|\mathbf{x}_i) =& \frac{\exp(\mathbf{x}_i'\boldsymbol{\beta})}{1 + \exp(\mathbf{x}_i'\boldsymbol{\beta})}\\ \Pr(y_i = 0|\mathbf{x}_i) =& \frac{1}{1 + \exp(\mathbf{x}_i'\boldsymbol{\beta})},\end{aligned}$$

and in the case of Probit model,

$$\begin{aligned}\Pr(y_i = 1|\mathbf{x}_i) =& \Phi(\mathbf{x}_i'\boldsymbol{\beta})\\ \Pr(y_i = 0|\mathbf{x}_i) =& 1 - \Phi\left(\mathbf{x}_i'\boldsymbol{\beta}\right),\end{aligned}$$

where $\Phi(x)$ denotes the standard normal *cumulative distribution function* (CDF). Like in the case of linear models as discussed in Chapter 1, the coefficient vector $\boldsymbol{\beta}$ can contain many zeros. In this case, the associated covariates do not affect the probability of y_i. An objective of a shrinkage estimator is to identify which elements in $\boldsymbol{\beta}$ are zeros.

While there exists variable selection procedures, such as the algorithm proposed in Benjamini and Hochberg (1995), the resulting forward and backward stepwise procedures are known to have several drawbacks. In the case of forward stepwise, it is sensitive to the initial choice of variable, while backward stepwise is often not possible if the number of variable is greater than the number of observations. Shrinkage estimators can alleviate some of these drawbacks. Readers are referred to Hastie, Tibshirani and Friedman (2009) for more detailed discussion on stepwise procedures.

Similarly to the nonlinear least squares case, shrinkage estimator is a variable selector only when the log-likelihood function can be expressed as a function of a linear combination of the covariates. That is $L(\boldsymbol{\beta},\boldsymbol{\gamma};\mathbf{y},\mathbf{X}) = L(\mathbf{X}'\boldsymbol{\beta},\boldsymbol{\gamma};\mathbf{y})$. Popular econometric models that fall within this category include the Logit and Probit models. For Logit model, or logistic regression, let

$$f(\mathbf{x};\boldsymbol{\beta}) = \log\left(\frac{\pi(\mathbf{x})}{1-\pi(\mathbf{x})}\right) = \mathbf{x}'\boldsymbol{\beta}, \tag{2.15}$$

where $\pi(\mathbf{x}) = \Pr(y = 1|\mathbf{x})$ denotes the probability of a binary outcome conditional on $\mathbf{x}$. The log-likelihood for the binary model is well-known and given this, the shrinkage estimator for the Logit model can be written as

$$\hat{\boldsymbol{\beta}} = \arg\max_{\boldsymbol{\beta}} \left[\sum_{i=1}^{n} \{y_i f(\mathbf{x}_i;\boldsymbol{\beta}) - \log(1+\exp f(\mathbf{x}_i;\boldsymbol{\beta}))\} - \lambda p(\boldsymbol{\beta};\boldsymbol{\alpha})\right]. \tag{2.16}$$

The shrinkage estimator for the Probit model follows a similar formulation. The log-likelihood function for binary Probit model is

$$\sum_{i=1}^{n} y_i \log\Phi\left(\mathbf{x}_i'\boldsymbol{\beta}\right) + (1-y_i)\log\left[1-\Phi(\mathbf{x}_i'\boldsymbol{\beta})\right]$$

and therefore shrinkage estimator for the binary Probit model can be written as

$$\hat{\boldsymbol{\beta}} = \arg\max_{\boldsymbol{\beta}} \sum_{i=1}^{n} y_i \log\Phi\left(\mathbf{x}_i'\boldsymbol{\beta}\right) + (1-y_i)\log\left[1-\Phi(\mathbf{x}_i'\boldsymbol{\beta})\right] - \lambda p(\boldsymbol{\beta};\boldsymbol{\alpha}). \tag{2.17}$$

Again, the type of shrinkage estimator depends on the regularizer. In principle, all regularizers introduced in Chapter 1 can also be used in this case, but their theoretical properties and finite sample performance are often unknown. This is discussed further in Section 2.2.3.

The binary choice framework can be extended to incorporate multiple choices. Based on the above definitions, once the log-likelihood function for a model is known, it can be substituted into Equation (2.5) with a specific regularizer to obtain the desired shrinkage estimator.

The second case for a discrete random response variable is when $\mathcal{D}$ is an infinite and countable set, e.g., $\mathcal{D} = \mathbb{Z}^+ = \{0,1,2,\ldots\}$. This type of response variables are often appeared as *count data* in econometrics. One popular model for count data is the Poisson model. This model can be written as

$$\Pr(y|\mathbf{x}_i) = \frac{\exp(-\mu_i)\mu_i^y}{y!}$$

where $y!$ denotes the factorial of y and $\mu_i = \exp\left(\mathbf{x}_i'\boldsymbol{\beta}\right)$. The log-likelihood function now is

$$\sum_{i=1}^{n} -\exp(\mathbf{x}_i'\boldsymbol{\beta}) + y_i(\mathbf{x}_i'\boldsymbol{\beta}) + y_i!,$$

where the last term $y_i!$ is a constant and is often omitted for purposes of estimation as it does not affect the computation of the estimator. The shrinkage estimator with a particular regularizer for Poisson model is defined as

$$\hat{\boldsymbol{\beta}} = \arg\max_{\boldsymbol{\beta}} \sum_{i=1}^{n} -\exp(\mathbf{x}_i'\boldsymbol{\beta}) + y_i(\mathbf{x}_i'\boldsymbol{\beta}) - \lambda p(\boldsymbol{\beta};\boldsymbol{\alpha}). \tag{2.18}$$

2.2.3 Estimation, Tuning Parameter and Asymptotic Properties

This section discusses the estimation, determination of the tuning parameter, λ and the asymptotic properties of shrinkage estimators with the least squares and maximum likelihood objectives. The overall observation is that theoretical properties of shrinkage estimators with the least squares and likelihood objectives are sparse. In fact, the computation of the shrinkage estimators can itself be a challenging problem given the present knowledge in solving constrained optimisation. Asymptotic results are also rare and concepts, such as selection consistency, often require additional care in their interpretations for nonlinear models. This section covers some of these issues and identify the existing gaps for future research.

Estimation

To the best of our knowledge, the computation of shrinkage estimators with the least squares or likelihood objectives is still a challenging problem in general. This is due to the fact that constrained optimisation for nonlinear function is a difficult numerical problem.

There are specific cases, such as when the log-likelihood function is concave or when the nonlinear least squares function is convex, efficient algorithms do exist to solve the associated optimisation problems for certain regularizers, see for example Kwangmoo, Seung-Jean and Boyd (2007). Another exception is the family of models that falls under the *Generalised Linear Model* (GLM) framework. This includes Logit, Probit and Poisson models as special cases. For these models, efficient algorithms do exist for regularizers that are convex, such as the Bridge regularizers $p(\boldsymbol{\beta}) = ||\boldsymbol{\beta}||_{\gamma}^{\gamma}$ for $\gamma \geq 1$ as well as Elastic Net and SCAD. These estimators are readily available in open source languages such as R[1], Python[2] and Julia[3].

[1] https://cran.r-project.org/web/packages/glmnet/index.html

[2] https://scikit-learn.org/stable/index.html

[3] https://www.juliapackages.com/p/glmnet

Tuning Parameter and Cross-Validation

Like in the linear case, the tuning parameter, λ, plays an important role in terms of the performance of the shrinkage estimators for nonlinear model in practice. In the case when the response variable is discrete, the cross validation as introduced in Chapter 1 needs to be modified for shrinkage estimator with likelihood objective. The main reason for this modification is that the calculation of 'residuals', which the least squares seek to minimise, is not obvious in the context of nonlinear models. This is particularly true when the response variable is discrete. A different measurement of 'errors' in the form of *deviance* is required for purposes of identifying the optimal tuning parameter. One such modification can be found below:

1. For each λ value in a sequence of values $(\lambda_1 > \lambda_2 > \cdots > \lambda_T)$, estimate the model for each fold, leaving one fold out at a time. This produces a vector of estimates for each λ value: $\hat{\boldsymbol{\beta}}_1 \ldots \hat{\boldsymbol{\beta}}_T$.
2. For each set of estimates, calculate the deviance based on the left out fold/testing dataset. The deviance is defined as

$$e_t^k = -\frac{2}{n_k} \sum_i \log p(y_i | \boldsymbol{x}_i' \hat{\boldsymbol{\beta}}_T^k), \tag{2.19}$$

   where n_k is the number of observations in the fold k. The quantity computed for each fold and across all λ values.
3. Compute the average and standard deviation of the error/deviance resulting from K folds.

$$\bar{e}_t = \frac{1}{K} \sum_k e_t^k. \tag{2.20}$$

   The above average represents the average error/deviance for each λ.

$$sd(\bar{e}_t) = \sqrt{\frac{1}{K-1} \sum_k (e_t^k - \bar{e}_t)^2}. \tag{2.21}$$

   This is the standard deviation of average error associated with each λ.
4. Choose the *best* λ_t value based on the measures provided.

Given the above procedure, it is quite often useful to plot the average error rates for each λ value. Figure 2.1 plots the average misclassification error corresponding to a sequence of log λ_t values.[4] Based on this, we can identify the value of log λ that minimises the average error.[5] In order to be conservative (apply a slightly higher penalty), a log λ value of one standard deviation higher from the minimum value can also be selected. The vertical dotted lines in Figure 2.1 show both these values.

---

[4] This figure is reproduced from Friedman, Hastie and Tibshirani (2010).

[5] The process discussed here is known to be unstable for moderate sample sizes. In order to ensure robustness, the K-fold process can be repeated $n-1$ times and the λ can be obtained as the average of these repeated k-fold cross validations.

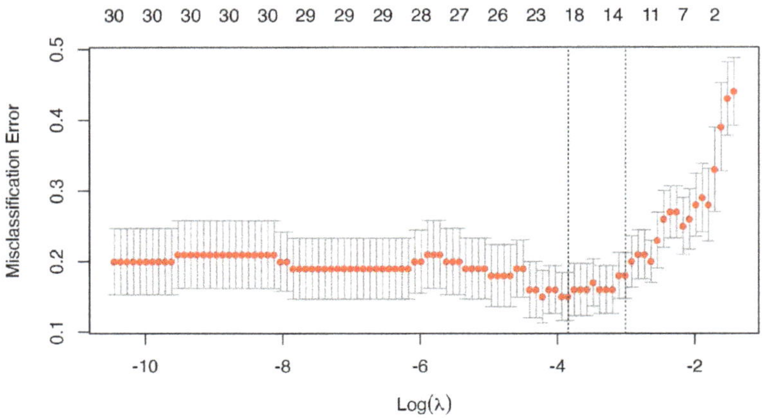

Fig. 2.1: LASSO CV Plot

Asymptotic Properties and Statistical Inference

The development of asymptotic properties, such as the Oracle properties as defined in Chapter 1, for shrinkage estimators with nonlinear least squares or likelihood objectives is still in its infancy. To the best of the authors' knowledge, the most general result to-date is provided in, Fan, Xue and Zou (2014) where they obtained Oracle properties for shrinkage estimators with convex objectives and concave regularizers. This covers the case of LASSO, Ridge, SCAD and Adaptive LASSO with likelihood objectives for the Logit, Probit and Poisson models.

However, as discussed in Chapter 1, the practical usefulness of the Oracle properties has been scrutinised by Leeb and Pötscher (2005) and Leeb and Pötscher (2008). The use of point-wise convergence in developing these results means the number of observations required for the Oracle properties to be relevant depends on the true parameters, which are unknown in practice. As such, it is unclear if valid inference can be obtained purely based on the Oracle properties.

For linear model, Chapter 1 shows that it is possible to obtain valid inference for *Partially Penalised Estimator*, at least for the subset of the parameter vector that is not subject to regularization. Shi, Song, Chen and Li (2019) show that the same idea can also apply to shrinkage estimators with likelihood objective for Generalised Linear Model with canonical link. Specifically, their results apply to response variables with probability distribution function of the form

$$f(y_i; \mathbf{x}_i) = \exp\left[\frac{y_i \mathbf{x}_i' \boldsymbol{\beta} - \psi(\mathbf{x}_i' \boldsymbol{\beta})}{\phi_0}\right] \delta(y_i) \tag{2.22}$$

for some smooth functions ψ and δ. This specification includes the Logit and Poisson models as special cases, but not the Probit model.

Given Equation (2.22), the corresponding log-likelihood function can be derived in the usual manner. Let $\boldsymbol{\beta} = (\boldsymbol{\beta}_1', \boldsymbol{\beta}_2')'$ where $\boldsymbol{\beta}_1$ and $\boldsymbol{\beta}_2$ are $p_1 \times 1$ and $p_2 \times 1$ sub-vectors with $p = p_1 + p_2$. Assume one wishes to test the hypothesis that $\mathbf{B}\boldsymbol{\beta}_1 = \mathbf{C}$, where $\mathbf{B}$ and $\mathbf{C}$ are a $r \times p_1$ matrix and a $p_1 \times 1$ vector, respectively. Consider the following Partially Penalised Estimator

$$\left(\hat{\boldsymbol{\beta}}_{PR}, \hat{\boldsymbol{\gamma}}_{PR}\right) = \underset{\boldsymbol{\beta}, \boldsymbol{\gamma}}{\arg\max} \log L(\mathbf{X}\boldsymbol{\beta}, \boldsymbol{\gamma}; \mathbf{y}) - \lambda p(\boldsymbol{\beta}_2; \boldsymbol{\alpha}) \tag{2.23}$$

and the *restricted* Partially Penalised Estimator, where $\boldsymbol{\beta}_1$ is assumed to satisfy the restriction $\mathbf{B}\boldsymbol{\beta}_1 = \mathbf{C}$,

$$\left(\hat{\boldsymbol{\beta}}_{RPR}, \hat{\boldsymbol{\gamma}}_{RPR}\right) = \underset{\boldsymbol{\beta}, \boldsymbol{\gamma}}{\arg\max} \log L(\mathbf{X}\boldsymbol{\beta}, \boldsymbol{\gamma}; \mathbf{y}) - \lambda p(\boldsymbol{\beta}_2; \boldsymbol{\alpha}) \tag{2.24}$$

$$\text{s.t.} \qquad \mathbf{B}\boldsymbol{\beta}_1 = \mathbf{C}, \tag{2.25}$$

where $p(\boldsymbol{\beta}; \boldsymbol{\alpha})$ is a *folded concave* regularizer, which includes Bridge with $\gamma > 1$ and SCAD as special cases, then the likelihood ratio test statistics

$$2\left[\log L(\hat{\boldsymbol{\beta}}_{PR}, \hat{\boldsymbol{\gamma}}_{PR}) - \log L(\hat{\boldsymbol{\beta}}_{RPR}, \hat{\boldsymbol{\gamma}}_{RPR})\right] \overset{d}{\sim} \chi^2(r). \tag{2.26}$$

The importance of the formulation above is that $\boldsymbol{\beta}_1$, the subset of parameters that is subject to the restriction/hypothesis, $\mathbf{B}\boldsymbol{\beta}_1 = \mathbf{C}$, is not part of the regularization. This is similar to the linear case, where the subset of the parameters that are to be tested are not part of the regularization. In this case, Shi et al. (2019) show that log-ratio test as defined in Equation (2.26) has an asymptotic χ^2 distribution.

The result as stated in Equation (2.26), however, only applies when the response variable has the distribution function in the form of Equation (2.22) with a regularizer that belongs to the *folded concave* family. While SCAD belongs to this family, other popular regularizers, such as LASSO, adaptive LASSO and Bridge with $\gamma \leq 1$, are not part of this family. Thus, from the perspective of econometrics, the result above is quite limited as it is relevant only to Logit and Poisson models with a SCAD regularizer. It does not cover the Probit model or other popular models and regularizers that are relevant in econometric analysis. Thus, hypothesis testing using shrinkage estimators for the cases that are relevant in econometrics is still an open problem.

However, combining the approach as proposed in Chernozhukov, Hansen and Spindler (2015) as discussed in Chapter 1 with those considered in Shi et al. (2019) may prove to be useful in progressing this line of research. For example, it is possible to derive the asymptotic distribution for the Partially Penalised Estimator for the models considered in Shi et al. (2019) using the *Immunization Condition* approach in, Chernozhukov et al. (2015) as shown in Propositions 2.1 and 2.2.

Proposition 2.1 *Let y_i be a random variable with the conditional distribution as defined in Equation (2.22) and consider the following Partially Penalised Estimator with likelihood objective,*

$$\hat{\boldsymbol{\beta}} = \arg\max_{\boldsymbol{\beta}} S(\boldsymbol{\beta}) = l(\boldsymbol{\beta}) - \lambda p(\boldsymbol{\beta}_2) \tag{2.27}$$

$$l(\boldsymbol{\beta}) = \sum_{i=1}^{n} y_i \mathbf{x}_i'\boldsymbol{\beta} - \psi(\mathbf{x}_i'\boldsymbol{\beta}), \tag{2.28}$$

where $\boldsymbol{\beta} = (\boldsymbol{\beta}_1'\boldsymbol{\beta}_2')$ and $\mathbf{x}_i = (\mathbf{x}_{1i}, \mathbf{x}_{2i})$, such that $\mathbf{x}_i'\boldsymbol{\beta} = \mathbf{x}_{1i}'\boldsymbol{\beta}_1 + \mathbf{x}_{2i}'\boldsymbol{\beta}_2$, and $p(\boldsymbol{\beta})$ denotes the regularizer, such that $\dfrac{\partial^2 p}{\partial \boldsymbol{\beta}_2, \partial \boldsymbol{\beta}_2'}$ exists.

Under the assumptions that $\mathbf{x}_i$ is stationary with finite mean and variance and there exists a well-defined $\boldsymbol{\mu}$ for all $\boldsymbol{\beta}$, such that

$$\boldsymbol{\mu} = -\frac{\partial \psi}{\partial z_i} \mathbf{x}_{1i}\mathbf{x}_{2i}' \left[\frac{\partial \psi}{\partial z_i} \mathbf{x}_{2i}\mathbf{x}_{2i}' + \lambda \frac{\partial^2 p}{\partial \boldsymbol{\beta}_2 \partial \boldsymbol{\beta}_2'} \right]^{-1}, \tag{2.29}$$

where $z_i = \mathbf{x}_i'\boldsymbol{\beta}$, then

$$\sqrt{n}\left(\hat{\boldsymbol{\beta}}_1 - \boldsymbol{\beta}_0\right) \xrightarrow{d} N(0, \boldsymbol{\Gamma}^{-1}\boldsymbol{\Omega}\boldsymbol{\Gamma}^{-1}), \tag{2.30}$$

where

$$\boldsymbol{\Gamma} = -\frac{\partial^2 \psi}{\partial z_i^2} \mathbf{x}_{1i}\mathbf{x}_{1i}' + \left(\frac{\partial^2 \psi}{\partial z_i^2}\right)^2 \left[\mathbf{x}_{1i}\mathbf{x}_{2i}' \left(\frac{\partial^2 \psi}{\partial z_i^2} \mathbf{x}_{2i}\mathbf{x}_{2i}' - \lambda \frac{\partial^2 p}{\partial \boldsymbol{\beta}_2 \partial \boldsymbol{\beta}_2'} \right)^{-1} \mathbf{x}_{2i}\mathbf{x}_{1i}' \right]$$

$$\boldsymbol{\Omega} = Var\left[\sqrt{n} S(\hat{\boldsymbol{\beta}})\right].$$

Proof See Appendix. □

The result above is quite general and covers Logit model when $\psi(\mathbf{x}_i'\boldsymbol{\beta}) = \log\left[1 + \exp(\mathbf{x}_i'\boldsymbol{\beta})\right]$ and Poisson model when $\psi(\mathbf{x}_i'\boldsymbol{\beta}) = \exp(\mathbf{x}_i'\boldsymbol{\beta})$ with various regularizers including Bridge and SCAD. But Proposition 2.1 does not cover Probit model. The same approach can be used to derive the asymptotic distribution of its shrinkage estimators with likelihood objective as shown in Proposition 2.2

Proposition 2.2 *Consider the following Partially Penalised Estimators for a Probit model*

$$\hat{\boldsymbol{\beta}} = \arg\max_{\boldsymbol{\beta}} S(\boldsymbol{\beta}) = l(\boldsymbol{\beta}) - \lambda p(\boldsymbol{\beta}_2) \tag{2.31}$$

$$l(\boldsymbol{\beta}) = \sum_{i=1}^{n} y_i \log \Phi(\mathbf{x}_i'\boldsymbol{\beta}) - (1 - y_i) \log\left[1 - \Phi(\mathbf{x}_i'\boldsymbol{\beta})\right], \tag{2.32}$$

where $\boldsymbol{\beta} = (\boldsymbol{\beta}_1'\boldsymbol{\beta}_2')$ *and* $\mathbf{x}_i = (\mathbf{x}_{1i}, \mathbf{x}_{2i})$, *such that* $\mathbf{x}_i'\boldsymbol{\beta} = \mathbf{x}_{1i}'\boldsymbol{\beta}_1 + \mathbf{x}_{2i}'\boldsymbol{\beta}_2$ *and* $p(\boldsymbol{\beta})$ *denotes the regularizer, such that* $\frac{\partial^2 p}{\partial \boldsymbol{\beta}_2, \partial \boldsymbol{\beta}_2'}$ *exists.*

Under the assumptions that $\mathbf{x}_i$ *is stationary with finite mean and variance and there exists a well-defined* $\boldsymbol{\mu}$ *for all* $\boldsymbol{\beta}$, *such that*

$$\boldsymbol{\mu} = -\boldsymbol{\Lambda}(\boldsymbol{\beta})\boldsymbol{\Theta}(\boldsymbol{\beta}), \tag{2.33}$$

where

$$\begin{aligned}
\boldsymbol{\Lambda}(\boldsymbol{\beta}) &= \sum_{i=1}^{n} \{-y_i\,[z_i + \phi(z_i)]\,\eta(z_i) + [\,(1-y_i)\,[\phi(z_i) - z_i]\,\xi(z_i)\}\mathbf{x}_{1i}\mathbf{x}_{2i}', \\
\boldsymbol{\Theta}(\boldsymbol{\beta}) &= \sum_{i=1}^{n} \{-y_i\,[z_i + \phi(z_i)]\,\eta(z_i) + [\,(1-y_i)\,[\phi(z_i) - z_i]\,\xi(z_i)\}\mathbf{x}_i\mathbf{x}_{2i}' - \lambda \frac{\partial^2 p}{\partial \boldsymbol{\beta}_2 \partial \boldsymbol{\beta}_2'}, \\
\eta(z_i) &= \frac{\phi(z_i)}{\Phi(z_i)}, \\
\xi(z_i) &= \frac{\phi(z_i)}{1 - \Phi(z_i)},
\end{aligned}$$

where $z_i = \mathbf{x}_i'\boldsymbol{\beta}$, $\phi(x)$ *and* $\Phi(x)$ *denote the probability density and cumulative distribution functions for a standard normal distribution respectively. Then*

$$\sqrt{n}\left(\hat{\boldsymbol{\beta}}_1 - \boldsymbol{\beta}_0\right) \xrightarrow{d} N(0, \boldsymbol{\Gamma}^{-1}\boldsymbol{\Omega}\boldsymbol{\Gamma}^{-1}), \tag{2.34}$$

where

$$\begin{aligned}
\boldsymbol{\Gamma} &= \mathbf{A}_1 + \mathbf{A}_2\mathbf{B}^{-1}\mathbf{A}_2', \\
\mathbf{A}_1 &= \sum_{i=1}^{n} -y_i\,[z_i + \phi(z_i)]\,\eta(z_i)\mathbf{x}_{1i}\mathbf{x}_{1i}' + (1-y_i)\,[\phi(z_i) - z_i]\,\xi(z_i)\mathbf{x}_{1i}\mathbf{x}_{1i}', \\
\mathbf{A}_2 &= \sum_{i=1}^{n} -y_i\,[z_i + \phi(z_i)]\,\eta(z_i)\mathbf{x}_{1i}\mathbf{x}_{1i}' + (1-y_i)\,[\phi(z_i) - z_i]\,\xi(z_i)\mathbf{x}_{1i}\mathbf{x}_{2i}', \\
\mathbf{B} &= \sum_{i=1}^{n} -y_i\,[z_i + \phi(z_i)]\,\eta(z_i)\mathbf{x}_{1i}\mathbf{x}_{2i}' + (1-y_i)\,[\phi(z_i) - z_i]\,\xi(z_i)\mathbf{x}_{1i}\mathbf{x}_{2i}' - \lambda \frac{\partial^2 p}{\partial \boldsymbol{\beta}_2 \partial \boldsymbol{\beta}_2'}, \\
\boldsymbol{\Omega} &= Var\left[\sqrt{n}S(\hat{\boldsymbol{\beta}})\right].
\end{aligned}$$

Proof See Appendix. □

Given the results in Propositions 2.1 and 2.2, it is now possible to obtain the asymptotic distribution for the Partially Penalised Estimators for the Logit, Probit and Poisson models under Bridge and SCAD. This should facilitate valid inferences for the subset of parameters that are not part of the regularizations for these models. These results should also lead to similar results as those derived in Shi et al. (2019), which would further facilitate inferences on parameter restrictions using conventional techniques such as the log-ratio tests. The formal proof of these is left for further research.

2.2.4 Monte Carlo Experiments – Binary Model with shrinkage

This section contains a small set of Monte Carlo experiments examining the finite sample performance of shrinkage estimators for a binary model[6]. The aim of this experiment is to assess both selection and estimation consistency. The experiments are carried out under two scenarios; with and without the presence of correlation between the covariates. The shrinkage estimators chosen for this exercise are the LASSO and Elastic net, and their performance is compared to the standard maximum likelihood estimator. The details of the data generation process are outlined below.

1. Generate an n x 10 matrix of normally distributed covariates with means and standard deviations, provided in Table 2.1.
2. In the first scenario, there is no correlation between the covariates. This scenario serves as benchmark for the second scenario where correlation is present in the covariates. The correlation coefficient is calculated as

$$\rho(X_i, X_j) = (0.5)^{|i-j|}, \tag{2.35}$$

   where X_i is the ith covariate.
3. The true model parameters $\boldsymbol{\beta}_0$ are provided in Table 2.2. A range of values are selected for $\boldsymbol{\beta}_0$ and given the relative magnitudes of $\boldsymbol{\beta}_0$, variables X01 and X02 can be considered strongly relevant variables; X03 a weakly irrelevant variable and remaining variables are considered to be irrelevant.
4. The response variable is generated as follows:

   Step 1. Generate π (as per a Logit model specification) as:

$$\pi_i = \frac{\exp(\mathbf{x}_i'\boldsymbol{\beta}_0)}{1+\exp(\mathbf{x}_i'\boldsymbol{\beta}_0)}.$$

   Step 2. Draw $u_i \sim U[0,1]$, where $U[0,1]$ denotes a uniform random distribution with a 0 to 1 domain.
   Step 3. If $u_i < \pi_i$ set, $y_i = 1$ else set $y_i = 0$.

The results from 4000 repetitions on the selection consistency for the three estimators; LASSO, Elastic net and the standard maximum likelihood estimator of a Logit model are provided. The analysis is carried out over five sample sizes of 100, 250, 500, 1000 and 2000. For the LASSO and Elastic net, selection consistency is defined having a nonzero value for a coefficient. For the maximum likelihood

[6] The code used to carry out this Monte Carlo experiment is available in the electronic supplementary materials.

Table 2.1: Means and standard deviations of covariates

Variables	X01	X02	X03	X04	X05	X06	X07	X08	X09	X10
Means	0	0	0	0	0	0	0	0	0	0
Std Dev	2	2.5	1.5	2	1.25	3	5	2.75	4	1

Table 2.2: True values of the parameters

	β_1	β_2	β_3	β_4	β_5	β_6	β_7	β_8	β_9	β_10
Coefficient Values	2	-2	0.05	0	0	0	0	0	0	0

case, selection is defined by a t-value with an absolute value more than 1.96.[7] The selection consistency results across both scenarios (with and without correlation) are presented as graphs for variables X01, X03 and X05 as they represent a strongly relevant, weakly relevant and irrelevant variable respectively.

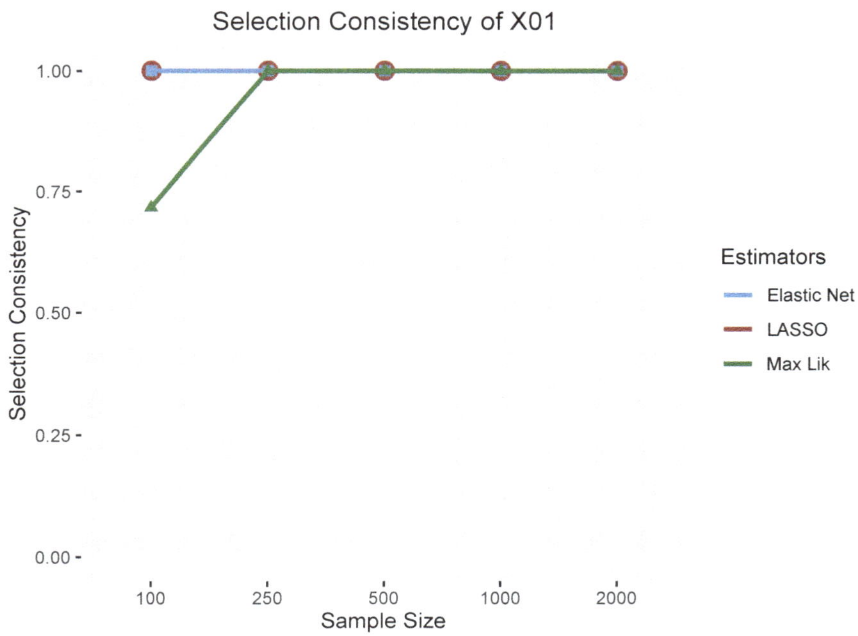

Fig. 2.2: X01 Selection Consistency

[7] Results where convergence was not achieved were excluded.

When no correlation is present, the strongly relevant variable X01 is selected consistently by all estimators for sample sizes equal to and above 250 (Figure 2.2). Both Elastic net and LASSO selected X01 across all sample sizes. However, for a sample size of 100, the maximum likelihood estimator had a 75% chance of selecting X01. Given a strongly relevant variable, this demonstrates that shrinkage estimators outperform the maximum likelihood estimator at smaller sample sizes. This result remains relatively unchanged when correlation is introduced (Figure 2.3). This implies that for a strongly relevant variable, the correlation does not affect selection consistency. For situations where a strongly relevant variable exits and is correlated with other variables, shrinkage estimators can be used with confidence.

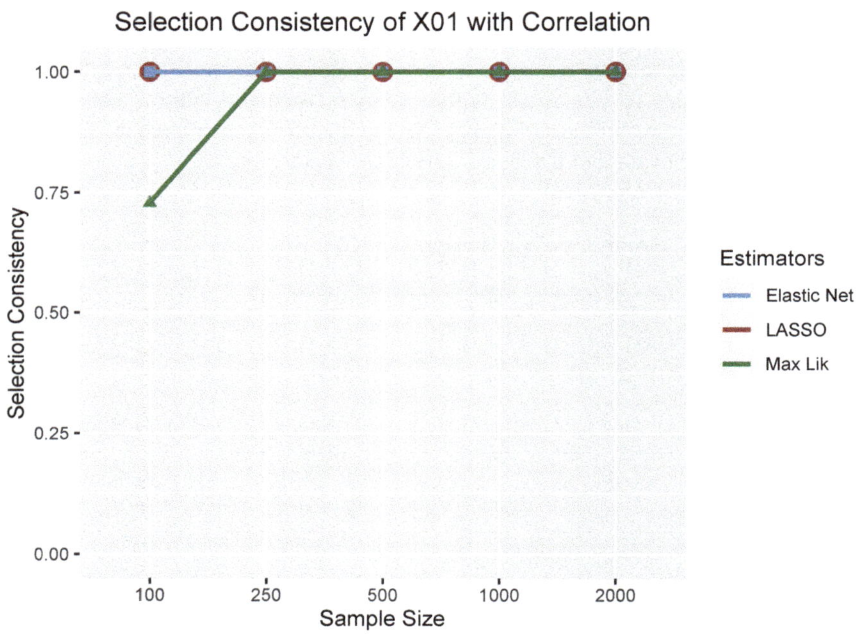

Fig. 2.3: X01 Selection Consistency with Correlation

When detecting a weakly relevant variable, X03, with no correlations present (Figure 2.4), both shrinkage estimators outperform the maximum likelihood estimator in all sample sizes. The Elastic net estimator is the best performer by a large margin. In fact, its selection consistency improves significantly as the sample size increases, moving from 50% to 75%. Both the LASSO and the maximum likelihood estimator have a selection consistency of less than 25% across all sample sizes. The LASSO estimator's selection consistency remains relatively constant across all sample sizes, whereas the maximum likelihood estimator does demonstrate a slight improvement in selection consistency as the sample size increases.

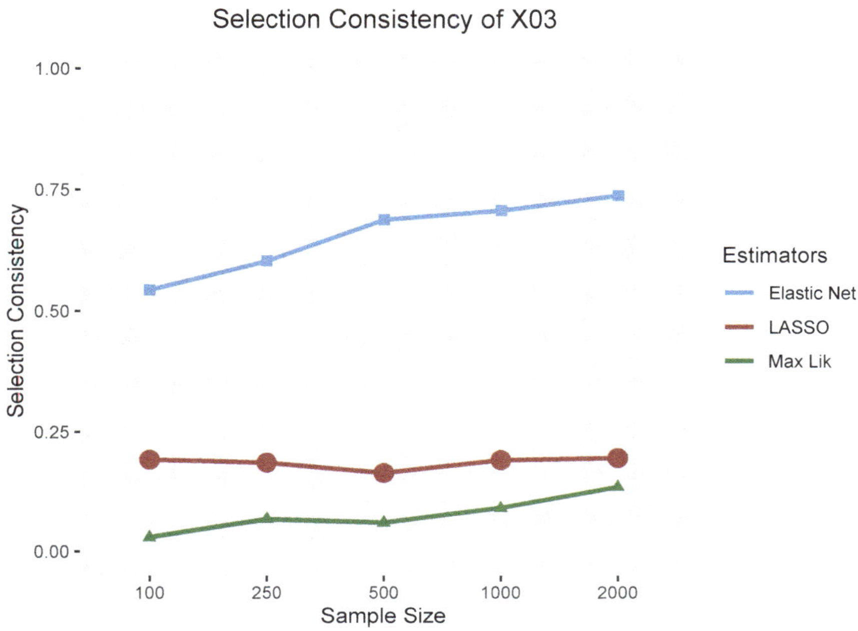

Fig. 2.4: X03 Selection Consistency

The results for selection consistency change significantly once correlation is introduced (Figure 2.5). The overall selection consistency decreases for both the shrinkage estimators and the maximum likelihood estimator. Unlike the first scenario, the Elastic net's selection consistency remains relatively unchanged across all sample sizes (approximately 50%). However, it still remains the best performer with regard to selection consistency. Interestingly, the LASSO's selection consistency decreases marginally as the sample size increases. Hence, for large sample sizes, the selection consistency of the maximum likelihood estimator is similar to or slightly better than LASSO. These results indicate that for weakly relevant variables, the presence of correlation does affect the selection consistency of shrinkage estimators. In such circumstances, using Elastic net may be a good choice.

When detecting the irrelevant variable, X05, with no correlations present (Figure 2.6), the maximum likelihood estimator outperforms the two shrinkage estimators. However, as the sample size increases, LASSO's selection consistency almost matches the maximum likelihood estimator. Elastic net performs the worst in this case, selecting the completely irrelevant variable more than 50% of the time for small sample sizes and almost 70% of the time for large sample sizes. The results remain largely unchanged when correlation is introduced (Figure 2.7). This implies that correlation does not affect the selection consistency of irrelevant variables. In such

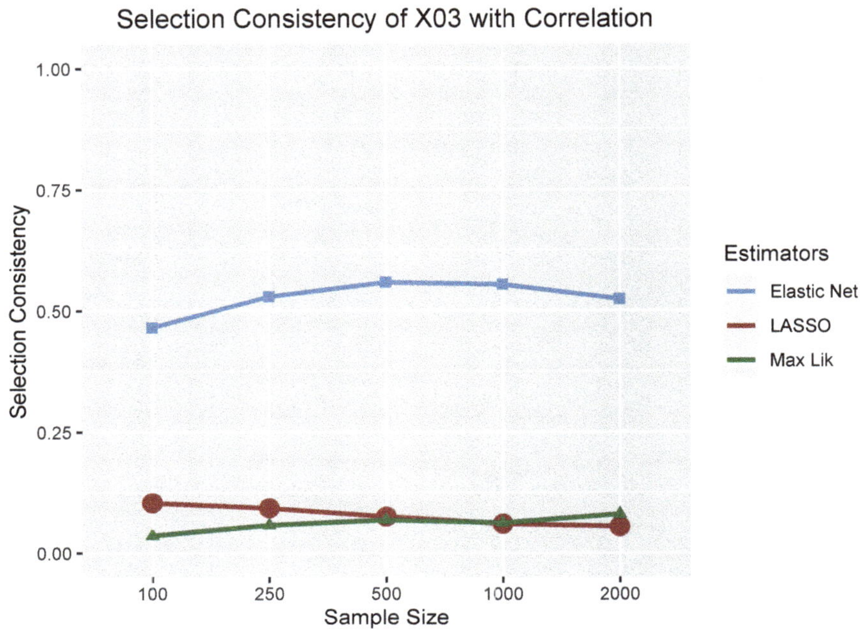

Fig. 2.5: X03 Selection Consistency with Correlation

situations, it may be best not to use shrinkage estimators. Although for large samples, the difference between the maximum likelihood estimator and LASSO is small.

The overall results on selection consistency indicate that correlation has an impact on weakly relevant variables. The selection consistency results for strongly relevant and irrelevant variables are not affected by the presence of correlation. In addition, shrinkage estimators have a higher selection consistency for both strongly relevant and weakly relevant variables. But, perform poorly for irrelevant variables. Given these preliminary results, it may be best to be careful and to carry out further experiments such as, for example, examining selection consistency for Partial Penalised Estimations.

With regard to estimation, when no correlation is present, the maximum likelihood estimator for β_1 corresponding to the strongly relevant variable X01, has a mean approximately equal to the true value (2). However, both the shrinkage estimators produce estimates with a significant negative bias (Figure 2.8). This is to be expected given the role of the penalty functions. In practise, shrinkage estimators are used for selection purposes and subsequently the selected variables are used in the final model i.e., post model estimation. The results for the correlated case do not differ significantly.

When no correlation is present, the maximum likelihood estimator for, β_5 corresponding to irrelevant variable X05, has a mean approximately equal to the true value

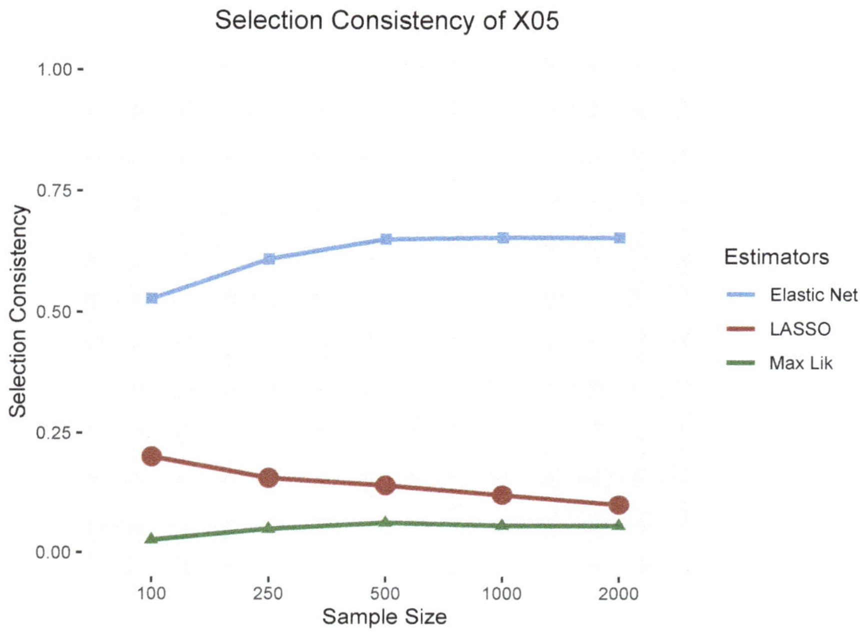

Fig. 2.6: X05 Selection Consistency

(0). The means of both shrinkage estimators are close to their true values. However, the main difference is in the variance between the maximum likelihood estimator and the shrinkage estimators. Figure 2.9 shows that the shrinkage estimators have much lower variance compared to the maximum likelihood estimator. As in the previous case, the results for the correlated case are identical.

The overall results for estimation imply that shrinkage estimators do have negative bias and as such should only be used for selection purposes. In addition, the maximum likelihood estimator has a lower variance compared to shrinkage estimators.

2.2.5 Applications to Econometrics

A general conclusion from the discussion thus far is that while applications of shrinkage estimators in nonlinear econometric models is promising, some important ingredients are still missing in order for these estimators to be widely used. The answers to the questions below may help to facilitate the use of shrinkage estimators for nonlinear models and thus, could be the focus of future research.

1. Asymptotic and finite sample properties, such as asymptotic distributions and Oracle Properties, of shrinkage estimators for nonlinear models with likelihood

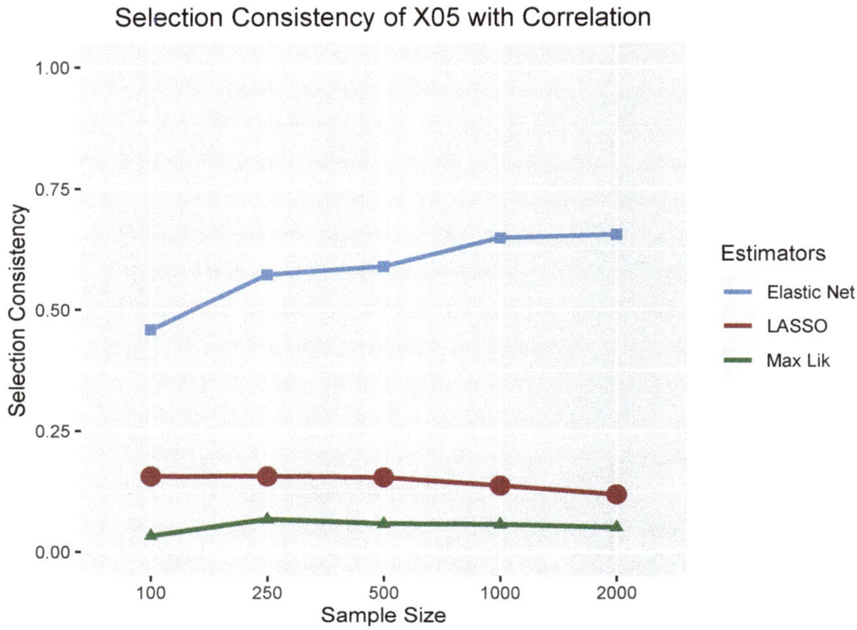

Fig. 2.7: X05 Selection Consistency with Correlation

objectives. The contributions thus far do not cover a sufficient set of models used in econometric analysis.

2. Are the asymptotic properties subject to the criticism of Leeb and Pötscher (2005) and Leeb and Pötscher (2008) and thus render Oracle properties irrelevant from the practical viewpoint?
3. What are the properties of Partially Penalised Estimators as introduced in Chapter 1 in the case of nonlinear models in general? In the event that the criticism of Leeb and Pötscher (2005) and Leeb and Pötscher (2008) hold for nonlinear models, can Partially Penalised Estimators help to facilitate valid inference? While the results presented in this chapter look promising, more research in this direction is required.
4. In the case when $p > N$, shrinkage estimators have good performance in terms of selection consistency in the linear case, but their finite sample performance for nonlinear models is not entirely clear. This is crucial since this is one area where shrinkage estimators offer a feasible solution over conventional estimators.

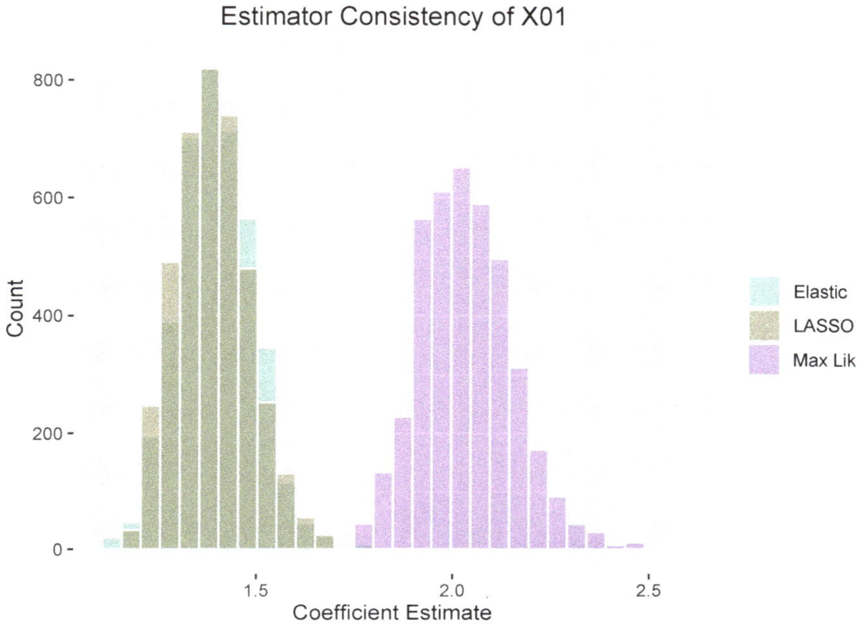

Fig. 2.8: Estimation of strongly relevant variable, X01

2.3 Overview of Tree-based Methods - Classification Trees and Random Forest

In this section, tree-based methods for classification and regression are covered. In contrast to the previous section, which covered parametric methods, tree based methods are nonparametric. As such, they assume very little about the data generating process, which has led to their use in a variety of different applications.

Tree based methods are similar to decision trees, where the hierarchical structure allows users to make a series of sequential decisions to arrive at a conclusion. Given this structure, the decisions can be thought of as inputs x, and the conclusion as output, y. Interestingly, one of the earliest papers on the automatic construction of decision trees – Morgan and Sonquist (1963) – was coauthored by an economist. However, it was only since the publication of Breiman, Friedman, Olshen and Stone (1984) that tree based methods gained prominence. Based on this seminal work, two types of trees – Classification and Regression Trees (CART) – have become popular. Both types of trees have a similar framework/structure, and the main difference is the nature of the output variable, y. In the case of classification trees, y is categorical or nominal, whereas for regression trees, y is numerical. In the ML literature, this distinction is referred to as qualitative *versus* quantitative, see James, Witten, Hastie and Tibshirani (2013) for further discussion.

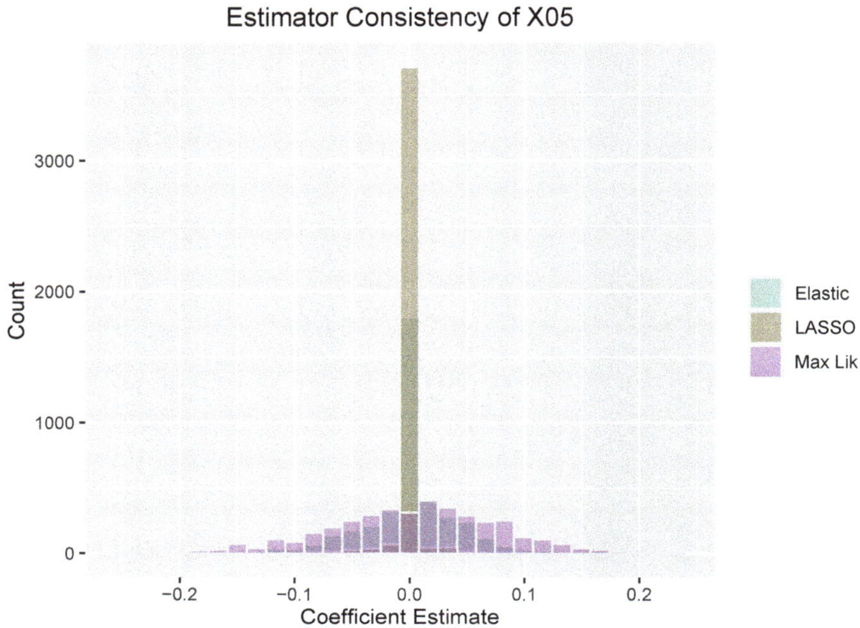

Fig. 2.9: Estimation of relevant variable, XO5

The popularity of trees is primarily due to their decision-centric framework, i.e., by generating a set of rules based inputs x, trees are able to predict an outcome y. The resulting structure of a tree is easy to understand, interpret and visualise. The most significant inputs x are readily identifiable from a tree. This is useful with regard to data exploration, and requires no or limited statistical knowledge to interpret the results. In most cases, the data requires relatively less preprocessing to generate a tree compared to other regression and classification methods. Given these advantages, trees are used to model numerical and categorical data. Next, an outline of the tree building process is provided.

The process of building a tree involves recursively partitioning the regressor space (X_i) to predict y_i. The partitioning yields J distinct and non-overlapping regions $R_1, R_2, \ldots R_J$. For all observations that fall into a region R_j, the same prediction applies to all of them. In the case of regression trees (numerical y), this is simply the mean of y for that region. For classification trees, the prediction is the most occurring case/category in that region. A natural question is 'how is the partitioning/splitting of the variable carried out?'. The approach used is known as *recursive binary splitting*. The tree begins by selecting a variable X_j and a corresponding threshold value s such that splitting the regressor space into regions $\{X|X_j < s\}$ and $\{X|X_j \geq s\}$ leads to the largest possible reduction in error, where 'error' is a measure of how well the predictions match actual values. This process of splitting into regions by considering all possible thresholds across all variables which minimises the error continues until a stopping criterion is reached; for example, a minimum number of observations in a particular region. In the case of regression trees (numerical y), the residual sum of squares (RSS) is used, defined as

$$\text{RSS} = \sum_{j=1}^{J} \sum_{i \in R_j} (y_i - \hat{y}_{R_j})^2, \tag{2.36}$$

where $\hat{y}_{R_j}$ is the mean of y for region R_j. Minimising RSS implies that all squared deviations for a given region are minimised across all regions. In a classification setting, RSS cannot be used as a criterion for splitting variables, since we cannot use the mean values of y, as it is categorical. There are however a number of alternative measures, each with its own strengths and limitations. The classification error rate (CER) is defined as the fraction of observations that do not belong to the most common class/category,

$$CER = 1 - \max_k(\hat{p}_{mk}), \tag{2.37}$$

where $\hat{p}_{mk}$ denotes the proportion of observations in the mth region that are from the kth class. In addition to CER, there are two other measures: the Gini index (also known as Gini impurity) and Cross Entropy. The Gini index is defined as

$$G = \sum_{k=1}^{K} \hat{p}_{mk}(1 - \hat{p}_{mk}). \tag{2.38}$$

Small values of the Gini index indicate that the region predominantly contains observations from a single category.[8] The cross-entropy measure is defined as

$$CEnt = -\sum_{k=1}^{K} \hat{p}_{mk} \log \hat{p}_{mk}. \tag{2.39}$$

This is based on Shannon's Entropy (Shannon (1948))) which forms the basis of Information theory and is closely related to the Gini Index.

Based on the above, the tree building process is as follows:

1. Start with the complete data, consider splitting variables X_j at threshold point s in order to obtain the first two regions:

$$R_1(j,s) = \{X|X_j \leq s\} \text{ and } R_2(j,s) = \{X|X_j > s\}. \tag{2.40}$$

2. Evaluate the following in order to determine the splitting variable X_j and threshold point s

[8] Even though both the Gini index in classification and that used in inequality analyses, are measures of variation, they are not equivalent.

$$\min_{j,s}\left[\sum_{x_i\in R_1(j,s)}(y_i-\bar{y}_{R_1})^2+\sum_{x_i\in R_2(j,s)}(y_i-\bar{y}_{R_2})^2\right], \tag{2.41}$$

where $\bar{y}_{R_m}$ is the average of y in region m. Note that this formulation relates to a regression tree as it is based on RSS; for a classification tree, RSS is replaced with either the Gini index or Cross-Entropy measure.

3. Having found the first split, repeat this process on the two resulting region s (R_1 and R_2). Continue to apply the above minimisation step on all resulting regions.
4. Stop if there is a low number of observations in a region (say 5 or less).

For a large set of variables, the process described above most likely yields an overfitted model. As such, it is important to take into consideration the size of a tree $|T|$. This is defined as the number of sub-samples/regions that are not split any further, or the number of terminal nodes (leaves) of a tree (see section below). A large tree may overfit the training data, and a small tree may not be able to provide an accurate fit. Given this trade-off, the act of reducing tree size is done in order to avoid overfitting. For a given tree, a measure that captures both accuracy and tree size is defined as

$$\text{RSS}+\alpha\ |T|, \tag{2.42}$$

where α is the tuning parameter/penalty. The tuning parameter is selected using cross validation. Note, for classification trees, we substitute the Gini Index with the RSS. Unlike, the case for shrinkage estimators, this tuning parameter does not have any theoretical foundations.

2.3.1 Conceptual Example of a Tree

This section introduces some specific CART terms and describes how a small tree might be constructed. It also provides a brief overview of an evaluation measure for classification problems. The very top of a tree where the first split takes place is known as the *root node*. A *branch* is a region resulting from a partition of a variable. If a node does not have branches, it is labelled as a *terminal* node or *leaf*. The process of building or *growing* a tree consists of increasing the number of branches. The process of decreasing/cutting the number of branches (Equation 2.42) is known as *pruning*.

For ease of presentation, define $\wedge$ and $\vee$ as the 'and' and 'or' operator, respectively, and let the usual indicator function $I(.) = 1$ if its argument is true, and zero otherwise.

Consider the following model where the response variable depends on two covariates, x_1 and x_2

$$\begin{aligned} y_i =&\beta_1 I(x_{1i} < c_1 \wedge x_{2i} < c_2) + \beta_2 I(x_{1i} < c_1 \wedge x_{2i} \geq c_2) \\ &+\beta_3 I(x_{1i} \geq c_1 \wedge x_{2i} < c_2) + \beta_4 I(x_{1i} \geq c_1 \wedge x_{2i} \geq c_2) + u_i. \end{aligned} \tag{2.43}$$

As shown in Figure 2.10, the tree regression divides the (x_1, x_2) space into four regions (branches). Within each region, the prediction of the response variable is the same for all values of x_1 and x_2 belonging to that region. In this particular example, if $c_1 = 2$ and $c_2 = 3$ the four regions/branches are:

$$\{(x_1, x_2) : x_1 < 3 \wedge x_2 < 2\}$$
$$\{(x_1, x_2) : x_1 < 3 \wedge x_2 \geq 2\}$$
$$\{(x_1, x_2) : x_1 \geq 3 \wedge x_2 < 2\}$$
$$\{(x_1, x_2) : x_1 \geq 3 \wedge x_2 \geq 2\}.$$

Based on this setup, the predicted y values in regions is given the corresponding β. Figure 2.11 displays the same splits using a hierarchical tree, which is how most of the tree based models are displayed.

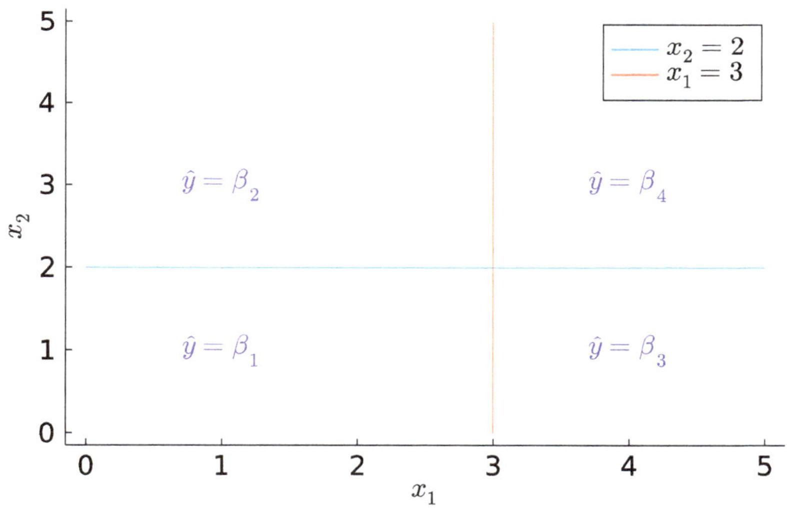

Fig. 2.10: Tree Based Regression

One popular way to evaluate the performance of a binary classification tree is by using a confusion matrix (Figure 2.12). This matrix directly compares each of the predicted values to their corresponding actual values. As such, this matrix classifies all observations into one of four categories. The first two being True Positives (TP) and True Negatives (TN). These categories count all the observations where the predictions are consistent with actual/observed value. The remaining two categories False Positives (FP) and False Negative (FN) count all the observations where the predictions are not consistent with the actual/observed values. Based on the counts across the four categories, a variety of metrics can be calculated to measure different facets of the model, such as overall accuracy. Note that the confusion matrix can be

Fig. 2.11: Tree view of the Regression Tree

calculated in any classification model. This includes the Logit or Probit model used in econometrics. This allows the user to not only compare different versions of the same model, but also compare performance across different types of models. For example, a confusion matrix based on a Logit model can be compared to a confusion matrix resulting from a classification tree.

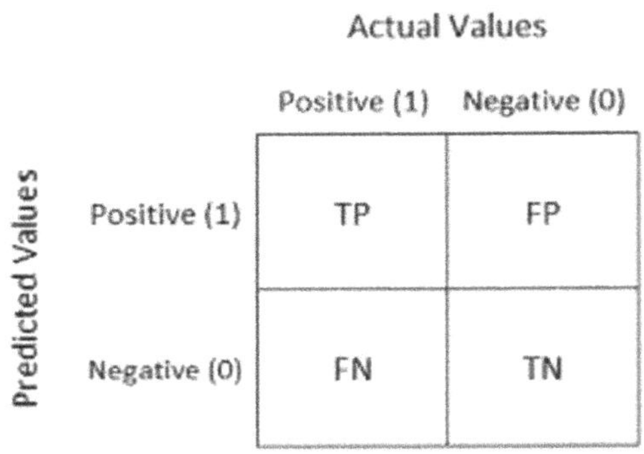

Fig. 2.12: Confusion Matrix

2.3.2 Bagging and Random Forests

As noted by Breiman (1996b), trees are well-known to be unstable as a small change in the training sample can cause the resulting tree structure to change drastically. This change implies that the predictions will also change, leading to high variance in predictions. Given this, it is not surprising that trees are generally out-performed by other classification and regression methods with regard to predictive accuracy. This is due to the fact that single trees are prone to overfitting, and pruning does not guarantee stability. Two other popular methods that can be applied to trees to improve their predictive performance are *Bagging* and *Random Forests*. Both methods are based on aggregation, whereby the overall prediction is obtained by combining predictions from many trees. Brief details of both these approaches are discussed below.

Bagging (Bootstrap Aggregation) consists of repeatedly taking samples and constructing trees from each sample, and subsequently combining the predictions from each tree in order to obtain the final prediction. The following steps outline this process:

1. Generate B different bootstrapped samples of size n (with replacement).
2. Build a tree on each sample and obtain a prediction for a given $\boldsymbol{x}$, $\hat{f}_b(\boldsymbol{x})$.
3. Compute the average of all B predictions to get the final bagging prediction,

$$\hat{f}_{\text{bagging}}(\boldsymbol{x}) = \frac{1}{B}\sum_{b=1}^{B} \hat{f}_b(\boldsymbol{x}). \tag{2.44}$$

4. The above steps work fine for regression trees. With regard to classification trees, the final prediction consists of selecting the most occurring category/class among the B predictions.

As argued in Breiman (1996a), the bagging procedure described above improves predictions by reducing their variance. The random forests concept extends the bagging approach by constraining the number of covariates in each bootstrapped sample. In other words, only a subset of all covariates are used to build each tree. As a rule of thumb, if there are a total of p covariates, then only $\sqrt{p}$ covariates are selected for each bootstrapped sample. The following steps outline the random forests approach:

1. Draw a random sample (with replacement) of size n.
2. Randomly select m covariates from the full set of p covariates (where $m \approx \sqrt{p}$).
3. Build a tree using the selected m covariates and obtain a prediction for a given $\boldsymbol{x}$, $\hat{f}_{b,m}(\boldsymbol{x})$. This is the prediction of the bth tree based on m selected covariates.
4. Repeat steps 1 to 3 for all B bootstrapped samples.
5. Compute the average of individual tree predictions in order to obtain the random forest prediction, $\hat{f}_{rf}$

$$\hat{f}_{rf}(\boldsymbol{x}) = \frac{1}{B}\sum_{b=1}^{B} \hat{f}_{b,m}(\boldsymbol{x}). \tag{2.45}$$

6. In the classification setting, each bootstrapped tree predicts the most commonly occurring category/class and the random forest predictor selects the most commonly occurring class across all B trees.

Note that bagging is a special case of random forest, when $m = p$ (step 2). Similar to the Bagging approach, the random forests approach reduces the variance, but unlike bagging, the random forest approach also reduces the correlation across the bootstrapped trees (due to the fact that not all trees have the same covariates). By allowing a random selection of covariates, the splitting process is not the same for each tree, and this seems to improve the overall prediction. Several numerical studies show that random forests perform well across many applications. These aggregation methods are related to the concept of forecast combinations, which is popular in mainstream econometrics. The next section contains further information on the applications and connection of trees to econometrics.

2.3.3 Applications and Connections to Econometrics

For econometricians, both regression and classification trees potentially offer some advantages compared to traditional regression-based methods. For a start, they are displayed graphically. This illustrates the decision-making process, and as such the results are relatively easy to interpret. The tree based approach allows the user to explore the variables in order to gauge their *partitioning* ability, i.e., how well a given variable is able to classify observations correctly using covariates. This relates to variable importance, which is discussed below. Trees can handle qualitative variables directly without the need to create dummy variables, and are relatively robust to outliers. Given enough data, a tree can estimate nonlinear means and interaction effects without the econometrician having to specify these in advance. Furthermore, non-constant variance (heteroskedasticity) is also accommodated by both classification and regression trees. Many easily accessible and tested packages can be used to build trees. Examples of such packages in R include rpart (Therneau, Atkinson and Ripley (2015)), caret (Kuhn (2008)) and tree (Ripley (2021)). For Python, refer to the scikit-learn library (Pedregosa et al. (2011)).

It is also important to consider some limitations of tree based methods. Trees do not produce any regression coefficients. Hence, it would be challenging to quantify the relation between the response variable and the covariates. Another consideration is computational time of trees, especially as the number of covariates increases. In addition, repeated sampling and tree building (bagging and random forests) are also computationally demanding. However, in such instances, parallel processing can help lower computational time. As mentioned earlier, single trees also have a tendency to overfit and as such offer lower predictive accuracy relative to other methods. But, this too can be improved by the use of bagging or random forests.

There are two connections that can be made between trees and topics in econometrics: nonparametric regression and threshold regression. With regard to the first, trees can be seen as a form nonparametric regression. Consider a simple nonparametric model:

$$y_i = m(x_i) + \varepsilon_i, \tag{2.46}$$

where $m(x_i) = E[y_i|x_i]$ is the conditional mean and ε_i are error terms. Here, m does not have a parametric form and its estimation occurs at particular values of x (local estimation). The most popular method of nonparametric regression is to use a local average of y. Compute the average of all y values with some window of x. By sliding this window along the domain, an estimate of the entire regression function can be obtained. In order to improve the estimation, a weighted local average is used, where the weights are applied to the y values. This is achieved using a kernel function. The size of the estimation window also known as bandwidth is part of the kernel function itself. Given this, a nonparametric regression estimator which uses local weighted averaging can be defined as:

$$\hat{m}_{NW}(x) = \sum_{i=1}^{n} w_i y_i \tag{2.47}$$

$$= \frac{\sum_{i=1}^{n} K_h(x - x_i) y_i}{\sum_{j=1}^{n} K_h(x - x_j)}, \tag{2.48}$$

where $K_h(x)$ denotes the scaled kernel function. Note that the sum of the weights is equal to 1 (the denominator is the normalising constant). The subscript NW credits the developers of this estimator: Nadaraya (1964) and Watson (1964).

With an appropriate choice of a kernel function and some modifications on the bandwidth selection, this nonparametric framework can replicate CART. Begin with a fixed bandwidth and create a series of non-overlapping x intervals. For each one of these intervals, choose a kernel function that computes the sample mean of y values that lie in that interval. This leads to the estimated regression function being a step function with regular intervals. The final step is to combine the adjacent intervals where the difference in the sample means (y) is small. The result is a step function with varying x intervals. Each of the end-points of these intervals represents a variable, splits between the difference in the mean value of y when it is significant. This is similar to the variable splits in the CART.

Replacing kernels with splines also leads to the CART framework. In equation (2.46) let $m(x_i) = s(x_i)$, then

$$y_i = s(x_i) + \varepsilon_i, \tag{2.49}$$

where $s(x)$ is a pth order spline and ε_i is the error term, which is assumed to have zero mean. The pth order spline can be written as

$$s(x) = \beta_0 + \beta_1 x + \cdots + \beta_p x^p + \sum_{j=1}^{J} b_j (x - \kappa_j)_+^p, \tag{2.50}$$

where κ_j denotes the jth threshold value and $(\alpha)_+ = \max(0, \alpha)$.

Based on the above, trees can be thought of as zero order ($p = 0$) splines. These are essentially piece-wise constant functions (step functions). Using a sufficiently large number of split points, a step function can approximate most functions. Similar to trees, splines are prone to overfitting, which can result in a jagged and noisy fit. However, unlike trees, the covariates need to be processed prior to entering the nonparametric model. For example, categorical variables need to be transformed to a set of dummy variables.

It is worth noting that the nonparametric regression can also be extended to accommodate y as a categorical or count variable. This is achieved by applying a suitable link function to the left-hand side of Equation (2.49). Although, the zero-spline is flexible, it is still a discrete step function, that is, an approximation of a nonlinear continuous function. In order to obtain smoother fit, shrinkage can be applied to reduce the magnitude of b_j.

$$(\hat{\beta}, \hat{b}) = \min_{\beta, b} \sum_i \{y_i - s(x_i)\}^2 + \alpha \sum_{j=1}^{J} b_j^2. \tag{2.51}$$

Another option would be to use the LASSO penalty, which would allow certain b_j to be shrunk to zero. This analogous to *pruning* a tree.

It is important to note that splines of order greater than zero can model non-constant relationships between threshold points. As such, this provides greater flexibility compared to trees. However, unlike trees, the splines require pre-specified thresholds.[9] As such, this does not accurately reflect the variable splitting in CART. In order to accommodate this shortcoming, the Multivariate Adaptive Regression Splines (MARS) was introduced. This method is based on linear splines and its adaptive ability consists of selecting thresholds in order to optimise fit to the data, see Hazelton (2015) for further explanation. Given this, MARS can be regarded a generalisation of trees. MARS can also handle continuous numerical variables better than trees. In the CART framework, midpoints of numerical variable were assigned as potential thresholds, which is limiting.

Recall that trees are zero-order splines. This is similar to a simplified version of threshold regression, where the only intercepts (between thresholds) are used to model the relationship between the response and the covariates. As an example, a single threshold model is written as

$$Y = \beta_{01} I(X \leq \kappa) + \beta_{02} I(X > \kappa) + \varepsilon, \tag{2.52}$$

where κ is the threshold value, and ε has a mean of zero. Based on this simple model, if X is less than or equal to the threshold value, the estimated value of Y is equal to the constant β_{01}. This is similar to a split in a regression tree, where the predicted value is equal to the average value of Y in that region. This is also evident in the conceptual example provided earlier, see Equation (2.43). As in the case for splines, threshold regression can easily accommodate non-constant relationships, between thresholds. Similarly to the MARS method, this also represents a generalisation of trees.

Given the two connections, econometricians who have used to nonparametric and/or threshold regressions are able to relate to the proposed CART methods. The estimation methods for both nonparametric and threshold regression are well-known in the econometrics literature. In addition, econometricians can also take advantage of the rich theoretical developments in both of these areas. This includes concepts such as convergence rates of estimators and their asymptotic properties. These developments represent an advantage over the CART methods, which to-date have no or little theoretical developments (see below). This is especially important to econometric work which covers inference and/or causality. However, if the primary purpose is illustrating the decision-making process or for exploratory analysis, then implementing CART may be advantageous. There are tradeoffs to consider when selecting a method of analysis. For more details on nonparametric and threshold regression, see Li and Racine (2006) and Hansen (2000) and references within. Based

[9] These are also called knots.

on this discussion, the section below covers the limited (and recent) developments for inference in trees.

Inference

For a single small tree, it may be possible to visually see the role that each variable plays in producing partitions. In a loose sense, a variable's partitioning ability corresponds to its importance/significance. However, as the number of variables increases, the visualisation aspect may not be so appealing. Furthermore, when the bagging and/or random forests approach is implemented, it is not possible to represent all the results into a single tree. This makes interpretability even more challenging, despite the improvement in predictions. Nevertheless, the importance/significance of a variable can still be measured by quantifying the increase in RSS (or Gini index) if the variable is excluded from the tree. Repeating this for B trees, the variable importance score is then computed as the average increase in RSS (or Gini Index). This process is repeated for all variables and the variable importance score is plotted. A large score indicates that a removing this variable leads to large increases in RSS (Gini index) on average, and hence the variable is considered 'important'.

Figure 2.13 shows a sample variable importance plot reproduced using Kuhn (2008). Based on the importance score, variable V11 is considered the most important variable: removing this variable from the tree/s leads to the greatest increase in the Gini index, on average. The second most important variable is V12 and so on. Practitioners often use the variable importance to select variables: keeping the ones with the largest scores and discarding those with lower ones. However, there seems no agreed cut-off score for this purpose. In-fact, there are no known theoretical properties of variable importance. As such, variable importance scores are to be considered with caution. Based on this, variable importance is of limited use with regard to inference.

More recently, trees have found applications in causal inference settings. This would be of interest to applied econometricians who are interested in modelling heterogeneous treatment effects. Athey and Imbens (2016) introduces methods for constructing trees in order to study causal effects, also providing valid inferences for such. For further details, see Chapter 3.

Given a single binary treatment, the conditional average treatment effect $y(\boldsymbol{x})$ is given by

$$y(\boldsymbol{x}) = \mathbb{E}[y|d=1,\boldsymbol{x}] - \mathbb{E}[y|d=0,\boldsymbol{x}], \tag{2.53}$$

which is the difference between the conditional expected response for the treated group ($d = 1$) and the control group ($d = 0$), given a set of covariates ($\boldsymbol{x}$). Athey and Imbens (2016) propose a causal tree framework to estimate $y(\boldsymbol{x})$. It is an extension of the classification and regression tree methods described above. Unlike, trees where the variable split is chosen based on minimising the RSS or Gini Index, causal trees choose a variable split (left and right) that maximises the squared difference between the estimated treatment effects; that is, maximise

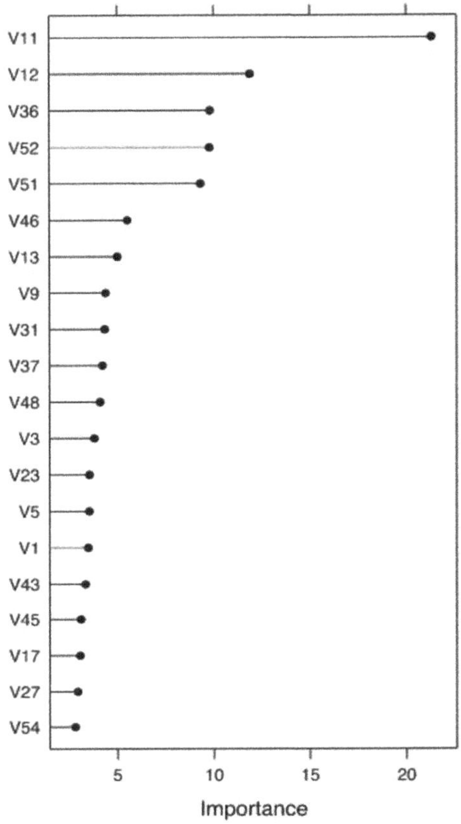

Fig. 2.13: Variable Importance Chart

$$\sum_{\text{left}} (\bar{y}_1 - \bar{y}_0)^2 + \sum_{\text{right}} (\bar{y}_1 - \bar{y}_0)^2, \tag{2.54}$$

where $\bar{y}_d$ is the sample mean of observations with treatment d. Furthermore, Athey and Imbens (2016) uses two samples to build a tree. The first sample determines the variable splits, and the second one is used to re-estimate the treatment effects conditional on the splits. Using this setup, Athey and Imbens (2016) derive the approximately Normal sampling distributions for $y(\boldsymbol{x})$. An R package called *causalTree* is available for implementing Causal Trees.

2.4 Concluding Remarks

The first section of this chapter provided a brief overview on regularization for nonlinear econometric models. This included regularization with both nonlinear least squares and the likelihood function. Furthermore, the estimation, tuning parameters and asymptotic properties were discussed in some detail. One of the important takeaways is that for shrinkage estimators of nonlinear models, the selection consistency and oracle properties do not simply carry over from shrinkage estimators for linear models. For example, it is not always the case that shrinkage estimators of nonlinear models can be used for variable selection. This is due to the functional form of nonlinear models. The exception is when nonlinear model have a single index form. Examples of this include both the Logit and Probit model. Another point of difference compared to the linear models' case is when using the maximum likelihood function as an objective function for nonlinear models, there are additional parameters, such as variance, that are part of the shrinkage estimators. If a given model can be estimated using both least squares or maximum likelihood, it is possible to compare the resulting tuning parameters across both shrinkage estimators. Based on this comparison, it is evident that the tuning parameters are not always equal to each other. Although, commonly used for empirical work, the theoretical properties of shrinkage estimators for nonlinear models are often unknown. In addition to this, the computational aspects of these estimators which involve constrained optimisation are challenging in general. With regard to asymptotic results, there has been some progress in laying the theoretical foundations. However, given the wide scope of nonlinear models, much work remains to be done.

This chapter also provided a brief overview of the tree based methods in the machine learning literature. This consisted of introducing both regression and classification trees, as well as methods of building these trees. An advantage of trees is that they are easy to understand, interpret and visualise. However, trees are prone to overfitting, which makes them less attractive compared to other regression and classification methods. Additionally, trees tend to unstable, i.e., small changes in the training data lead to drastic changes in predictions. To overcome this limitation, bagging and/or random forests approach can be used. There appears to be no or limited development with regard to inference for tree based techniques. A pseudo inference measure, variable importance does not have any theoretical basis and as such cannot be used with confidence. Causal trees, a recent development, offers some inference capabilities. However, the asymptotic results are provided for the response variable, i.e., not directly related to the estimation aspect for trees. Given the links between trees and nonparametric regression, inference procedures from the nonparametric framework may extend to tree based techniques. Perhaps, these links could aid in the development of inference for trees. This is a potential area for further research.

Appendix

Proof of Proposition 2.1

The Partially Penalised Estimator satisfies the following First Order Necessary Conditions

$$\frac{\partial S}{\partial \boldsymbol{\beta}_1} = 0$$
$$\frac{\partial S}{\partial \boldsymbol{\beta}_2} = 0.$$

Given μ as defined in Equation (2.29), define

$$M(\boldsymbol{\beta}_1, \boldsymbol{\beta}_2) = \frac{\partial S}{\partial \boldsymbol{\beta}_1} + \mu \frac{\partial S}{\partial \boldsymbol{\beta}_2}.$$

Given the definition of μ, it is straightforward to show that $M(\hat{\boldsymbol{\beta}}_1, \hat{\boldsymbol{\beta}}_2) = 0$ if and only if the First Order Conditions are satisfied. Moreover, it is also straightforward to show that

$$\frac{\partial M}{\partial \boldsymbol{\beta}_2} = 0$$

after some tedious algebra. Given the immunization condition, the result follows directly from Proposition 4 in Chernozhukov et al. (2015). This completes the proof.

Proof of Proposition 2.2

The proof of Proposition 2.2 follows the same argument as Proposition 2.1.

References

Athey, S. & Imbens, G. (2016). Recursive Partitioning for Heterogeneous Causal effects. *Proceedings of the National Academy of Sciences, 113*(27), 7353–7360. doi: 10.1073/pnas.1510489113

Benjamini, Y. & Hochberg, Y. (1995). Controlling the False Discovery Rate: A Practical and Powerful Approach to Multiple Testing. *Journal of the Royal Statistical Society. Series B (Methodological), 57*(1), 289–300.

Breiman, L. (1996a). Bagging Predictors. *Machine Learning, 24*(2), 123-140.

Breiman, L. (1996b). *Bias, Variance, and Arcing classifiers* (Tech. Rep. No. 460). Berkeley, CA: Statistics Department, University of California at Berkeley.

Breiman, L., Friedman, J., Olshen, R. & Stone, C. (1984). *Classification and Regression Trees*. Wadsworth and Brooks/Cole.

Chernozhukov, V., Hansen, C. & Spindler, M. (2015). Valid Post-Selection and Post-Regularization Inference: An Elementary, General Approach. *Annual Review of Economics*, *7*, 649–688.

Fan, J., Xue, L. & Zou, H. (2014). Strong Oracle Optimality of Folded Concave Penalized Estimation. *Annals of Statistics*, *42*(3), 819–849.

Friedman, J., Hastie, T. & Tibshirani, R. (2010). Regularization Paths for Generalized Linear Models via Coordinate Descent. *Journal of Statistical Software*, *33*(1), 1–22. Retrieved from https://www.jstatsoft.org/v33/i01/

Hansen, B. E. (2000). Sample Splitting and Threshold Estimation. *Econometrica*, *68*(3), 575–603.

Hastie, T., Tibshirani, R. & Friedman, J. (2009). *The Elements of Statistical Learning: Data Mining, Inference and Prediction*. Springer.

Hazelton, M. L. (2015). Nonparametric regression. In J. D. Wright (Ed.), *International Encyclopedia of the Social & Behavioral Sciences (Second Edition)* (Second Edition ed., p. 867-877). Oxford: Elsevier. doi: https://doi.org/10.1016/B978-0-08-097086-8.42124-0

James, G., Witten, D., Hastie, T. & Tibshirani, R. (2013). *An Introduction to Statistical Learning: with Applications in R*. Springer. Retrieved from https://faculty.marshall.usc.edu/gareth-james/ISL/

Jansen, D. & Oh, W. (1999). Modeling Nonlinearity of Business Cycles: Choosing between the CDR and Star Models. *The Review of Economics and Statistics*, *81*, 344-349.

Kuhn, M. (2008). Building Predictive Models in R Using the caret Package. *Journal of Statistical Software, Articles*, *28*(5), 1–26. Retrieved from https://www.jstatsoft.org/v028/i05 doi: 10.18637/jss.v028.i05

Kwangmoo, K., Seung-Jean, K. & Boyd, S. (2007). An Interior-Point Method for Large-Scale Logistic Regression. *Journal of Machine Learning Research*, *8*, 1519–1555.

Leeb, H. & Pötscher, B. M. (2005). Model Selection and Inference: Facts and Fiction. *Econometric Theory*, *21*(1), 21–59.

Leeb, H. & Pötscher, B. M. (2008). Sparse Estimators and the Oracle Property, or the Return of Hodges' Estimator. *Journal of Econometrics*, *142*(1), 201–211. doi: 10.1016/j.jeconom.2007.05.017

Li, Q. & Racine, J. S. (2006). *Nonparametric Econometrics: Theory and Practice* (No. 8355). Princeton University Press.

Morgan, J. N. & Sonquist, J. A. (1963). Problems in the Analysis of Survey Data, and a Proposal. *Journal of the American Statistical Association*, *58*(302), 415–434. Retrieved from http://www.jstor.org/stable/2283276

Nadaraya, E. A. (1964). On estimating Regression. *Theory of Probability and its Applications*, *9*, 141–142.

Pedregosa, F., Varoquaux, G., Gramfort, A., Michel, V., Thirion, B., Grisel, O., . . . Duchesnay, E. (2011). Scikit-learn: Machine Learning in Python. *Journal of Machine Learning Research*, *12*, 2825–2830.

Ripley, B. (2021). Tree: Classification and Regression Trees [Computer software manual]. Retrieved from https://cran.r-project.org/web/packages/tree/index .html (R pacakge version 1.0-41)

Shannon, C. (1948). The mathematical theory of communication. *Bell Systems Technical Journal*, *27*, 349–423.

Shi, C., Song, R., Chen, Z. & Li, R. (2019). Linear Hypothesis Testing for High Dimensional Generalized Linear Models. *The Annals of Statistics*, *47*(5). doi: 10.1214/18-AOS1761

Teräsvirta, T. & Anderson, H. (1992). Characterizing Nonlinearities in Business Cycles using Smooth Transition Autoregressive Models. *Journal of Applied Econometrics*, *7*, S119-S136.

Therneau, T., Atkinson, B. & Ripley, B. (2015). rpart: Recursive Partitioning and Regression Trees [Computer software manual]. Retrieved from http:// CRAN.R-project.org/package=rpart (R package version 4.1-9)

Tong, H. (2003). *Non-linear Time Series: A Dynamical System Approach.* Oxford University Press.

Watson, G. S. (1964). Smooth Regression Analysis. *Sankhyā Ser.*, *26*, 359–372.

Chapter 3
The Use of Machine Learning in Treatment Effect Estimation

Robert P. Lieli, Yu-Chin Hsu and Ágoston Reguly

Abstract Treatment effect estimation from observational data relies on auxiliary prediction exercises. This chapter presents recent developments in the econometrics literature showing that machine learning methods can be fruitfully applied for this purpose. The double machine learning (DML) approach is concerned primarily with selecting the relevant control variables and functional forms necessary for the consistent estimation of an average treatment effect. We explain why the use of orthogonal moment conditions is crucial in this setting. Another, somewhat distinct, strand of the literature focuses on treatment effect heterogeneity through the discovery of the conditional average treatment effect (CATE) function. Here we distinguish between methods aimed at estimating the entire function and those that project it on a pre-specified coordinate. We also present an empirical application that illustrates some of the methods.

3.1 Introduction

It is widely understood in the econometrics community that machine learning (ML) methods are geared toward solving prediction tasks (see Mullainathan & Spiess, 2017). Nevertheless, most applied work in economics goes beyond prediction and is often concerned with estimating the average effect of some policy or treatment on an outcome of interest in a given population. A question of first order importance

Robert P. Lieli ✉
Central European University, Budapest, Hungary and Vienna, Austria. e-mail: lielir@ceu.edu

Yu-Chin Hsu
Academia Sinica, Taipei, Taiwan; National Central University and National Chengchi University, Taipei, Taiwan. e-mail: ychsu@econ.sinica.edu.tw

Ágoston Reguly
Central European University, Budapest, Hungary and Vienna, Austria. e-mail: reguly_agoston@phd.ceu.edu

F. Chan, L. Mátyás (eds.), *Econometrics with Machine Learning*, Advanced Studies in Theoretical and Applied Econometrics 53, https://doi.org/10.1007/978-3-031-15149-1_3

is therefore identification — comparing the average outcome among participants and non-participants, can the difference be attributed to the treatment? If treatment status is randomly assigned across population units, then the answer is basically yes. However, if the researcher works with observational data, meaning that one simply observes what population units *chose* to do, then participants may be systematically different from non-participants and the difference in average outcomes is generally tainted by selection bias.

There are standard frameworks in econometrics to address the identification of treatment effects. In the present chapter we adopt one of the most traditional and fundamental settings in which the researcher has at their disposal a cross-sectional data set that includes a large set of pre-treatment measurements (covariates) assumed to be sufficient for adjusting for selection bias. The formalization of this idea is the widely used unconfoundedness assumption, also known as selection-on-observables, ignorability, conditional independence, etc. This assumption forms the basis of many standard methods used for average treatment effect estimation such as matching, regression, inverse propensity weighting, and their combinations (see e.g., Imbens & Wooldridge, 2009).

Assessing whether identification conditions hold in a certain setting is necessary for estimating treatment effects even if ML methods are to be employed. For example, in case of the unconfoundedness assumption, one has to rely on subject matter theory to argue that there are no unobserved confounders (omitted variables) affecting the outcome and the treatment status at the same time.[1] If the identification of the treatment effect is problematic in this sense, the use of machine learning will not fix it.

Nevertheless, economic theory has its own limitations. It may well suggest missing control variables, but given a large pool of *available* controls, it is usually not specific enough to select the most relevant ones or help decide the functional form with which they should enter a regression model. This is where machine learning methods can and do help. More generally, ML is useful in treatment effect estimation because the problem involves implicit or explicit predictive tasks. For example, the first stage of the two-stage least squares estimator is a predictive relationship, and the quality of the prediction affects the precision of the estimator. It is perhaps less obvious that including control variables in a regression is also a prediction exercise, both of the treatment status and the outcome. So is the specification and the estimation of the propensity score function, which is used by many methods to adjust for selection bias.

Another (related) reason for the proliferation of ML methods in econometrics is their ability to handle high dimensional problems in which the number of variables is comparable to or even exceeds the sample size. Traditional nonparametric estimators are well known to break down in such settings ('the curse of dimensionaliy') but even parametric estimators such as ordinary least squares (OLS) cannot handle a high variable to sample size ratio.

In this chapter we present two subsets of the literature on machine learning aided causal inference. The first came to be known as double or debiased machine learning

[1] There are rare circumstances in which the assumption is testable by statistical methods (see Donald, Hsu & Lieli, 2014), but in general it requires a substantive theory-based argument.

(DML) and has its roots in an extensive statistical literature on semiparametric estimation with orthogonal moment conditions. In the typical DML setup the model (e.g., a regression equation) contains a parameter that captures the treatment effect under the maintained identifying assumptions as well as nuisance functions (e.g., a control function) that facilitate consistent estimation of the parameter of interest. The nuisance functions are typically unknown conditional expectations; these are the objects that can be targeted by flexible ML methods. The applicability of ML turns out to depend on whether the nuisance functions enter the model in a way that satisfies an orthogonality condition. It is often possible to transform a model to satisfy this condition, usually at the cost of introducing an additional nuisance function (or functions) to be estimated.

The second literature that we engage with in this chapter is aimed at estimating heterogeneous treatment effects using ML methods. More specifically, the object of interest is the conditional average treatment effect (CATE) function that describes how the average treatment effect changes for various values of the covariates. This is equivalent to asking if the average treatment effect differs across subgroups such as males vs. females or as a function of age, income, etc. The availability of such information makes it possible to design better targeted policies and concentrate resources to where the expected effect is higher. One strand of this literature uses 'causal trees' to discover the CATE function in as much detail as possible without prior information, i.e., it tries to identify which variables govern heterogeneity and how they interact. Another strand relies on dimension reduction to estimate heterogeneous effects as a function of a *given* variable as flexibly as possible while averaging out the rest of the variables.

With machine learning methods becoming ubiquitous in treatment effect estimation, the number of papers that summarize, interpret and empirically illustrate the research frontier to broader audiences is also steadily increasing (see e.g., Athey & Imbens, 2019, Kreif & DiazOrdaz, 2019, Knaus, 2021, Knaus, Lechner & Strittmatter, 2021, Huber, 2021). So, what is the value added of this chapter? We do not claim to provide a comprehensive review; ML aided causal inference is a rapidly growing field and it is scarcely possible to cover all recent developments in this space. Rather, we choose to include fewer papers and focus on conveying the fundamental *ideas* that underlie the literature introduced above in an intuitive way. Our hope is that someone who reads this chapter will be able to approach the vast technical literature with a good general understanding and find their way around it much easier.

While the chapter also foregoes the discussion of most implementation issues, we provide an empirical illustration concerned with estimating the effect of a mother's smoking during pregnancy on the baby's birthweight. We use DML to provide an estimate of the average effect (a well-studied problem) and construct a causal tree to discover treatment effect heterogeneity in a data-driven way. The latter exercise, to our knowledge, has not been undertaken in this setting. The DML estimates of the average 'smoking effect' we obtain are in line with previous results, and are virtually identical to the OLS benchmark and naive applications of the Lasso estimator. Part of the reason for this is that the number of observations is large relative to the number of controls, even though the set of controls is also large. The heterogeneity analysis confirms the important role of mother's age already documented in the literature but also points to other variables that may be predictive of the magnitude of the treatment effect.

The rest of the chapter is organized as follows. In Section 3.2 we outline a standard treatment effect estimation framework under unconfoundedness and show more formally where the DML and the CATE literature fits in, and what the most important papers are. Section 3.3 is then devoted to the ideas and procedures that define the DML method with particular attention to why direct ML is not suitable for treatment effect estimation. Section 3.4 presents ML methods for estimating heterogeneous effects. We distinguish between the causal tree approach (aimed at discovering the entire CATE function) and dimension reduction approaches (aimed at discovering heterogeneity along a given coordinate). Section 3.5 presents the empirical applications and Section 3.6 concludes.

3.2 The Role of Machine Learning in Treatment Effect Estimation: a Selection-on-Observables Setup

Let $Y(1)$ and $Y(0)$ be the potential outcomes associated with a binary treatment $D \in \{0,1\}$ and let X stand for a vector of predetermined covariates.[2] The (hypothetical) treatment effect for a population unit is given by $Y(1)-Y(0)$. Modern econometric analysis typically allows for unrestricted individual treatment effect heterogeneity and focuses on identifying the average treatment effect (ATE) or the conditional average treatment effect (CATE) given the possible values of the full covariate vector X. These parameters are formally defined as $\tau = E[Y(1)-Y(0)]$ and $\tau(X) = E[Y(1)-Y(0)|\,X]$, respectively.

The fundamental identification problem is that for any individual unit only the potential outcome corresponding to their actual treatment status is observed—the counterfactual outcome is unknown. More formally, the available data consists of a random sample $\{(Y_i, D_i, X_i)\}_{i=1}^{n}$, where $Y = Y(0) + D[Y(1)-Y(0)]$. In order to identify ATE and CATE from the joint distribution of the observed variables, we make the selection-on-observables (unconfoundedness) assumption, which states that the potential outcomes are independent of the treatment status D conditional on X:

$$(Y(1), Y(0)) \perp D \,\big|\, X. \tag{3.1}$$

Condition (3.1) can be used to derive various identification results and corresponding estimation strategies for τ and $\tau(X)$.[3] Here we use a regression framework as our starting point. We can decompose the conditional mean of $Y(j)$ given X as

[2] Chapter 5 of this volume provides a more detailed account of the potential outcome framework.

[3] For these strategies (such as regression, matching, inverse probability weighting) to work in practice, one also needs the overlap assumption to hold. This ensures that in large samples there is a sufficient number of treated and untreated observations in the neighborhood of any point x in the support of X(see Imbens & Wooldridge, 2009).

$$E[Y(j)|X] = \mu_j + g_j(X),\ j = 0,1,$$

where $\mu_j = E[Y(j)]$ and $g_j(X)$ is a real-valued function of X with zero mean. Then we can write $\tau = \mu_1 - \mu_0$ for the average treatment effect and $\tau(X) = \tau + g_1(X) - g_0(X)$ for the conditional average treatment effect function. Given assumption (3.1), one can express the outcome Y as the partially linear regression model

$$Y = \mu_0 + g_0(X) + \tau D + [g_1(X) - g_0(X)]D + U, \tag{3.2}$$

where $E[U|D,X] = 0$. One can obtain a textbook linear regression model from (3.2) by further assuming that $g_0(X) = g_1(X) = [X - E(X)]'\beta$ or, more generally, $g_0(X) = [X - E(X)]'\beta_0$ and $g_1(X) = [X - E(X)]'\beta_1$.

The scope for employing machine learning methods in estimating model (3.2) arises at least in two different ways. First, the credibility of the unconfoundedness assumption (3.1) hinges on the researcher's ability to collect a rich set of observed covariates that are predictive of both the treatment status D and the potential outcomes $(Y(1),Y(0))$. Therefore, the vector X of potential controls may already be high-dimensional in that the number of variables is comparable to the sample size. In this case theory-based variable selection or ad-hoc comparisons across candidate models become inevitable even if the researcher only wants to estimate simple linear versions of (3.2) by OLS. Machine learning methods such as the Lasso or L_2-boosting (as proposed by Kueck, Luo, Spindler & Wang, 2022) offer a more principled and data-driven way to conduct variable selection. Nevertheless, as we will shortly see, *how* exactly the chosen ML estimator is used for this purpose matters a great deal.

Second, the precise form of the control functions $g_0(X)$ and $g_1(X)$ is unknown. Linearity is a convenient assumption, but it is just that — an assumption. Misspecifying the control function(s) can cause severe bias in the estimated value of the (conditional) average treatment effect (see e.g., Imbens & Wooldridge, 2009). The Lasso handles the discovery of the relevant functional form by reducing it to a variable selection exercise from an extended covariate pool or 'dictionary' $b(X)$, which contains a set of basis functions constructed from the raw covariates X. Thus, it can simultaneously address the problem of finding the relevant components of X and determining the functional form with which they should enter the model (3.2).

The previous two points are the primary motivations for what came to be known as the 'double' or 'debiased' machine learning (DML) literature in econometrics. Early works include Belloni, Chen, Chernozhukov and Hansen (2012), Belloni, Chernozhukov and Hansen (2013) and Belloni, Chernozhukov and Hansen (2014b). Belloni, Chernozhukov and Hansen (2014a) provides a very accessible and intuitive synopsis of these papers. The research program was developed further by Belloni, Chernozhukov, Fernández-Val and Hansen (2017), Chernozhukov et al. (2017) and Chernozhukov et al. (2018). The last paper implements the double machine learning method in a general moment-based estimation framework.

A third, somewhat distinct, task where machine learning methods facilitate causal inference is the discovery of treatment effect heterogeneity, i.e., the conditional average treatment effect function $\tau(X)$. An influential early paper in this area is by

Athey and Imbens (2016). While maintaining the unconfoundedness assumption, they move away from the model-based regression framework represented by equation (3.2), and employ a regression tree algorithm to provide a step-function approximation to $\tau(X)$. Several improvements and generalizations are now available, e.g., Wager and Athey (2018), Athey, Tibshirani and Wager (2019). Other approaches to estimating treatment effect heterogeneity involve reducing the dimension of $\tau(X)$ such as in Semenova and Chernozhukov (2020), Fan, Hsu, Lieli and Zhang (2020), and Zimmert and Lechner (2019).

We now present a more detailed review of these two strands of the causal machine learning literature — double machine learning and methods aimed at discovering treatment effect heterogeneity. These are very rapidly growing fields, so our goal is not to present every available paper but rather the central ideas.

3.3 Using Machine Learning to Estimate Average Treatment Effects

3.3.1 Direct versus Double Machine Learning

As discussed in Section 3.2, DML is primarily concerned with variable selection and the proper specification of the control function rather than the discovery of treatment effect heterogeneity. In order to focus on the first two tasks, we actually restrict treatment effect heterogeneity in model (3.2) by assuming $g_0(X) = g_1(X)$.[4] This yields:

$$Y = \mu_0 + g_0(X) + \tau D + U. \tag{3.3}$$

The need for DML is best understood if we contrast it with the natural (though ultimately flawed) idea of using, say, the Lasso to estimate model (3.3) directly. In order to employ the Lasso, one must first set up a dictionary $b(X)$ consisting of suitable transformations of the components of X. In most applications this means constructing powers and interactions between the raw variables up to a certain order.[5] We let p denote the dimension of $b(X)$, which can be comparable to or even larger than the sample size n. The Lasso then uses a linear combination $b(X)'\beta$ to approximate $\mu_0 + g_0(X)$ and the model (3.3) is estimated by solving

$$\min_{\tau,\beta} \sum_{i=1}^{n} [Y_i - \tau D_i - b(X_i)'\beta]^2 + \lambda \sum_{k=1}^{p} |\beta_k|$$

[4] This condition does not mean that $Y(1) - Y(0)$ is the same for all units; however, it does mean that the distribution of $Y(1) - Y(0)$ is mean-independent of X, i.e., that the CATE function is constant.

[5] For example, if $X = (X_1, X_2, X_3)$, then the dictionary that contains up to second order polynomial terms is given by $b(X) = (1, X_1, X_2, X_3, X_1^2, X_2^2, X_3^2, X_1X_2, X_1X_3, X_2X_3)'$.

for some $\lambda > 0$. Let $\hat{\tau}^{dir}$ and $\hat{\beta}^{dir}$ denote the solution, where the superscript *dir* stands for 'direct'. For sufficiently large values of the penalty λ, many components of $\hat{\beta}^{dir}$ are exact zeros, which is the reason why Lasso acts as selection operator. The coefficient on D is left out of the penalty term to ensure that one obtains a non-trivial treatment effect estimate for any value of λ. In practice, λ may be chosen by cross-validation (Chapter 1 of this volume provides a detailed discussion of the Lasso estimator).

While this direct procedure for estimating τ may seem reasonable at first glance, $\hat{\tau}^{dir}$ has poor statistical properties. As demonstrated by Belloni et al. (2014b), $\hat{\tau}^{dir}$ can be severely biased, and its asymptotic distribution is non-Gaussian in general (it has a thick right tail with an extra mode in their Figure 1). Thus, inference about τ based on $\hat{\tau}^{dir}$ is very problematic. Using some other regularized estimator or selection method instead of Lasso would run into similar problems (see again Chapter 1 for some alternatives).

The double (or debiased) machine learning procedure approaches the problem of estimating (3.3) in multiple steps, mimicking the classic econometric literature on the semiparametric estimation of a partially linear regression model (e.g., Pagan & Ullah, 1999, Ch. 5). To motivate this approach, we take the conditional expectation of (3.3) with respect to X to obtain

$$E(Y|X) = \mu_0 + \tau E(D|X) + g_0(X), \tag{3.4}$$

given that $E(U|X) = 0$. Subtracting equation (3.4) from (3.3) yields an estimating equation for τ that is free from the control function $g_0(X)$ but involves two other unknown conditional expectations $\xi_0(X) = E(Y|X)$ and the propensity score $m_0(X) = E(D|X) = P(D = 1|X)$:

$$Y - \xi_0(X) = \tau(D - m_0(X)) + U. \tag{3.5}$$

The treatment effect parameter τ is then estimated in two stages:

(i) One uses a machine learning method—for example, the Lasso—to estimate the two 'nuisance functions' $\xi_0(X)$ and $m_0(X)$ in a flexible way. It is this twofold application of machine learning that justifies the use of the adjective 'double' in the terminology.
(ii) One then estimates τ simply by regressing the residuals of the dependent variable, $Y - \hat{\xi}_0(X)$, on the residuals of the treatment dummy, $D - \hat{m}_0(X)$.

There are several variants of the DML procedure outlined above depending on how the available sample data is employed in executing stages (i) and (ii). In the early literature (reviewed by Belloni et al., 2014a), the full sample is used in both steps, i.e., the residuals $Y_i - \hat{\xi}_0(X_i)$ and $D_i - \hat{m}_0(X_i)$, are constructed 'in-sample,' for each observation used in estimating ξ_0 and m_0.

By contrast, the more recent practice involves employing different subsamples in stages (i) and (ii) as in Chernozhukov et al. (2018). More specifically, we can partition the full set of observations $I = \{1, \ldots, n\}$ into K folds (subsamples) of size n/K each, denoted I_k, $k = 1, \ldots K$. Furthermore, let $I_k^c = I \setminus I_k$, i.e., I_k^c is the set of all observations not contained in I_k. Setting aside I_1, one can execute stage (i), the machine learning estimators, on I_1^c. The resulting estimates are denoted as $\hat{m}_{0,I_1^c}$ and $\hat{\xi}_{0,I_1^c}$, respectively.

In stage (ii), the residuals are then constructed for the observations in I_1, i.e., one computes $Y_i - \hat{\xi}_{0,I_1^c}(X_i)$ and $D_i - \hat{m}_{0,I_1^c}(X_i)$ for $i \in I_1$. Instead of estimating τ right away, steps (i) and (ii) are repeated with I_2 taking over the role of I_1 and I_2^c taking over the role of I_1^c. Thus, we obtain another set of 'out-of-sample' residuals $Y_i - \hat{\xi}_{0,I_2^c}(X_i)$ and $D_i - \hat{m}_{0,I_2^c}(X_i)$ for $i \in I_2$. We iterate in this way until all folds $I_1, I_2, \ldots, I_K$ are used up and each observation $i \in I$ has a Y-residual and a D-residual associated with it. Stage (ii) is then completed by running an OLS regression of the full set of Y-residuals on the full set of D-residuals to obtain the DML estimate $\hat{\tau}^{DML}$. This sample splitting procedure is called the cross-fitting approach to double machine learning (see Chernozhukov et al., 2017, Chernozhukov et al., 2018).[6]

Regardless of which variant of $\hat{\tau}^{DML}$ is used, the estimator has well-behaved statistical properties: it is root-n consistent and asymptotically normal provided that the nuisance functions ξ_0 and m_0 satisfy some additional regularity conditions. Thus, one can use $\hat{\tau}^{DML}$ and its OLS standard error to conduct inference about τ in an entirely standard way. This is a remarkable result because machine learning algorithms involve a thorough search for the proper specification of the nuisance function estimators $\hat{m}_0$ and $\hat{\xi}_0$. Despite this feature of machine learning estimators, post-selection inference is still possible.

As mentioned above, the nuisance functions ξ_0 and m_0 need to satisfy some additional regularity conditions for standard inference based on $\hat{\tau}^{DML}$ to be valid. Essentially, what needs to be ensured is that the first stage ML estimators converge sufficiently fast—faster than the rate $n^{-1/4}$. When this estimator is the Lasso, the required conditions are called sparsity assumptions. These assumptions describe how 'efficiently' one can approximate m_0 and ξ_0 using linear combinations of the form $b(X)'\beta$. In particular, the assumption is that one can achieve a small approximation error just by using a few important terms, i.e., a 'sparse' coefficient vector β (see also Chapter 1 of this volume for a discussion).

More technically, in the absence of strong functional form assumptions, uniformly consistent estimation of ξ_0 and m_0 requires the inclusion of higher and higher order polynomial terms into $b(X)$; hence, in theory, the dimension of $b(X)$ expands with the sample size n. For a corresponding coefficient vector $\beta = \beta_n$, let the 'sparsity index' s_n be the number of nonzero components. The sparsity assumption then states that there exists a sequence of approximations $b(X)'\beta_n$ to ξ_0 and m_0 such that s_n increases slowly but the approximation error still vanishes at a sufficiently fast rate (and is negligible relative to the estimation error).[7] For example, Belloni et al. (2017)

---

[6] Another variant of cross-fitting involves running K separate regressions, one over each fold I_k, using the nuisance function estimators $\hat{m}_{0,I_k^c}$ and $\hat{\xi}_{0,I_k^c}$. There are K resulting estimates of τ, which can be averaged to obtain the final estimate.

[7] A special case of the sparsity assumption is that the functions m_0 and ξ_0 obey parametric models linear in the coefficients, i.e., the approximation error $m_0(X) - b_n(X)'\beta_n$ can be made identically zero for a finite value of s_n.

specify rigorously the sparsity conditions needed for the full-sample DML estimation of τ.

The required sparsity assumptions also provide theoretical motivation for the cross-fitted version of the DML estimator. As pointed out by Chernozhukov et al. (2017), the sparsity assumptions required of the nuisance functions to estimate ATE are milder when the first stage estimates are constructed over an independent subsample. In particular, this induces a tradeoff between how strict the sparsity assumptions imposed on m_0 and ξ_0 need to be — if, say, m_0 is easy to approximate, then the estimation of ξ_0 can use a larger number of terms and vice versa. In the full-sample case the sparsity indices for estimating m_0 and ξ_0 have to satisfy strong restrictions individually.[8] It is generally true that the split sample approach naturally mitigates biases due to overfitting (estimating the nuisance functions and the parameter of interest using the same data), whereas the full sample approach requires more stringent complexity restrictions, such as entropy conditions, to do so (Chernozhukov et al., 2018). Despite the theoretical appeal of cross-fitting, we are at present not aware of applications or simulation studies, apart from illustrative examples in *ibid*., where a substantial difference arises between the results delivered by the two DML approaches.

3.3.2 Why Does Double Machine Learning Work and Direct Machine Learning Does Not?

Loosely speaking, direct estimation of (3.3) by Lasso (or some other ML estimator) fails because it stretches the method beyond its intended use – prediction – and expects it to produce an interpretable, ‘structural’ coefficient estimate. Mullainathan and Spiess (2017) provide several insights into why this expectation is misguided in general. By contrast, the double machine learning approach uses these methods for their intended purpose only — to approximate conditional expectation functions in a flexible way. It is however possible to provide a more formal explanation.

For the sake of argument we start by assuming that the control function $g_0(X)$ is known. In this case τ can be consistently estimated by an OLS regression of $Y - g_0(X)$ on D (and a constant). This estimation procedure is of course equivalent to the moment condition

$$E[UD] = E\{[Y - g_0(X) - \tau D]D\} = 0.$$

Now suppose that instead of the true control function $g_0(X)$, we are presented with a somewhat perturbed version, $g_0(X) + t[g(X) - g_0(X)]$, where $g(X) - g_0(X)$ is the direction of perturbation and $t > 0$ is a scaling factor. When the scalar t is sufficiently small, the deviation of the perturbed control function from g_0 is

[8] More formally, let $s_{m,n}$ and $s_{\xi,n}$ denote the sparsity indices of m_0 and ξ_0, respectively. The full-sample variant of DML requires that both $s_{m,n}$ and $s_{\xi,n}$ grow slower than $\sqrt{n}$. The cross-fitted variant only requires that $s_{m,n} \cdot s_{\xi,n}$ grow slower than n.

(uniformly) small. In practice, the perturbed function is the Lasso estimate, which is subject to approximation error (selection mistakes) as well as estimation error.

How does working with the perturbed control function affect our ability to estimate τ? To answer this question, we can compute the derivative

$$\partial_t E\{[Y - g_0(X) - t \cdot h(X) - \tau D]D\}\big|_{t=0} \tag{3.6}$$

where $h(X) = g(X) - g_0(X)$. This derivative expresses the change in the moment condition used to estimate τ as one perturbs $g_0(X)$ in the direction $h(X)$ by a small amount. It is easy to verify that (3.6) is equal to $-E[h(X)D]$, which is generally non-zero, given that the covariates X are predictive of treatment status.

Intuitively, we can interpret this result in the following way. When equation (3.3) is estimated by Lasso directly, the implicit perturbation to $g_0(X)$ is the combined estimation and approximation error $h(X) = b(X)'\hat{\beta}^{dir} - g_0(X)$. Due to the impact of regularization (using a nonzero value of λ), this error can be rather large in finite samples, even though it vanishes asymptotically under sparseness conditions. More specifically, Lasso may mistakenly drop components of X from the regression that enter $g_0(X)$ nontrivially and are also correlated with D. This causes the derivative (3.6) to differ from zero and hence induces first-order omitted variable bias in the estimation of τ.

The DML procedure guards against large biases by using a moment condition with more favorable properties to estimate τ. As explained in Section 3.3.1, DML first estimates the conditional mean functions $\xi_0(X) = E(Y|X)$ and $m_0(X) = E(D|X)$, and then the estimated value of τ is obtained by a regression of $Y - \xi_0(X)$ on $D - m_0(X)$. The second step is equivalent to estimating τ based on the moment condition

$$E\Big\{\big(Y - \xi_0(X)\big)\big(D - m_0(X)\big) - \tau\big(D - m_0(X)\big)^2\Big\} = 0. \tag{3.7}$$

Once again, consider a thought experiment where we replace ξ_0 and m_0 with perturbed versions $\xi_0 + t(\xi - \xi_0)$ and $m_0 + t(m - m_0)$, respectively. To gauge how these perturbations affect the moment condition used to estimate τ, we can compute the derivative

$$\partial_t E\Big\{\big(Y - \xi_0 - t \cdot h_\xi\big)\big(D - m_0 - t \cdot h_m\big) - \tau\big(D - m_0 - t \cdot h_m\big)^2\Big\}\Big|_{t=0} \tag{3.8}$$

where $h_m(X) = m(X) - m_0(X)$ and $h_\xi(X) = \xi(X) - \xi_0(X)$. Given that the residuals $Y - \xi_0(X)$ and $D - m_0(X)$ are uncorrelated with all functions of X, including the deviations $h_m(X)$ and $h_\xi(X)$, it is straightforward to verify that

$$\begin{aligned}
&E\Big\{\big(Y - \xi_0 - t \cdot h_\xi\big)\big(D - m_0 - t \cdot h_m\big) - \tau\big(D - m_0 - t \cdot h_m\big)^2\Big\} \\
&= E\big[(Y - \xi_0)(D - m_0)\big] + t^2 E\big[h_m(X)h_\xi(X)\big] - \tau \cdot t^2 E\big[h_m(X)^2\big]
\end{aligned}$$

so that the derivative (3.8) evaluated at $t = 0$ is clearly zero.

This result means that the moment condition (3.7) is robust to small perturbations around the true nuisance functions $\xi_0(X)$ and $m_0(X)$. For example, even if Lasso drops some relevant components of X, causing a large error in the approximation to $g_0(X)$, the resulting bias in $\hat{\tau}^{DML}$ is an order smaller. This property is a key reason why DML works for estimating τ.

3.3.3 DML in a Method of Moments Framework

We can now give a more general perspective on double machine learning that is not directly tied to a regression framework. In particular, stage (ii) of the procedure described in Section 3.3.1 can be replaced by any moment condition that identifies the parameter of interest and satisfies an orthogonality condition analogous to (3.8) with respect to the unknown nuisance functions involved. (The nuisance functions are typically conditional means such as the expectation of the outcome or the treatment indicator conditional on a large set of covariates.) Stage (i) of the general DML framework consists of the machine learning estimation of the nuisance functions using a suitable method (such as the Lasso, random forest, etc.). The cross-fitted version of the estimator is recommended for practical use in general. That is, the first and the second stages should be conducted over non-overlapping subsamples and the roles of the subsamples should be rotated.

Chernozhukov et al. (2018) provides a detailed exposition of the general moment based framework with several applications. The general theory involves stating high level conditions on the nuisance functions and their first stage machine learning estimators so that the selection and estimation errors have a negligible impact on the second stage moment condition. In typical problems, the minimum convergence rate that is required of the first stage ML estimators is faster than $n^{-1/4}$.

As an example, we follow Chernozhukov et al. (2017) and consider the estimation of τ using a moment condition that is, in a sense, even more robust than (3.7). In particular, let us define $\xi_{0|D=0}(X) = E(Y|D=0,X)$, $\xi_{0|D=1}(X) = E(Y|D=1,X)$, $\xi_0 = (\xi_{0|D=0}, \xi_{0|D=1})$, and

$$\begin{aligned}\psi(W,m_0,\xi_0) =& \frac{D(Y-\xi_{0|D=1}(X))}{m_0(X)} + \xi_{0|D=1}(X) \\ &- \frac{(1-D)(Y-\xi_{0|D=0}(X))}{1-m_0(X)} - \xi_{0|D=0}(X),\end{aligned}$$

where $W = (Y,D,X)$. The true value of τ is identified by the orthogonal moment condition

$$E[\psi(W,m_0,\xi_0) - \tau] = 0. \tag{3.9}$$

Splitting the sample into, say, two parts I_1 and I_2, let $\hat{m}_{0,I_k}$ and $\hat{\xi}_{0,I_k}$ denote the first stage ML estimators over the subsample I_k. Define $\hat{\psi}_i$ as $\psi(W_i, \hat{m}_{0,I_2}, \hat{\xi}_{0,I_2})$ for $i \in I_1$ and as $\psi(W_i, \hat{m}_{0,I_1}, \hat{\xi}_{0,I_1})$ for $i \in I_2$. Then the DML estimator of τ is simply given by the sample average

$$\hat{\tau}^{DML} = \frac{1}{n}\sum_{i=1}^{n}\hat{\psi}_i.$$

Under regularity conditions mentioned above, the distribution of $\sqrt{n}(\hat{\tau}^{DML} - \tau)$ is asymptotically normal with mean zero and variance $E[(\psi - \tau)^2]$, which is consistently estimated by $n^{-1}\sum_{i=1}^{n}(\hat{\psi}_i - \hat{\tau}^{DML})^2$.

Equation (3.9) satisfies the previously introduced orthogonality condition

$$\partial_t E\left[\psi(W, m_0 + th_m, \xi_0 + th_\xi) - \tau\right]\big|_{t=0} = 0,$$

where $h_m(X)$ and $h_\xi(X)$ are perturbations to m_0 and ξ_0. However, an even stronger property is true. It is easy to verify that

$$E\left[\psi(W, m_0, \xi_0 + th_\xi) - \tau\right] = 0\ \forall t \quad \text{and} \quad E\left[\psi(W, m_0 + th_m, \xi_0) - \tau\right] = 0\ \forall t. \tag{3.10}$$

This means that if m_0 is reasonably well approximated, the moment condition for estimating τ is completely robust to specification (selection) errors in modeling ξ_0 and, conversely, if ξ_0 is well approximated, then specification errors in m_0 do not affect the estimation of τ. For example, if Lasso mistakenly drops an important component of X in estimating $\xi_{0|D=1}$ or $\xi_{0|D=0}$, this error is inconsequential as long as this variable (and all other relevant variables) are included in the approximation of m_0.

In settings in which the nuisance functions ξ_0 and m_0 are estimated based on finite dimensional parametric models, the moment condition (3.9) is often said to be 'doubly robust.' This is because property (3.10) implies that if fixed dimensional parametric models are postulated for ξ_0 and m_0, then consistent estimation of τ is still possible even when one of the models is misspecified.

The notion of an orthogonal moment condition goes back to Neyman (1959). There is an extensive statistical literature spanning several decades on the use of such moment conditions in various estimation problems. We cannot possibly do justice to this literature in this space; the interested reader could consult references in Chernozhukov et al. (2018) and Fan et al. (2020) for some guidance in this direction.

3.3.4 Extensions and Recent Developments in DML

There are practically important recent extensions of the DML framework that consider treatment effect estimation with continuous treatments or in the presence of mediating variables. For example, Colangelo and Lee (2022) study the average dose-response function (the mean of the potential outcome) as a function of treatment intensity. Let T denote a continuous treatment taking values in the set $\mathcal{T}$ and let $Y(t)$ be the potential outcome associated with treatment intensity $t \in \mathcal{T}$. The average dose-response function is given by $\mu_t = E[Y(t)]$. Under the weak unconfoundedness assumption $Y(t) \perp T | X$ and suitable continuity conditions, μ_t is identified by

$$\mu_t = \lim_{h \to 0} E\left[\frac{1}{h}K\left(\frac{T-t}{h}\right)\frac{Y}{f_{T|X}(t|X)}\right],$$

where $f_{T|X}(t|X)$ is the conditional density function of T given X (also called the generalized propensity score), $K(\cdot)$ is a kernel function and h is a bandwidth parameter. Colangelo and Lee (2022) show that the associated orthogonal moment condition is

$$\mu_t = \lim_{h \to 0} E\left[\gamma(t,X) + \frac{1}{h}K\left(\frac{T-t}{h}\right)\frac{Y-\gamma(t,X)}{f_{T|X}(t|X)}\right], \quad (3.11)$$

where $\gamma(t,x) = E[Y|T=t, X=x]$. They apply the DML method to estimate μ_t based on (3.11) and show that their kernel-based estimator is asymptotically normal but converges at a nonparametric rate.

Using the same framework, Hsu, Huber, Lee and Liu (2022) propose a Cramer-von Mises-type test for testing whether μ_t has a weakly monotonic relationship with the treatment dose t. They first transform the null hypothesis of a monotonic relationship to countably many moment inequalities where each of the moments can be identified by an orthogonal moment condition, and the DML method can be applied to obtain estimators converging at the parametric ($\sqrt{n}$) rate. They propose a multiplier bootstrap procedure to construct critical values and show that their test controls asymptotic size and is consistent against any fixed alternative.

Regarding causal mediation, Farbmacher, Huber, Laffers, Langen and Spindler (2022) study the average direct and indirect effect of a binary treatment operating through an intermediate variable that lies on the causal path between the treatment and the outcome. They provide orthogonal moment conditions for these quantities under the unconfoundedness assumption and show that the associated DML estimators are $\sqrt{n}$-consistent and asymptotically normal (see *ibid.* for further details).

Nevertheless, the use of DML extends beyond the estimation of various average treatment effects and is also applicable to related problems such as learning the optimal policy (assignment rule) from observational data (Athey & Wager, 2021). For example, officials may need to choose who should be assigned to a job training program using a set of observed characteristics and data on past programs. At the same time, the program may need to operate within a budget constraint and/or the assignment rule may need to satisfy other restrictions.

To set up the problem more formally, let π be a policy that maps a subject's characteristics to a 0-1 binary decision (such as admission to a program). The policy π is assumed to belong to a class of policies Π, which incorporates problem-specific constraints pertaining to budget, functional form, fairness, etc. The utilitarian welfare of deploying the policy π relative to treating no one is defined as

$$V(\pi) = E[Y(1)\pi(X) + Y(0)(1-\pi(X))] - E[Y(0)] = E[\pi(X)(Y(1)-Y(0))].$$

The policy maker is interested in finding the treatment rule π with the highest welfare in the class Π, i.e., solving $V^* = \max_{\pi \in \Pi} V(\pi)$. A π^* satisfying $V(\pi^*) = V^*$ or equivalently $\pi^* \in \arg\max_{\pi \in \Pi} V(\pi)$ is called an optimal treatment rule (the optimal treatment rule might not be unique). Under the unconfoundedness assumption, $V(\pi)$ is identified as

$$V(\pi) = E\left[\pi(X)\Big(\frac{DY}{m_0(X)} - \frac{(1-D)Y}{1-m_0(X)}\Big)\right], \tag{3.12}$$

or, alternatively, as

$$V(\pi) = E\left[\pi(X)\Big(\frac{D(Y-\xi_{0|D=1}(X))}{m_0(X)} + \xi_{0|D=1}(X) - \frac{(1-D)(Y-\xi_{0|D=0}(X))}{1-m_0(X)} - \xi_{0|D=0}(X)\Big)\right], \tag{3.13}$$

where $\xi_{0|D=0}$, $\xi_{0|D=1}$ and m_0 are defined as in Section 3.3.3. In studying the policy learning problem, Kitagawa and Tetenov (2018) assume that the propensity score function m_0 is known and estimate $V(\pi)$ based on (3.12) using an inverse probability weighted estimator. They show that the difference between the estimated optimal welfare and the true optimal welfare decays at the rate of $1/\sqrt{n}$ under suitable control over the complexity of the class of decision rules. Athey and Wager (2021) extend Kitagawa and Tetenov (2018) in two aspects. First, Athey and Wager (2021) allow for the case in which the propensity score is unknown and estimate $V(\pi)$ based on the orthogonal moment condition (3.13) using the DML approach. They show that the difference between the estimated optimal welfare and the true optimal welfare continues to converge to zero at the rate of $1/\sqrt{n}$ under suitable conditions. Second, in addition to binary treatments with unconfounded assignment, Athey and Wager (2021) also allow for endogenous and continuous treatments.

3.4 Using Machine Learning to Discover Treatment Effect Heterogeneity

3.4.1 The Problem of Estimating the CATE Function

Under the unconfoundedness assumption (3.1), it is the full dimensional conditional average treatment effect (CATE) function that provides the finest breakdown of the average treatment effect across all the subpopulations defined by the possible values of X. Without any assumptions restricting individual treatment effect heterogeneity, it is given by

$$\tau(X) = E[Y(1) - Y(0)|\, X] = \tau + g_1(X) - g_0(X).$$

For example, in the context of the empirical application presented in Section 3.5, $\tau(X)$ describes the average effect on birthweight of smoking during pregnancy given the mother's age, education, various medical conditions, the pattern of prenatal care received, etc. Of course, some of these variables may be more important in capturing heterogeneity than others.

While under the treatment effect homogeneity assumption $g_0(X)$ was simply a nuisance object, now $\tau + g_1(X) - g_0(X)$ has become an infinite dimensional parameter of interest. Following the same steps as in Section 3.3.1, equation (3.5) generalizes to

$$Y - \xi_0(X) = \tau(X)(D - m_0(X)) + U$$

In addition to pre-estimating ξ_0 and m_0, the generalization of the DML approach requires the introduction of some type of structured approximation to $\tau(X)$ to make this equation estimable. For example, one could specify a parametric model $\tau(X) = b(X)'\beta$, where the dimension of $b(X)$ is fixed and its relevant components are selected based on domain-specific theory and expert judgement. Then β can still be estimated by OLS, but the procedure is subject to misspecification bias. Using instead the Lasso to estimate β brings back the previously described problems associated with direct machine learning estimation of causal parameters.

But there is an even more fundamental problem—when X is high dimensional, the CATE function may be quite complex and hard to describe, let alone visualize. While $\tau(X)$ could be completely flat along some coordinates of X, it can be highly nonlinear in others with complex interactions between multiple components. Suppose, for example, that one could somehow obtain a debiased series approximation to $\tau(X)$. It might contain terms such as $2.4X_3 - 1.45X_2^2X_5^2 + 0.32X_1X_3^2X_5 + \ldots$. Understanding and analyzing this estimate is already a rather formidable task.

There are two strands of the econometrics literature offering different solutions to this problem. The first does not give up the goal of discovering $\tau(X)$ in its entirety and uses a suitably modified regression tree algorithm to provide a step-function approximation to $\tau(X)$. Estimated regression trees are capable of capturing and presenting complex interactions in a relatively straightforward way that is often amenable to interpretation. A pioneering paper of this approach is Athey and Imbens (2016) with several follow-ups and extensions such as Wager and Athey (2018) and Athey et al. (2019). The second idea is to reduce the dimensionality of $\tau(X)$ to a coordinate (or a handful of coordinates) of interest and integrate out the 'unneeded' components. More formally, let X_1 be a component or a small subvector of X. Abrevaya, Hsu and Lieli (2015) introduce the reduced dimensional CATE function as

$$\tau(X_1) = E[Y(1) - Y(0)|X_1] = E[\tau(X)|X_1],$$

where the second equality follows from the law of iterated expectations.[9] As proposed by Fan et al. (2020) and Semenova and Chernozhukov (2021), the reduced dimensional CATE function can be estimated using an extension of the moment-based DML method, where the second stage of the procedure, which is no longer a high dimensional problem, employs a traditional nonparametric estimator.

[9] It is a slight abuse of notation to denote the functions $\tau(X)$ and $\tau(X_1)$ with the same letter. We do so for simplicity; the arguments of the two function will distinguish between them.

The two approaches to estimating heterogeneous average treatment effects complement each other. The regression tree algorithm allows the researcher to discover in an automatic and data-driven way which variables, if any, are the most relevant drivers of treatment effect heterogeneity. The downside is that despite the use of regression trees, the results can still be too detailed and hard to present and interpret. By contrast, the dimension-reduction methods ask the researcher to pre-specify a variable of interest and the heterogeneity of the average treatment effect is explored in a flexible way focusing on this direction. The downside, of course, is that other relevant predictors of treatment effect heterogeneity may remain undiscovered.

3.4.2 The Causal Tree Approach

Regression tree basics. Regression trees are algorithmically constructed step functions used to approximate conditional expectations such as $E(Y|X)$. More specifically, let $\Pi = \{\ell_1, \ell_2, \ldots, \ell_{\#\Pi}\}$ be a partition of $\mathcal{X} = support(X)$ and define

$$\bar{Y}_{\ell_j} = \frac{\sum_{i=1}^{n} Y_i 1_{\ell_j}(X_i)}{\sum_{i=1}^{n} 1_{\ell_j}(X_i)}, \quad j = 1, \ldots, \#\Pi,$$

to be the average outcome for those observations i for which $X_i \in \ell_j$. A regression tree estimates $E(Y|X = x)$ using a step function

$$\hat{\mu}(x;\Pi) = \sum_{j=1}^{\#\Pi} \bar{Y}_{\ell_j} 1_{\ell_j}(x). \tag{3.14}$$

The regression tree algorithm considers partitions Π that are constructed based on recursive splits of the support of the components of X. Thus, the subsets ℓ_j, which are called the leaves of the tree, are given by intersections of sets of the form $\{X_k \leq c\}$ or $\{X_k > c\}$, where X_k denotes the kth component of X. In building the regression tree, candidate partitions are evaluated through a mean squared error (MSE) criterion with an added term that penalizes the number of splits to avoid overfitting. Chapter 2 of this volume provides a more in-depth look at classification and regression trees.

From a regression tree to a causal tree. Under the unconfoundendess assumption one way to estimate ATE consistently is to use the inverse probability weighted estimator proposed by Hirano, Imbens and Ridder (2003):

$$\hat{\tau} = \frac{1}{\#S} \sum_{i \in S} \left[\frac{Y_i D_i}{m(X_i)} - \frac{Y_i(1 - D_i)}{1 - m(X_i)} \right], \tag{3.15}$$

where $\mathcal{S}$ is the sample of observations on (Y_i, D_i, X_i), $\#\mathcal{S}$ is the sample size, and $m(\cdot)$ is the propensity score function, i.e., the conditional probability $m(X) = P(D = 1|X)$. To simplify the exposition, we will assume that the function $m(X)$ is known.[10]

Given a subset $\ell \subset \mathcal{X}$, one can also implement the estimator (3.15) in the subsample of observations for which $X_i \in \ell$, yielding an estimate of the conditional average treatment effect $\tau(\ell) = E[Y(1) - Y(0)|X_i \in \ell]$. More specifically, we define

$$\hat{\tau}_{\mathcal{S}}(\ell) = \frac{1}{\#\ell} \sum_{i \in \mathcal{S}, X_i \in \ell} \left[\frac{Y_i D_i}{m(X_i)} - \frac{Y_i(1 - D_i)}{1 - m(X_i)} \right],$$

where $\#\ell$ is the number of observations that fall in ℓ.

Computing $\hat{\tau}_{\mathcal{S}}(\ell)$ for various choices of ℓ is called subgroup analysis. Based on subject matter theory or policy considerations, a researcher may pre-specify some subgroups of interest. For example, theory may predict that the effect changes as a function of age or income. Nevertheless, theory is rarely detailed enough to say exactly how to specify the relevant age or income groups and may not be able to say whether the two variables interact with each other or other variables. There may be situations, especially if the dimension of X is high, where the relevant subsets need to be discovered completely empirically by conducting as detailed a search as possible. However, it is now well understood that mining the data 'by hand' for relevant subgroups is problematic for two reasons. First, some of these groups may be complex and hard to discover, e.g., they may involve interactions between several variables. Moreover, if X is large there are simply too many possibilities to consider. Second, it is not clear how to conduct inference for groups uncovered by data mining. The (asymptotic) distribution of $\hat{\tau}(\ell)$ is well understood for fixed ℓ. But if the search procedure picks a group $\hat{\ell}$ because, say, $\hat{\tau}(\hat{\ell}) - \hat{\tau}$ is large, then the distribution of $\hat{\tau}(\hat{\ell})$ will of course differ from the fixed ℓ case.

In their influential paper Athey and Imbens (2016) propose the use of the regression tree algorithm to search for treatment effect heterogeneity, i.e., to discover the relevant subsets $\hat{\ell}$ from the data itself. The resulting partition $\hat{\Pi}$ and the estimates $\hat{\tau}(\hat{\ell})$, $\hat{\ell} \in \hat{\Pi}$ are called a causal tree. Their key contribution is to modify to the standard regression tree algorithm in a way that accommodates treatment effect estimation (as opposed to prediction) and addresses the statistical inference problem discussed above.

The first proposed modification is what Athey and Imbens (2016) call an 'honest' approach. This consist of partitioning the available data $\mathcal{S} = \{(Y_i, D_i, X_i)\}_{i=1}^n$ into an estimation sample $\mathcal{S}^{est}$ and a training sample $\mathcal{S}^{tr}$. The search for hetereogeneity, i.e., the partitioning of $\mathcal{X}$ into relevant subsets $\hat{\ell}$, is conducted entirely over the training sample $\mathcal{S}^{tr}$. Once a suitable partition of $\mathcal{X}$ is identified, it is taken as given and the group-specific treatment effects are re-estimated over the independent estimation sample that has been completely set aside up to that point. More formally, the eventual conditional average treatment effect estimates are of the form $\hat{\tau}_{\mathcal{S}^{est}}(\hat{\ell}_{\mathcal{S}^{tr}})$, where the notation emphasizes the use of the two samples for different purposes. Inference based

[10] It is not hard to extend the following discussions to the more realistic case in which $m(X)$ needs to be estimated. We also note that even when $m(X)$ is known, it is more efficient to work with an estimated counterpart; see Hirano et al. (2003).

on $\hat{\tau}_{\mathcal{S}^{est}}(\hat{\ell}_{\mathcal{S}^{tr}})$ can then proceed as if $\hat{\ell}_{\mathcal{S}^{tr}}$ were fixed. While in the DML literature sample splitting is an enhancement, it is absolutely crucial in this setting.

The second modification concerns the MSE criterion used in the tree-building algorithm. The criterion function is used to compare candidate partitions, i.e., it is used to decide whether it is worth imposing additional splits on the data to estimate the CATE function in more detail. The proposed changes account for the fact that (i) instead of approximating a conditional expectation function the goal is to estimate treatment effects; and (ii) for any partition $\hat{\Pi}$ constructed from $\mathcal{S}^{tr}$, the corresponding conditional average treatment effects will be re-estimated using $\mathcal{S}^{est}$.

Technical discussion of the modified criterion. We now formally describe the proposed criterion function. Given a partition $\Pi = \{\ell_1, \ldots, \ell_{\#\Pi}\}$ of $\mathcal{X}$ and a sample $\mathcal{S}$, let

$$\hat{\tau}_{\mathcal{S}}(x;\Pi) = \sum_{j=1}^{\#\Pi} \hat{\tau}_{\mathcal{S}}(\ell_j) 1_{\ell_j}(x)$$

be the corresponding step function estimator of the CATE function $\tau(x)$, where the value of $\hat{\tau}_{\mathcal{S}}(x;\Pi)$ is the constant $\hat{\tau}_{\mathcal{S}}(\ell_j)$ for $x \in \ell_j$. For a given $x \in \mathcal{X}$, the MSE of the CATE estimator is $E[(\tau(x) - \hat{\tau}_{\mathcal{S}}(x;\Pi))^2]$; the proposed criterion function is based on the expected (average) MSE

$$\text{EMSE}(\Pi) = E_{X_t, \mathcal{S}^{est}}\left[\left(\tau(X_t) - \hat{\tau}_{\mathcal{S}^{est}}(X_t;\Pi)\right)^2\right],$$

where X_t is a new, independently drawn,'test' observation. Thus, the goal is to choose the partition Π in a way so that $\hat{\tau}_{\mathcal{S}^{est}}(x;\Pi)$ provides a good approximation to $\tau(x)$ on average, where the averaging is with respect to the marginal distribution of X.

While EMSE(Π) cannot be evaluated analytically, it can still be estimated. To this end, one can rewrite EMSE(Π) as[11]

$$\text{EMSE}(\Pi) = E_{X_t}\left\{V_{\mathcal{S}^{est}}[\hat{\tau}_{\mathcal{S}^{est}}(X_t;\Pi)]\right\} - E[\tau(X_t;\Pi)^2] + E[\tau(X_t)^2], \qquad (3.16)$$

where $V_{\mathcal{S}^{est}}(\cdot)$ denotes the variance operator with respect to the distribution of the sample $\mathcal{S}^{est}$ and

$$\tau(x;\Pi) = \sum_{j=1}^{\#\Pi} \tau(\ell_j) 1_{\ell_j}(x).$$

As the last term in (3.16) does not depend on Π, it does not affect the choice of the optimal partition. We will henceforth drop this term from (3.16) and denote the remaining two terms as EMSE—a convenient and inconsequential abuse of notation. Recall that the key idea is to complete the tree building process (i.e., the choice of the partition Π) on the basis of the training sample alone. Therefore, EMSE(Π) will be

[11] Equation (3.16) is derived in the Electronic Online Supplement, *Section 3.1*. The derivations assume that $E(\hat{\tau}_{\mathcal{S}}(\ell_j)) = \tau(\ell_j)$, i.e., that the leaf-specific average treatment effect estimator is unbiased. This is true if the propensity score function is known but only approximately true otherwise.

estimated using $\mathcal{S}^{tr}$; the only information used from $\mathcal{S}^{est}$ is the sample size, denoted as $\#\mathcal{S}^{est}$.

We start with the expected variance term. It is given by

$$E_{X_t}\left\{V_{\mathcal{S}^{est}}[\hat{\tau}_{\mathcal{S}^{est}}(X_t;\Pi)]\right\} = \sum_{j=1}^{\#\Pi} V_{\mathcal{S}^{est}}[\hat{\tau}_{\mathcal{S}^{est}}(\ell_j)]P(X_t \in \ell_j).$$

The variance of $\hat{\tau}_{\mathcal{S}^{est}}(\ell_j)$ is of the form $\sigma_j^2/\#\ell_j^{est}$, where $\#\ell_j^{est}$ is the number of observations in the estimation sample falling in leaf ℓ_j and

$$\sigma_j^2 = V\left[Y_i D_i/p(X_i) - Y_i(1-D_i)/(1-p(X_i))\middle| X_i \in \ell_j\right].$$

Substituting $V_{\mathcal{S}^{est}}[\hat{\tau}_{\mathcal{S}^{est}}(\ell_j)] = \sigma_j^2/\#\ell_j^{est}$ into the expected variance equation yields

$$\begin{aligned} E_{X_t}\left\{V_{\mathcal{S}^{est}}[\hat{\tau}_{\mathcal{S}^{est}}(X_t;\Pi)]\right\} &= \sum_{j=1}^{\#\Pi} \frac{\sigma_j^2}{\#\ell_j^{est}} P(X_t \in \ell_j) \\ &= \frac{1}{\#\mathcal{S}^{est}} \sum_{j=1}^{\#\Pi} \sigma_j^2 P(X_t \in \ell_j) \frac{\#\mathcal{S}^{est}}{\#\ell_j^{est}}. \end{aligned}$$

As $\#\mathcal{S}^{est}/\#\ell_j^{est} \approx 1/P(X_i \in \ell_j)$ we can simply estimate the expected variance term by

$$\hat{E}_{X_t}\left\{V_{\mathcal{S}^{est}}[\hat{\tau}_{\mathcal{S}^{est}}(X_t;\Pi)]\right\} = \frac{1}{\#\mathcal{S}^{est}} \sum_{j=1}^{\#\Pi} \hat{\sigma}_{j,\mathcal{S}^{tr}}^2, \tag{3.17}$$

where $\hat{\sigma}_{j,\mathcal{S}^{tr}}^2$ is a suitable (approximately unbiased) estimator of σ_j^2 over the *training sample*.

Turning to the second moment term in (3.16), note that for any sample $\mathcal{S}$ and an independent observation X_t,

$$E_X[\tau(X_t;\Pi)^2] = E_X E_{\mathcal{S}}[\hat{\tau}_{\mathcal{S}}(X_t;\Pi)^2] - E_X\{V_{\mathcal{S}}[\hat{\tau}_{\mathcal{S}}(X_t;\Pi)]\}$$

because $E_{\mathcal{S}}[\hat{\tau}_{\mathcal{S}}(x;\Pi)] = \tau(x;\Pi)$ for any fixed point x. Thus, an unbiased estimator of $E_X[\tau(X_t;\Pi)^2]$ can be constructed from the training sample $\mathcal{S}^{tr}$ as

$$\hat{E}_X[\tau(X_t;\Pi)^2] = \frac{1}{\#\mathcal{S}^{tr}} \sum_{i \in \mathcal{S}^{tr}} \hat{\tau}_{\mathcal{S}^{tr},-i}(X_i;\Pi)^2 - \frac{1}{\#\mathcal{S}^{tr}} \sum_{j=1}^{\#\Pi} \hat{\sigma}_{j,\mathcal{S}^{tr}}^2, \tag{3.18}$$

where $\hat{\tau}_{\mathcal{S}^{tr},-i}$ is the leave-one-out version of $\hat{\tau}_{\mathcal{S}^{tr}}$ and we use the analog of (3.17) to estimate the expected variance of $\hat{\tau}_{\mathcal{S}^{tr}}(X_t;\Pi)$. Combining the estimators (3.17) and (3.18) with the decomposition (3.16) gives the estimated EMSE criterion function

$$\widehat{\text{EMSE}}(\Pi) = \left(\frac{1}{\#\mathcal{S}^{est}} + \frac{1}{\#\mathcal{S}^{tr}}\right)\sum_{j=1}^{\#\Pi} \hat{\sigma}^2_{j,\mathcal{S}^{tr}} - \frac{1}{\#\mathcal{S}^{tr}} \sum_{i \in \mathcal{S}^{tr}} \hat{\tau}_{\mathcal{S}^{tr},-i}(X_i;\Pi)^2. \tag{3.19}$$

The criterion function (3.19) has almost exactly the same form as in Athey and Imbens (2016), except that they do not use a leave-one-out estimator in the second term.

The intuition about how the criterion (3.19) works is straightforward. Say that Π and Π' are two partitions where Π' is finer in the sense that there is an additional split along a given X coordinate. If the two estimated treatment effects are not equal across this extra split, then the second term in the difference will decrease in absolute value, making $\widehat{\text{EMSE}}(\Pi')$ ceteris paribus lower. However, the first term of the criterion also takes into account the fact that an additional split will result in leaves with fewer observations, increasing the variance of the ultimate CATE estimate. In other words, the first term will generally increase with an additional split and it is the net effect that determines whether Π or Π' is deemed as a better fit to the data.

The criterion function (3.19) is generalizable to other estimators. In fact, the precise structure of $\hat{\tau}_{\mathcal{S}}(\ell)$ played almost no role in deriving (3.19); the only properties we made use of was unbiasedness ($E_{\mathcal{S}}(\hat{\tau}_{\mathcal{S}}(\ell_j)) = \tau(\ell_j)$) and that the variance of $\hat{\tau}_{\mathcal{S}}(\ell_j)$ is of the form $\sigma_j^2/\#\ell_j$. Hence, other types of estimators could be implemented in each leaf; for example, Reguly (2021) uses this insight to extend the framework to the parametric sharp regression discontinuity design.

3.4.3 Extensions and Technical Variations on the Causal Tree Approach

Wager and Athey (2018) extend the causal tree approach of Athey and Imbens (2016) to causal forests, which are composed of causal trees with a conditional average treatment effect estimate in each leaf. To build a causal forest estimator, one first generates a large number of random subsamples and grows a causal tree on each subsample using a given procedure. The causal forest estimate of the CATE function is then obtained by averaging the estimates produced by the individual trees. Practically, a forest approach can reduce variance and smooth the estimate.

Wager and Athey (2018) propose two honest causal forest algorithms. One is based on double-sample trees and the other is based on propensity score trees. The construction of a double-sample tree involves a procedure similar to the one described in Section 4.2. First one draws without replacement B subsamples of size $s = O(n^\rho)$ for some $0 < \rho < 1$ from the original data. Let these artificial samples be denoted as $\mathcal{S}_b$, $b = 1, \dots, B$. Then one splits each $\mathcal{S}_b$ into two parts with size $\#\mathcal{S}_b^{tr} = \#\mathcal{S}_b^{est} = s/2$.[12] The tree is grown using the $\mathcal{S}_b^{tr}$ data and the leafwise treatment effects are estimated using the $\mathcal{S}_b^{est}$ data, generating an individual estimator $\hat{\tau}_b(x)$. The splits of the tree are chosen by minimizing an expected MSE criterion analogous to (3.19), where $\mathcal{S}_b^{tr}$ takes the role of $\mathcal{S}^{tr}$ and $\mathcal{S}_b^{est}$ takes the role of $\mathcal{S}^{est}$. For any fixed value x, the causal forest CATE estimator $\hat{\tau}(x)$ is obtained by averaging the individual estimates

[12] We assume s is an even number.

$\hat{\tau}_b(x)$, i.e., $\hat{\tau}(x) = B^{-1}\sum_{b=1}^{B}\hat{\tau}_b(x)$. The propensity score tree procedure is similar except that one grows the tree using the whole subsample $\mathcal{S}_b$ and uses the treatment assignment as the outcome variable. Wager and Athey (2018) show that the random forest estimates are asymptotically normally distributed and the asymptotic variance can be consistently estimated by infinitesimal jackknife so valid statistical inference is available.

Athey et al. (2019) propose a generalized random forest method that transforms the traditional random forest into a flexible procedure for estimating any unknown parameter identified via local moment conditions. The main idea is to use forest-based algorithms to learn the problem specific weights so as to be able to solve for the parameter of interest via a weighted local M-estimator. For each tree, the splitting decisions are based on a gradient tree algorithm; see Athey et al. (2019) for further details.

There are other methods proposed in the literature for the estimation of treatment effect heterogeneity such as the neural network based approaches by Yao et al. (2018) and Alaa, Weisz and van der Schaar (2017). These papers do not provide asymptotic theory and valid statistical inference is not available at the moment, so this is an interesting direction for future research.

3.4.4 The Dimension Reduction Approach

As discussed in Section 3.4.1, the target of this approach is the reduced dimensional CATE function $\tau(X_1) = E[Y(1) - Y(0)|X_1]$, where X_1 is a component of X.[13] This variable is a priori chosen (rather than discovered from the data) and the goal is to estimate $\tau(X_1)$ in a flexible way. This can be accomplished by an extension of the DML framework.

Let $\psi(W, m_0, \xi_0)$ be defined as in Section 3.4.1. Then it is not hard to show the reduced dimensional CATE function is identified by the orthogonal moment condition

$$E[\psi(W, m_0, \xi_0) - \tau(X_1)|X_1] = 0 \Longleftrightarrow \tau(X_1) = E[\psi(W, m_0, \xi_0)|X_1].$$

While ATE is given by the unconditional expectation of the ψ function, the reduced dimensional CATE function $\tau(X_1)$ is the conditional expectation $E(\psi|X_1)$. As X_1 is a scalar, it is perfectly feasible to estimate this regression function by a traditional kernel-based nonparametric method such as a local constant (Nadaraya-Watson) or a local linear regression estimator.

To be more concrete, let us consider the same illustrative setup as in Section 3.3.3. Splitting the sample into, say, two parts I_1 and I_2, let $\hat{m}_{0,I_k}$ and $\hat{\xi}_{0,I_k}$ denote the first stage ML estimators over the subsample I_k. Define $\hat{\psi}_i$ as $\psi(W_i, \hat{m}_{0,I_2}, \hat{\xi}_{0,I_2})$ for $i \in I_1$ and as $\psi(W_i, \hat{m}_{0,I_1}, \hat{\xi}_{0,I_1})$ for $i \in I_2$. Then, for a given $x_1 \in support(X_1)$, the DML estimator of $\tau(x_1)$ with a Nadaraya-Watson second stage is given by

[13] The theory allows for X_1 to be a small vector, but $\tau(X_1)$ is easiest to visualize, which is perhaps its main attraction, when X_1 is a scalar. So this is the case we will focus on here.

$$\hat{\tau}^{DML}(x_1) = \frac{\sum_{i=1}^{n} \hat{\psi}_i K\left(\frac{X_{1i}-x_1}{h}\right)}{\sum_{i=1}^{n} K\left(\frac{X_{1i}-x_1}{h}\right)},$$

where $h = h_n$ is a bandwidth sequence (satisfying $h \to 0$ and $nh \to \infty$) and $K(\cdot)$ is a kernel function. Thus, instead of simply taking the average of the $\hat{\psi}_i$ values, one performs a nonparametric regression of $\hat{\psi}_i$ on X_{1i}. An estimator of this form was already proposed by Abrevaya et al. (2015), except their identification was based on a non-orthogonal inverse probability weighted moment condition, and their first stage estimator of the propensity score was a traditional parametric or nonparametric estimator.

The asymptotic theory of $\hat{\tau}^{DML}(x_1)$ is developed by Fan et al. (2020).[14] The central econometric issue is again finding the appropriate conditions under which the first stage model selection and estimation leaves the asymptotic distribution of the second stage estimator unchanged in the sense that $\hat{\tau}^{DML}(x_1)$ is first order asymptotically equivalent to the infeasible estimator in which $\hat{\psi}_i$ is replaced by ψ_i. In this case

$$\sqrt{nh}[\hat{\tau}^{DML}(x_1) - \tau(x_1)] \to_d N(0, \sigma^2(x_1)), \text{ for all } x_1 \in support(X_1), \quad (3.20)$$

provided that the undersmoothing condition $nh^5 \to 0$ holds to eliminate asymptotic bias. However, one is often interested in the properties of the entire function $x_1 \mapsto \tau(x_1)$ rather than just its value evaluated at a fixed point. *Ibid.* state a uniform representation result for $\hat{\tau}^{DML}(x_1)$ that implies (3.20) and but also permits the construction of *uniform* confidence bands that contain the whole function $x_1 \mapsto \tau(x_1)$ with a pre-specified probability. They propose the use of a multiplier bootstrap procedure for this purpose, which requires only a single estimate $\hat{\tau}^{DML}(x_1)$.

The high level conditions used by Fan et al. (2020) to derive these results account for the interaction between the required convergence rate of the first stage ML estimators and the second stage bandwidth h. (A simplified general sufficient condition is that the ML estimators converge faster than $h^{1/4}n^{-1/4}$.) They also show that the high level conditions are satisfied in case of a Lasso first stage and strengthened sparsity assumptions on the nuisance functions relative to ATE estimation; see *ibid.* for details.[15]

The paper by Semenova and Chernozhukov (2020) extends the general DML framework of Section 3.3.3 to cases in which the parameter of interest is a function identified by a low-dimensional conditional moment condition. This includes the reduced dimensional CATE function discussed thus far but other estimands as well. They use series estimation in the second stage and provide asymptotic results that

[14] To be exact, ibid. use local linear regression in the second stage; the local constant version of the estimator is considered in an earlier working paper. The choice makes no difference in the main elements of the theory and the results.

[15] Using the notation of footnote 8, a sufficient sparsity assumption for the cross-fitted case considered here is that $s_m \cdot s_\xi$ grows slower than nh, where s_ξ is the larger sparsity index among the two components of ξ.

can be used for inference under general high-level conditions on the first stage ML estimators.

3.5 Empirical Illustration

We revisit the application in Abrevaya et al. (2015) and Fan et al. (2020) and study the effect of maternal smoking during pregnancy on the baby's birthweight. The data set comes from vital statistics records in North Carolina between 1988 and 2002 and is large both in terms of the number of observations and the available covariates. We restrict the sample to first time mothers and, following the literature, analyze the black and caucasian (white) subsamples separately.[16] The two sample sizes are 157,989 and 433,558, respectively.

The data set contains a rich set of individual-level covariates describing the mother's socioeconomic status and medical history, including the progress of the pregnancy. This is supplemented by zip-code level information corresponding to the mother's residence (e.g., per capita income, population density, etc.). Table 3.1 summarizes the definition of the dependent variable (Y), the treatment variable (D) and the control variables used in the analysis. We divide the latter variables into a primary set X_1, which includes the more important covariates and some of their powers and interactions, and a secondary set X_2, which includes auxiliary controls and their transformations.[17] Altogether, X_1 includes 26 variables while the union of X_1 and X_2 has 679 components for the black subsample and 743 for the caucasian.

[16] The reason for focusing on first time mothers is that in case of previous deliveries we cannot identify the data point that corresponds to it. Birth outcomes across the same mother at different points in time are more likely to be affected by unobserved heterogeneity (see Abrevaya et al., 2015 for a more detailed discussion).

[17] As X_1 and X_2 already include transformations of the raw covariates, these vectors correspond to the dictionary $b(X)$ in the notation of Section 3.3.

Table 3.1: The variables used in the empirical exercise

Y	bweight	birth weight of the baby (in grams)
D	smoke	if mother smoked during pregnancy (1 if yes)
X_1	mage	mother's age (in years)
	meduc	mother's education (in years)
	prenatal	month of first prenatal visit
	prenatal_visits	number of prenatal visits
	male	baby's gender (1 if male)
	married	if mother is married (1 if yes)
	drink	if mother used alcohol during pregnancy (1 if yes)
	diabetes	if mother has diabetes (1 if yes)
	hyperpr	if mother has high blood pressure (1 if yes)
	amnio	if amniocentesis test (1 if yes)
	ultra	if ultrasound during pregnancy (1 if yes)
	dterms	previous terminated pregnancies (1 if yes)
	fagemiss	if father's age is missing (1 if yes)
	polynomials:	mage^2
	interactions: mage ×	meduc, prenatal, prenatal_visits, male, married, drink, diabetes, hyperpr, amnio, ultra, dterms, fagemiss
X_2	mom_zip	zip code of mother's residence (as a series of dummies)
	byear	birth year 1988-2002 (as series of dummies)
	anemia	if mother had anemia (1 if yes)
	med_inc	median income in mother's zip code
	pc_inc	per capita income in mother's zip code
	popdens	population density in mother's zip code
	fage	father's age (in years)
	feduc	father's education (in years)
	feducmisss	if father's education is missing (1 if yes)
	polynomials:	fage^2, fage^3, mage^3

We present two exercises. First, we use the DML approach to estimate τ, i.e., the average effect (ATE) of smoking. This is a well-studied problem with several estimates available in the literature using various data sets and methods (see e.g., Abrevaya, 2006, Da Veiga & Wilder, 2008 and Walker, Tekin & Wallace, 2009). The point estimates range from about -120 to -250 grams with the magnitude of the effect being smaller for blacks than whites. Second, we use the causal tree approach to search for covariates that drive heterogeneity in the treatment effect and explore the full-dimensional CATE function $\tau(X)$. To our knowledge, this is a new exercise; both Abrevaya et al. (2015) and Fan et al. (2020) focus on the reduced dimensional CATE function with mother's age as the pre-specified variable of interest.

The findings from the first exercise for black mothers are presented in Table 3.2; the corresponding results for white mothers can be found in the Electronic Online Supplement, *Table 3.1.* The left panel shows a simple setup where we use X_1 as the

Table 3.2: Estimates of τ for black mothers

	Basic setup		Extended setup	
	Point-estimate	SE	Point-estimate	SE
OLS	-132.3635	6.3348	-130.2478	6.3603
Naive Lasso (λ^*)	-131.6438	-	-132.3003	-
Naive Lasso ($0.5\lambda^*$)	-131.6225	-	-130.7610	-
Naive Lasso ($2\lambda^*$)	-131.4569	-	-134.6669	-
Post-naive-Lasso (λ^*)	-132.3635	6.3348	-128.6592	6.2784
Post-naive-Lasso ($0.5\lambda^*$)	-132.3635	6.3348	-130.7227	6.2686
Post-naive-Lasso ($2\lambda^*$)	-132.3635	6.3348	-130.2032	6.3288
DML (λ^*)	-132.0897	6.3345	-129.9787	6.3439
DML ($0.5\lambda^*$)	-132.1474	6.3352	-128.8866	6.3413
DML ($2\lambda^*$)	-132.1361	6.3344	-131.5680	6.3436
DML-package	-132.0311	6.3348	-131.0801	6.3421

Notes: Sample size= $157,989$. λ^* denotes Lasso penalties obtained by 5-fold cross validation. The DML estimators are implemented by 2-fold cross-fitting. The row titled 'DML-package' contains the estimate obtained by using the 'official' DML code (`dml2`) available at https://docs.doubleml.org/r/stable/. All other estimators are programmed directly by the authors.

vector of controls, whereas the right panel works with the extended set $X_1 \cup X_2$. In addition to the DML estimates, we report several benchmarks: (i) the OLS estimate of ATE from a regression of Y on D and the set of controls; (ii) the corresponding direct/naive Lasso estimates of ATE (with the treatment dummy excluded from the penalty); and (iii) post-Lasso estimates where the model selected by the Lasso is re-estimated by OLS. The symbol λ^* denotes Lasso penalty levels chosen by 5-fold cross validation; we also report the results for the levels $\lambda^*/2$ and $2\lambda^*$. The DML estimators are implemented using 2-fold cross-fitting.

The two main takeaways from Table 3.2 are that the estimated 'smoking effect' is, on average, -130 grams for first time black mothers (which is consistent with previous estimates in the literature), and that this estimate is remarkably stable across the various methods. This includes the OLS benchmark with the basic and extended set of controls as well as the corresponding naive Lasso and post-Lasso estimators. We find that even at the penalty level $2\lambda^*$ the naive Lasso keeps most of the covariates and so do the first stage Lasso estimates from the DML procedure. This is due to the fact that even small coefficients are precisely estimated in such large samples, so one does not achieve a meaningful reduction in variance by setting them to zero. As the addition of virtually any covariate improves the cross-validated MSE by a small amount, the optimal penalty λ^* is small. With close to the full set of covariates utilized in either setup, and a limited amount of shrinkage due to λ^* being small, there is little difference across the methods.[18] In addition, using $X_1 \cup X_2$ versus X_1 alone affects the size of the point estimates only to a small degree—the post-Lasso and DML estimates based on the extended setup are 1 to 3 grams smaller in magnitude. Physically this is a very small difference, though still a considerable fraction of the standard error.

It is also noteworthy that the standard errors provided by the post-Lasso estimator are very similar to the DML standard errors. Hence, in the present situation naively conducting inference using the post-Lasso estimator would lead to the proper conclusions. Of course, relying on this estimator for inference is still bad practice as there are no *a priori* theoretical guarantees for it to be unbiased or asymptotically normal.

The results for white mothers (see the Electronic Online Supplement, *Table 3.1.*) follow precisely the same patterns as discussed above—the various methods are in agreement and the basic and extended model setups deliver the same results. The only difference is that the point estimate of the average smoking effect is about −208 grams.

The output from the second exercise is displayed in Figure 3.1, which shows the estimated CATE function for black mothers represented as a tree (the corresponding figure for white mothers is in the Electronic Online Supplement, *Figure 3.1*). The two most important leaves are in the bottom right corner, containing about 92% of the observations in total. These leaves originate from a parent node that splits the sample by mother's age falling below or above 19 years; the average smoking effect is then significantly larger in absolute value for the older group (−221.7 grams) than for the younger group (−120.1 grams).

This result is fully consistent with Abrevaya et al. (2015) and Fan et al. (2020), who estimate the reduced dimensional CATE function for mother's age and find that the smoking effect becomes more detrimental for older mothers. The conceptual difference between these two papers and the tree in Figure 3.1 is twofold. First, in building the tree, age is not designated *a priori* as a variable of interest; it is the tree building algorithm that recognized its relevance for treatment effect heterogeneity. Second, not all other control variables are averaged out in the two leaves highlighted above; in fact, the preceding two splits condition on normal blood pressure as well as the mother being older than 15. However, given the small size of the complementing leaves, we would caution against over-interpreting the rest of the tree. Whether or not blood pressure is related to the smoking effect in a stable way would require some further robustness checks. We nevertheless note that even within the high blood pressure group the age-related pattern is qualitatively similar—smoking has a larger negative effect for older mothers and is in fact statistically insignificant below age 22.

The results for white mothers (see the Electronic Online Supplement, *Figure 3.1.*) are qualitatively similar in that they confirm that the smoking effect becomes more

[18] Another factor that contributes to this result is that no individual covariate is strongly correlated with the treatment dummy. Hence, even if one were mistakenly dropped from the naive Lasso regression, it would not create substantial bias. Indeed, in exercises run on small subsamples (so that the $dim(X)/n$ ratio is much larger) we still find little difference between the naive (post) Lasso and the DML method in this application.

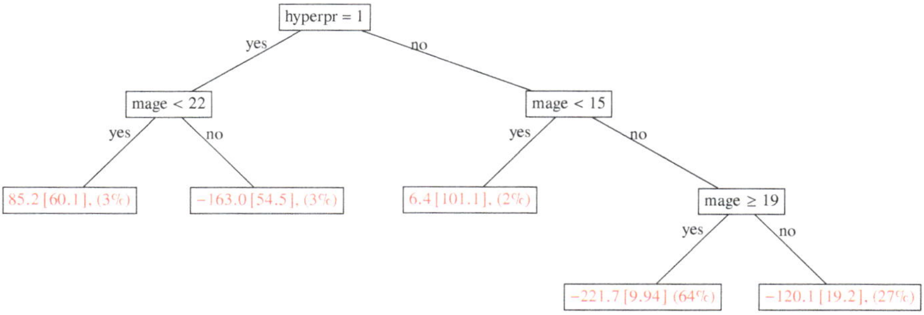

Fig. 3.1: A causal tree for the effect of smoking on birth weight (first time black mothers).
Notes: standard errors are in brackets and the percentages in parenthesis denote share of group in the sample. The total number of observations is $N = 157,989$ and the covariates used are X_1, except for the polynomial and interaction terms. To obtain a simpler model, we choose the pruning parameter using the 1SE-rule rather than the minimum cross-validated MSE.

negative with age (compare the two largest leaves on the $mage \geq 28$ and $mage < 28$ branches). Hypertension appears again as a potentially relevant variable, but the share of young women affected by this problem is small. We also note that the results are obtained over a subsample of $N = 150,000$ and robustness checks show sensitivity to the choice of the subsample. Nonetheless, the age pattern is qualitatively stable.

The preceding results illustrate both the strengths and weaknesses of using a causal tree for heterogeneity analysis. On the one hand, letting the data speak for itself is philosophically attractive and can certainly be useful. On the other hand, the estimation results may appear to be too complex or arbitrary and can be challenging to interpret. This problem is exacerbated by the fact that trees grown on different subsamples can show different patterns — suggesting that in practice it is prudent to construct a causal forest, i.e., use the average of multiple trees.

3.6 Conclusion

Until about the early 2010s econometrics and machine learning developed on parallel paths with only limited interaction between the two fields. This has changed considerably over the last decade and ML methods are now widely used in economic applications. In this chapter we reviewed their use in treatment effect estimation, focusing on two strands of the literature.

In applications of the *double or debiased machine learning* (DML) approach, the parameter of interest is typically some average treatment effect, and ML methods are employed in the first stage to estimate the unknown nuisance functions necessary for identification (such as the propensity score). A key insight that emerges from this literature is that machine learning can be fruitfully applied for this purpose if the nuisance functions enter the second stage estimating equations (more precisely, moment conditions) in a way that satisfies an orthogonality condition. This condition ensures that the parameter of interest is consistently estimable despite the selection and approximation errors introduced by the first stage machine learning procedure. Inference in the second stage can then proceed as usual. In practice a cross-fitting procedure is recommended, which involves splitting the sample between the first and second stages.

In applications of the *causal tree or forest* methodology, the parameter of interest is the full dimensional conditional average treatment effect (CATE) function, i.e., the focus is on treatment effect heterogeneity. The method permits near automatic and data-driven discovery of this function, and if the selected approximation is re-estimated on an independent subsample (the 'honest approach') then inference about group-specific effects can proceed as usual. Another strand of the heterogeneity literature estimates the projection of the CATE function on a given, pre-specified coordinate to facilitate presentation and interpretation. This can be accomplished by an extension of the DML framework where in the first stage the nuisance functions are estimated by an ML method and in the second stage a traditional nonparametric estimator is used (e.g., kernel-based or series regression).

In our empirical application (the effects of maternal smoking during pregnancy on the baby's birthweight) we illustrate the use of the DML estimator as well as causal trees. While the results confirm previous findings in the literature, they also highlight some limitations of these methods. In particular, with the number of observations orders of magnitude larger than the number of covariates, and the covariates not being very strong predictors of the treatment, DML virtually coincides with OLS and even the naive (direct) Lasso estimator. The causal tree approach successfully uncovers important patterns in treatment effect heterogeneity but also some that seem somewhat incidental and/or less straightforward to interpret. This suggests that in practice a causal forest should be used unless the computational cost is prohibitive.

In sum, machine learning methods, while geared toward prediction tasks in themselves, can be used to enhance treatment effect estimation in various ways. This is an active research area in econometrics at the moment, with a promise to supply exciting theoretical developments and a large number of empirical applications for years to come.

Acknowledgements We thank Qingliang Fan for his help in collecting literature. We are also grateful to Alice Kuegler, Henrika Langen and the editors for their constructive comments, which led to noticeable improvements in the exposition. The usual disclaimer applies.

References

Abrevaya, J. (2006). Estimating the effect of smoking on birth outcomes using a matched panel data approach. *Journal of Applied Econometrics*, *21*(4),

489–519.

Abrevaya, J., Hsu, Y.-C. & Lieli, R. P. (2015). Estimating conditional average treatment effects. *Journal of Business & Economic Statistics*, *33*(4), 485-505.

Alaa, A. M., Weisz, M. & van der Schaar, M. (2017). Deep counterfactual networks with propensity-dropout. *arXiv preprint arXiv:1706.05966, https://arxiv.org/abs/1706.05966*.

Athey, S. & Imbens, G. (2016). Recursive partitioning for heterogeneous causal effects. *Proceedings of the National Academy of Sciences*, *113*(27), 7353–7360.

Athey, S. & Imbens, G. W. (2019). Machine learning methods that economists should know about. *Annual Review of Economics*, *11*(1), 685-725.

Athey, S., Tibshirani, J. & Wager, S. (2019). Generalized random forests. *Annals of Statistics*, *47*(2), 1148–1178.

Athey, S. & Wager, S. (2021). Policy learning with observational data. *Econometrica*, *89*(1), 133-161.

Belloni, A., Chen, D., Chernozhukov, V. & Hansen, C. (2012). Sparse models and methods for optimal instruments with an application to eminent domain. *Econometrica*, *80*(6), 2369–2429.

Belloni, A., Chernozhukov, V., Fernández-Val, I. & Hansen, C. (2017). Program evaluation and causal inference with high-dimensional data. *Econometrica*, *85*(1), 233–298.

Belloni, A., Chernozhukov, V. & Hansen, C. (2013). Inference for high-dimensional sparse econometric models. In M. A. D. Acemoglu & E. Dekel (Eds.), *Advances in economics and econometrics. 10th world congress, vol. 3.* (pp. 245–95). Cambridge University Press.

Belloni, A., Chernozhukov, V. & Hansen, C. (2014a). High-dimensional methods and inference on structural and treatment effects. *Journal of Economic Perspectives*, *28*(2), 29-50.

Belloni, A., Chernozhukov, V. & Hansen, C. (2014b). Inference on treatment effects after selection among high-dimensional controls. *The Review of Economic Studies*, *81*(2), 608–650.

Chernozhukov, V., Chetverikov, D., Demirer, M., Duflo, E., Hansen, C. & Newey, W. (2017, May). Double/debiased/neyman machine learning of treatment effects. *American Economic Review*, *107*(5), 261-65.

Chernozhukov, V., Chetverikov, D., Demirer, M., Duflo, E., Hansen, C., Newey, W. & Robins, J. (2018). Double/debiased machine learning for treatment and structural parameters. *The Econometrics Journal*, *21*(1), C1-C68.

Colangelo, K. & Lee, Y.-Y. (2022). Double debiased machine learning nonparametric inference with continuous treatments. *arXiv preprint arXiv:2004.03036, https://arxiv.org/abs/2004.03036*.

Da Veiga, P. V. & Wilder, R. P. (2008). Maternal smoking during pregnancy and birthweight: a propensity score matching approach. *Maternal and Child Health Journal*, *12*(2), 194–203.

Donald, S. G., Hsu, Y.-C. & Lieli, R. P. (2014). Testing the unconfoundedness assumption via inverse probability weighted estimators of (l)att. *Journal of Business & Economic Statistics*, *32*(3), 395-415.

Electronic Online Supplement. (2022). *Online Supplement of the book Econometrics with Machine Learning.* https://sn.pub/0ObVSo.

Fan, Q., Hsu, Y.-C., Lieli, R. P. & Zhang, Y. (2020). Estimation of conditional average treatment effects with high-dimensional data. *Journal of Business and Economic Statistics.* (forthcoming)

Farbmacher, H., Huber, M., Laffers, L., Langen, H. & Spindler, M. (2022). Causal mediation analysis with double machine learning. *The Econometrics Journal.* (forthcoming)

Hirano, K., Imbens, G. W. & Ridder, G. (2003). Efficient estimation of average treatment effects using the estimated propensity score. *Econometrica, 71*(4), 1161–1189.

Hsu, Y.-C., Huber, M., Lee, Y.-Y. & Liu, C.-A. (2022). Testing monotonicity of mean potential outcomes in a continuous treatment with high-dimensional data. *arXiv preprint arXiv:2106.04237 ,https://arxiv.org/abs/2106.04237.*

Huber, M. (2021). *Causal analysis* (Working Paper). Fribourg, Switzerland: University of Fribourg.

Imbens, G. W. & Wooldridge, J. M. (2009). Recent developments in the econometrics of program evaluation. *Journal of economic literature, 47*(1), 5–86.

Kitagawa, T. & Tetenov, A. (2018). Who should be treated? empirical welfare maximization methods for treatment choice. *Econometrica, 86*(2), 591-616.

Knaus, M. (2021). Double machine learning based program evaluation under unconfoundedness. *arXiv preprint arXiv:2003.03191, https://arxiv.org/abs/2003.03191.*

Knaus, M., Lechner, M. & Strittmatter, A. (2021). Machine learning estimation of heterogeneous causal effects: Empirical monte carlo evidence. *The Econometrics Journal, 24*(1), 134-161.

Kreif, N. & DiazOrdaz, K. (2019). Machine learning in policy evaluation: New tools for causal inference. *arXiv preprint arXiv:1903.00402, https://arxiv.org/abs/1903.00402.*

Kueck, J., Luo, Y., Spindler, M. & Wang, Z. (2022). Estimation and inference of treatment effects with l2-boosting in high-dimensional settings. *Journal of Econometrics.* (forthcoming)

Mullainathan, S. & Spiess, J. (2017). Machine learning: An applied econometric approach. *Journal of Economic Perspectives, 31*(2), 87-106.

Neyman, J. (1959). Optimal asymptotic tests of composite statistical hypotheses. In U. Grenander (Ed.), *Probability and statistics* (p. 416-44).

Pagan, A. & Ullah, A. (1999). *Nonparametric econometrics.* Cambridge University Press.

Reguly, A. (2021). Heterogeneous treatment effects in regression discontinuity designs. *arXiv preprint arXiv:2106.11640, https://arxiv.org/abs/2106.11640.*

Semenova, V. & Chernozhukov, V. (2020). Debiased machine learning of conditional average treatment effects and other causal functions. *The Econometrics Journal, 24*(2), 264-289.

Semenova, V. & Chernozhukov, V. (2021). Debiased machine learning of conditional average treatment effects and other causal functions. *The Econometrics Journal,*

24(2), 264–289.

Wager, S. & Athey, S. (2018). Estimation and inference of heterogeneous treatment effects using random forests. *Journal of the American Statistical Association, 113*(523), 1228–1242.

Walker, M. B., Tekin, E. & Wallace, S. (2009). Teen smoking and birth outcomes. *Southern Economic Journal, 75*(3), 892–907.

Yao, L., Li, S., Li, Y., Huai, M., Gao, J. & Zhang, A. (2018). Representation learning for treatment effect estimation from observational data. *Proceedings of the 32nd International Conference on Neural Information Processing Systems*, 2638-2648.

Zimmert, M. & Lechner, M. (2019). Nonparametric estimation of causal heterogeneity under high-dimensional confounding. *arXiv preprint arXiv:1908.08779, https://arxiv.org/abs/1908.08779.*

Chapter 4
Forecasting with Machine Learning Methods

Marcelo C. Medeiros

Abstract This chapter surveys the use of supervised Machine Learning (ML) models to forecast time-series data. Our focus is on covariance stationary dependent data when a large set of predictors is available and the target variable is a scalar. We start by defining the forecasting scheme setup as well as different approaches to compare forecasts generated by different models/methods. More specifically, we review three important techniques to compare forecasts: the Diebold-Mariano (DM) and the Li-Liao-Quaedvlieg tests, and the Model Confidence Set (MCS) approach. Second, we discuss several linear and nonlinear commonly used ML models. Among linear models, we focus on factor (principal component)-based regressions, ensemble methods (bagging and complete subset regression), and the combination of factor models and penalized regression. With respect to nonlinear models, we pay special attention to neural networks and autoenconders. Third, we discuss some hybrid models where linear and nonlinear alternatives are combined.

4.1 Introduction

This chapter surveys the recent developments in the Machine Learning (ML) literature to forecast time-series data. ML methods have become an important estimation, model selection and forecasting tool for applied researchers in different areas, ranging from epidemiology to marketing, economics and finance. With the availability of vast datasets in the era of *Big Data*, producing reliable and robust forecasts is of great importance.

ML has gained a lot of popularity during the last few years and there are several definitions in the literature of what exactly the term means. Some of the most popular definitions yield the misconception that ML is a sort of magical framework where computers learn patterns in the data without being explicitly

Marcelo C. Medeiros ✉
Pontifical Catholic University of Rio de Janeiro, Rio de Janeiro, Brazil, e-mail: mcm@econ.puc-rio.br

F. Chan, L. Mátyás (eds.), *Econometrics with Machine Learning*, Advanced Studies in Theoretical and Applied Econometrics 53, https://doi.org/10.1007/978-3-031-15149-1_4

programmed; see, for example, Samuel (1959). With respect to the framework considered in this chapter, we define ML, or more specifically, Supervised ML, as set of powerful statistical models/methods combined with automated computer algorithms to learn hidden patterns between a target variable and a potentially very large set of explanatory variables. More specifically, in the forecasting framework, we want to learn the expectation of a random variable Y, which take values on $\mathbb{R}$, conditional on observations of a p-dimensional set of random variables $\boldsymbol{X}$. Therefore, our goal is to learn the function $f(\boldsymbol{x}) := \mathbb{E}(Y|\boldsymbol{X} = \boldsymbol{x})$, based on a sample of T observations of $\{Y_t, \boldsymbol{X}_t'\}$. We have a particular interest in the *Big Data* scenario where the number of predictors (p) is much larger than the sample size (T). We target the conditional expectation due to the well-known result that it gives the optimal prediction, in the mean-squared error sense, of Y based on observations of $\boldsymbol{X}$.

The supervised ML methods presented here can be roughly divided in three groups. The first one includes linear models, where the conditional expectation $\mathbb{E}(Y|\boldsymbol{X} = \boldsymbol{x})$ is assumed to be a linear function of the data. We consider three subclasses of linear alternatives. We start by surveying models based on factors, where the large set predictors $\boldsymbol{X}$ is represented by a small number of factors $\boldsymbol{F}$ taking values on $\mathbb{R}^k$, where k is much smaller than p; see for example, Stock and Watson (2002b, 2002a). We continue by discussing the combination of factor models and penalized regression. For a nice overview of penalized regression models see the Chapter 1. Finally, we present methods based on ensemble of forecasts, where a potentially large number of very simple models is estimated and the final forecast is a combination of the predictions from individual models. Examples of such techniques are Bagging (Breiman, 1996) and the Complete Subset Regression (Elliott, Gargano & Timmermann, 2013, 2015).

The second group of models consists of nonlinear approximations to $\mathbb{E}(Y|\boldsymbol{X} = \boldsymbol{x})$. In general, the models considered here assume that the unrestricted nonlinear relation between Y and $\boldsymbol{X}$ can be well approximated by a combination of simpler nonlinear (basis) functions. We start by presenting an unified framework based on sieve semiparametric approximation as in Grenander (1981). We continue by analysing specific models as special cases of our general setup. More specifically, we cover feedforward neural networks, both in their shallow and deep versions, and convolution and recurrent neural networks. Neural Networks (NN) are probably one of the most popular ML methods. The success is partly due to the, in our opinion, misguided analogy to the functioning of the human brain. Contrary of what has been boasted in the early literature, the empirical success of NN models comes from a mathematical fact that a linear combination of sufficiently many simple basis functions is able to approximate very complicated functions arbitrarily well in some specific choice of metric.

Finally, we discuss several alternatives to combine linear and nonlinear ML models. For example, we can write the conditional expectation as $f(x) := \mathbb{E}(Y|X = x) = \boldsymbol{x}'\boldsymbol{\beta} + g(\boldsymbol{x})$, where $g(\cdot)$ is a nonlinear function and $\boldsymbol{\beta}$ is a parameter to be estimated.

Before reviewing the ML methods discussed above, we start by defining the estimation and forecasting framework that is considered throughout this chapter. More specifically, we discuss the use of rolling versus expanding windows and direct versus indirect construction of multi-step-ahead forecasts. In the sequence, we review approaches to compare forecasts from different alternative models. Such tests are in the class of equal or superior predictive ability.

4.1.1 Notation

A quick word on notation: an uppercase letter as in X denotes a random quantity as opposed to a lowercase letter x which denotes a deterministic (non-random) quantity. Bold letters as in $\boldsymbol{X}$ and $\boldsymbol{x}$ are reserved for multivariate objects such as vector and matrices. The symbol $\|\cdot\|_q$ for $q \geq 1$ denotes the ℓ_q norm of a vector. For a set S we use $|S|$ to denote its cardinality.

4.1.2 Organization

In addition to the Introduction, this Chapter is organized as follows. In Section 4.2 we present the forecasting setup considered and we discuss the benefits of rolling window versus expanding window approach. We also analyze the pros and cons of direct forecasts as compared to indirect forecasts. Section 4.3 discuss different methods to compare forecasts from different models/methods. In Section 4.4 we present the linear ML models, while in Section 4.5 we consider the nonlinear alternatives. Hybrid approaches are reviewed in Section 4.5.5.

4.2 Modeling Framework and Forecast Construction

A typical exercise to construct forecasting models involve many decisions from the practitioner, which can be summarized as follows.

1. Definition of the type of data considered in the exercise. For example, is the target variable covariance-stationary or non-stationary? And the predictors? In this chapter we focus on covariance-stationary data. Moreover, our aim is to predict the mean of the target variable in the future conditioned on the information available at the moment the forecasts are made. We do not consider forecasts for higher moments, such as the conditional variance of the target.
2. How are the forecasts computed for each forecasting horizon? Are the forecasts direct or iterated?
3. Usually, to evaluate the generalization potential of a forecast model/method, running some sort of *backtesting* exercise is strongly advisable. The idea of backtesting models is to assess what the models' performance would have been if we had generated predictions over the past history of the data. This is also called a *pseudo-out-sample* exercise. The forecasting models are estimated in several

points of the past history of the data and forecasts for the 'future' are generated and compared to the realized values of the target variable. The question is when the models are estimated and based on which sample estimation is carried out. For example, are the models estimated in a rolling window or expanding window setup?
4. Which models are going to be considered? With the recent advances in the ML litearure and the availability of large and new datasets, the number of potential forecasting models has been increasing at a vertiginous pace.

In this section we review the points above and give some practical guidance to the forecaster.

4.2.1 Setup

Given a sample with $T > 0$ realizations of the random vector $\left(Y_t, \boldsymbol{W}_t'\right)'$, the goal is to predict Y_{T+h} for horizons $h = 1, \ldots, H$. For example, Y_t may represent the daily sales of a product and we want to forecast for each day of the upcoming week. Similarly, Y_t can be daily Covid-19 new cases or monthly inflation rates. In the later case, it is of extreme importance to policy makers to have precise forecasts of monthly inflation for the next 12 months, such that $h = 1, \ldots, 12$.

Throughout the chapter, we consider the following assumption:

Assumption Let $\{\boldsymbol{D}_t := (Y_t, \boldsymbol{W}_t')'\}_{t=1}^{\infty}$ be a sequence of zero-mean covariance-stationary stochastic process taking values on $\mathbb{R}^{d+1}$. Furthermore, $\mathbb{E}(\boldsymbol{D}_t \boldsymbol{D}_{t-j}') \longrightarrow \boldsymbol{0}$, as $|j| \longrightarrow \infty$. □

Therefore, we are excluding important processes that usually appear in time-series applications. In particular, unit-roots and long-memory processes are excluded by Assumption 1. The assumption of a zero mean is without lack of generality as we can always consider the data to be demeaned.

4.2.2 Forecasting Equation

For (usually predetermined) integers $r \geq 1$ and $s \geq 0$ define the p-dimensional vector of predictors $\boldsymbol{X}_t := \left(Y_t, \ldots, Y_{t-r+1}, \boldsymbol{W}_t', \ldots, \boldsymbol{W}_{t-s+1}'\right)'$ where $p = r + ds$ and consider the following assumption on the *data generating process* (DGP):

Assumption (Data Generating Process)

$$Y_{t+h} = f_h(\boldsymbol{X}_t) + U_{t+h}, \quad h = 1, \ldots, H, \quad t = 1, \ldots, T-h, \tag{4.1}$$

where $f_h(\cdot) : \mathbb{R}^n \to \mathbb{R}$ is an unknown (measurable) function and $\{U_{t+h}\}_{t=1}^{T-h}$ is a sequence of zero mean covariance-stationary stochastic process. In addition, $\mathbb{E}(U_t U_{t-j}') \longrightarrow \infty$, as $|j| \longrightarrow 0$. □

Model (4.1) is an example of a *direct forecasting* equation. In this case, the future value of the target variable, Y_{t+h}, is explicitly modeled as function of the data in time t. We are going to adopt this forecasting approach in this chapter.

An alternative to direct forecast is to write a model for $h = 1$ and iterate it in order to produce forecasts for longer horizons. This is the *iterated forecast approach*. This is trivial to be achieved for some linear specifications, but can be rather complicated in general as it requires a forecasting model to all variables in $\boldsymbol{W}$. Furthermore, for nonlinear models, the construction of iterated multi-step forecasts requites numerical evaluations of integrals via Monte Carlo techniques; see, for instance, Teräsvirta (2006).

Example (Autoregressive Model) Consider tat the DGP is an autoregressive model of order 1, AR(1), such that:

$$Y_{t+1} = \phi Y_t + V_{t+1}, \quad |\phi| < 1, \tag{4.2}$$

where ϕ is an unknown parameter and V_t is a uncorrelated zero-mean process.

By recursive iteration of (4.2), we can write:

$$\begin{aligned} Y_{t+h} &= \phi^h Y_t + \phi^{h-1} V_{t+1} + \cdots + V_{t+h}, \\ &= \theta Y_t + U_{t+h}, \quad h = 1, \ldots, H. \end{aligned} \tag{4.3}$$

where $\theta := \phi^h$ and $U_{t+h} := \phi^{h-1} V_{t+1} + \cdots + V_{t+h}$.

Note that the forecast for Y_{t+h} can be computed either by estimating model (4.2) and iterating it h-steps-ahead or by estimating (4.3) directly for each h, $h = 1, \ldots, H$.□

4.2.3 Backtesting

The next choice faced the applied researcher is how the models are going to be estimated and how they are going to be evaluated. Typically, in the time-series literature, forecasting models are evaluated by their (pseudo) out-of-sample (OOS) performance. Hence, the sample is commonly divided into two subsamples. The first one is used to estimate the parameters of the model. This is known as the in-sample (IS) estimation of the model. After estimation, forecasts are constructed for the OOS period. However, the conclusions about the (relative) quality of a model can be heavily dependent on how the sample is split, i.e., how many observations we use for estimation and how many we leave to test the model. The set of observations used to estimate the forecasting models is usually called the *estimation window*.

A common alternative is evaluate forecasts using different combinations of IS/OOS periods constructed in different ways as, for example:

Expanding window: the forecaster chooses an initial window size to estimate the models, say R. After estimation, the forecasts for $t = R+1, \ldots, R+h$ are computed. When a new observation arrives, the forecaster incorporate the new data in the estimation window, such that R is increased by one unit. The process is repeated until we reach the end of the sample. See the upper panel in Figure 4.1. In this case, the models are estimated with an increasing number of observations over time.

Rolling window: the forecaster chooses an initial window size to estimate the models, say R. After estimation, the forecasts for $t = R+1, \ldots, R+h$ are computed. When a new observation arrives, the first data point is dropped and the forecaster incorporate the new observation in the estimation window. Therefore, the window size is kept constant. See lower panel in Figure 4.1. In this case, all models are estimated in a sample with R observations.

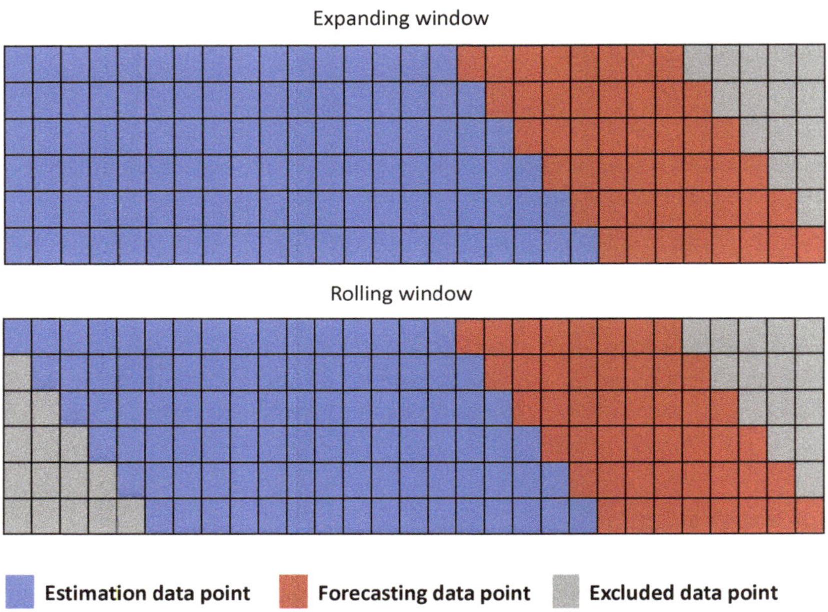

Fig. 4.1: Expanding versus rolling window framework

However, it is important to notice the actual number of observations used for estimation depend on the maximum lag order in the construction of the predictor vector $\boldsymbol{X}$, on the choice of forecasting horizon, and on the moment the forecasts are constructed. For example, suppose we are at the time period t and we want to estimate a model to forecast the target variable at $t+h$. If we consider an expanding window framework, the effective number of observations used to estimate the model is $T^* := T^*(t,r,s,h) = t - \max(s,r) - h$, for $t = R, R+1, \ldots, T-h$.[1] For the rolling window case we have

[1] We drop the dependence of T^* on t, r, s, h in order to simplify notation.

$$T^* = \begin{cases} R & \text{if } t > \max(r,s,h) \\ R - \max(r,s,h) & \text{otherwise,} \end{cases}$$

for $t = R, R+1, \ldots, T-h$. To avoid making the exercise too complicated, we can start the backtesting of the model for $t > \max(r,s,h)$.

The choice between expanding versus rolling window is not necessarily trivial. On one hand, with expanding windows, the number of observations increase over time, potentially yielding more precise estimators of the models. On the other hand, estimation with expanding windows is more susceptible to the presence of structural breaks and outliers, therefore, yielding less precise estimators. Forecasts based on rolling windows are influenced less by structural breaks and outliers.

4.2.4 Model Choice and Estimation

Let $\widehat{Y}_{t+h|t} := \widehat{f}_h(\boldsymbol{x}_t)$ be the forecast for Y_{t+h} based on information up to time t. In order to estimate a given forecasting model, we must choose a *loss function*, $\mathcal{L}(Y_{t+h}, \widehat{Y}_{t+h|h})$, which measures the discrepancy between Y_{t+h} and $\widehat{Y}_{t+h|t}$. We define the *risk function* as $\mathcal{R}(Y_{t+h}, \widehat{Y}_{t+h|h}) := \mathbb{E}_{Y_{t+h}|\boldsymbol{x}_t}\left[\mathcal{L}(Y_{t+h}, \widehat{Y}_{t+h|h})\right]$. The pseudo-true-model is given by

$$f_h^*(\boldsymbol{x}_t) = \arg\min_{f_h(\boldsymbol{x}_t)\in\mathcal{F}} \mathcal{R}(Y_{t+h}, \widehat{Y}_{t+h|h}), \tag{4.4}$$

where $\mathcal{F}$ is a generic function space.

In this chapter we set f_h as the *conditional expectation function*: $f_h(\boldsymbol{x}) := \mathbb{E}(Y_{t+h}|\boldsymbol{X}_t = \boldsymbol{x})$.[2] Therefore, the pseudo-true-model is the function $f_h^*(\boldsymbol{x}_t)$ that minimizes the expected value of the quadratic loss:

$$f_h^*(\boldsymbol{x}_t) = \arg\min_{f_h(\boldsymbol{X}_t)\in\mathcal{F}} \mathbb{E}_{Y_{t+h}|\boldsymbol{X}_t}\left\{[Y_{t+h} - f_h(\boldsymbol{X}_t)]^2|\boldsymbol{X}_t = \boldsymbol{x}_t\right\}, \quad h = 1, \ldots, H.$$

In practice, the model should be estimated based on a sample of data points. Hence, in the rolling window setup with length R and for $t > \max(r,s,h)$,

$$[\widehat{f}_h(\boldsymbol{X}_{t-R-h+1}), \ldots, \widehat{f}_h(\boldsymbol{X}_{t-h})]' = \arg\min_{f_h(\boldsymbol{X}_\tau)\in\mathcal{F}} \frac{1}{R} \sum_{\tau=t-R-h+1}^{t-h} [Y_{\tau+h} - f_h(\boldsymbol{X}_\tau)]^2.$$

However, the optimization problem stated above is infeasible when $\mathcal{F}$ is infinite dimensional, as there is no efficient technique to search over all $\mathcal{F}$. Of course, one solution is to restrict the function space, as for instance, imposing linearity or specific forms of parametric nonlinear models as in, for example, Teräsvirta (1994), Suarez-Fariñas, Pedreira and Medeiros (2004) or McAleer and Medeiros (2008); see also Teräsvirta, Tjøstheim and Granger (2010) for a recent review of such models.

[2] Therefore, whenever we write $\widehat{Y}_{t+h|t}$ in this Chapter, we mean an estimator of $\mathbb{E}(Y_{t+h}|\boldsymbol{X}_t = \boldsymbol{x})$.

Alternatively, we can replace $\mathcal{F}$ by a simpler and finite dimensional $\mathcal{F}_D$. The idea is to consider a sequence of finite dimensional spaces, the *sieve spaces*, $\mathcal{F}_D$, $D = 1, 2, 3, \ldots$, that converges to $\mathcal{F}$ in some norm. The approximating function $f_{h,D}(\boldsymbol{X}_t)$ is written as

$$f_{h,D}(\boldsymbol{X}_t) = \sum_{j=1}^{J} \beta_j g_{h,j}(\boldsymbol{X}_t), \tag{4.5}$$

where $g_{h,j}(\cdot)$ is the j-th *basis function* for $\mathcal{F}_D$ and can be either fully known or indexed by a vector of parameters, such that: $g_{h,j}(\boldsymbol{X}_t) := g_h(\boldsymbol{X}_t; \boldsymbol{\theta}_j)$. The number of basis functions $J := J_R$ depends on the sample size R. D is the dimension of the space and it also depends on the sample size: $D := D_R$.[3] Therefore, the optimization problem is then modified to

$$[\widehat{f}_{h,D}(\boldsymbol{X}_{t-R-h+1}), \ldots, \widehat{f}_{h,D}(\boldsymbol{X}_t)]' = \\ \arg \min_{f_{h,D}(\boldsymbol{X}_t) \in \mathcal{F}_D} \frac{1}{R} \sum_{\tau=t-R-h+1}^{t-h} \left[Y_{\tau+h} - f_{h,D}(\boldsymbol{X}_\tau) \right]^2 .$$

In terms of parameters, set $f_{h,D}(\boldsymbol{X}_t) := f_{h,D}(\boldsymbol{X}_t; \boldsymbol{\theta})$, where $\boldsymbol{\theta} = (\boldsymbol{\theta}_1', \ldots, \boldsymbol{\theta}_J')' := \boldsymbol{\theta}_{h,D}$. Therefore, the *pseudo-true parameter* $\boldsymbol{\theta}^*$ is defined as

$$\boldsymbol{\theta}^* := \arg \min_{\boldsymbol{\theta} \in \mathbb{R}^D} \mathbb{E}_{Y_{\tau+h}|\boldsymbol{X}_\tau} [\left(Y_{\tau+h} - f_{h,D}(\boldsymbol{X}_\tau; \boldsymbol{\theta}) \right)^2 | \boldsymbol{X}_\tau = \boldsymbol{x}_\tau]. \tag{4.6}$$

As a consequence, the estimator for $\boldsymbol{\theta}^*$ is

$$\widehat{\boldsymbol{\theta}} = \arg \min_{\boldsymbol{\theta} \in \mathbb{R}^D} \frac{1}{R} \sum_{\tau=t-R-h+1}^{t-h} \left[Y_{\tau+h} - f_{h,D}(\boldsymbol{X}_\tau; \boldsymbol{\theta}) \right]^2 . \tag{4.7}$$

The sequence of approximating spaces $\mathcal{F}_D$ is chosen by using the structure of the original underlying space $\mathcal{F}$ and the fundamental concept of *dense sets*.

Definition (Dense sets) If we have two sets A and $B \in \mathcal{X}$, $\mathcal{X}$ being a metric space, A is dense in B if for any $\epsilon > 0, \in \mathbb{R}$ and $x \in B$, there is a $y \in A$ such that $\|x - y\|_{\mathcal{X}} < \epsilon$.□

The approach to approximate $\mathcal{F}$ by a sequence of simpler spaces is called the *method of sieves*. For a comprehensive review of the method for time-series data, see Chen (2007). There are many examples of sieves in the literature. Examples are: polynomial series, Fourier series, trigonometric series, neural networks, etc.

When the basis functions are all known (*linear sieves*), the problem is linear in the parameters and methods like ordinary least squares (when $J \ll T^*$, where T^* is the size of the estimation sample) or penalized estimation can be used as we discuss later in this chapter.

Example (Linear Sieves) From the theory of approximating functions we know that the proper subset $\mathcal{P} \subset C$ of polynomials is dense in C, the space of continuous

[3] We are assuming here that the models are estimated with a sample of R observations.

functions. The set of polynomials is smaller and simpler than the set of all continuous functions. In this case, it is natural to define the sequence of approximating spaces $\mathcal{F}_D$, $D = 1,2,3,\ldots$ by making $\mathcal{F}_D$ the set of polynomials of degree smaller or equal to $D-1$ (including a constant in the parameter space). Note that $\dim(\mathcal{F}_D) = D < \infty$. In the limit this sequence of finite dimensional spaces converges to the infinite dimensional space of polynomials, which on its turn is dense in C.

Let $p = 1$ and pick a polynomial basis such that

$$f_D(X_t) = \beta_0 + \beta_1 X_t + \beta_2 X_t^2 + \beta_3 X_t^3 + \cdots + \beta_J X_t^J.$$

In this case, the dimension D of $\mathcal{F}_D$ is $J+1$, due to the presence of a constant term. If $J << T^*$, the vector of parameters $\boldsymbol{\beta} = (\beta_1, \ldots, \beta_J)'$ can be estimated by

$$\widehat{\boldsymbol{\beta}} = \left(\boldsymbol{X}_J' \boldsymbol{X}_J\right)^{-1} \boldsymbol{X}_J' \boldsymbol{Y},$$

where $\boldsymbol{X}_J$ is the $T^* \times (J+1)$ design matrix and $\boldsymbol{Y}$ is a T^* vector. When $J > T$, $\boldsymbol{\beta}$ can be estimated by penalized regression:

$$\widehat{\boldsymbol{\beta}} = \arg\min_{\boldsymbol{\beta} \in \mathbb{R}^D} \frac{1}{T^*} \|\boldsymbol{Y} - \boldsymbol{X}\boldsymbol{\beta}\|_2^2 + \lambda p(\boldsymbol{\beta}),$$

where $\lambda > 0$ and $p(\boldsymbol{\beta})$ is a penalty function as discussed in Chapter 1 of this book.□

When the basis functions are also indexed by parameters (*nonlinear sieves*), nonlinear least-squares methods should be used.

Example (Nonlinear Sieves) Let $p = 1$ and consider the case where

$$f_D(X_t) = \beta_0 + \sum_{j=1}^{J} \beta_f \frac{1}{1 + \exp\left[-\gamma_j (X_t - \gamma_{0j})\right]}.$$

This is an example of a single-hidden-layer feedforward neural network, one of the most popular machine learning models. We discuss such models later in this chapter. The vector of parameters is given by

$$\boldsymbol{\theta} = (\beta_0, \beta_1, \ldots, \beta_J, \gamma_1, \ldots, \gamma_j, \gamma_{01}, \ldots, \gamma_{0J})',$$

which should be estimated by nonlinear least squares. As the number of parameters can be very large compared to the sample size, some sort of regularization is necessary to estimate the model and avoid overfitting. □

The forecasting model to be estimated has the following general form:

$$Y_{t+h} = f_{h,D}(\boldsymbol{X}_t) + Z_{t+h},$$

where $Z_{t+h} = U_{t+h} + \left[f_h(\boldsymbol{X}_t) - f_{h,D}(\boldsymbol{X}_t)\right]$.

In a rolling window framework, the forecasts are computed as follows:

$$\widehat{Y}_{t+h|t} = \widehat{f}_{h,D,(t-R+1:t)}(\boldsymbol{x}_t), \quad t = R,\ldots,T-h, \tag{4.8}$$

where $\widehat{f}_{h,D,(t-R_h+1:t)}(\boldsymbol{x}_t)$ is the estimated approximating function based on data from time $t-R+1$ up to t and R is the window size. Therefore, for a given sample of T observations, the pseudo OOS exercise results in $P_h := T-h-R+1$ forecasts for each horizon.

In practice, it is hardly the case where there is only one forecasting model. The common scenario is to have a potentially large number of competing models. Even we the forecaster is restricted to the linear setup there are multiple potential alternatives. For example, different set of variables or different estimation techniques, specially in the high-dimensional environment. Therefore, it is important to compare the forecasts from set of alternatives available. This is what we review in the next section of this chapter.

4.3 Forecast Evaluation and Model Comparison

Let $\mathcal{M}_h$ be a set of alternative forecasting models for the target variable Y at horizon $t+h$, $h = 1,2,\ldots,H$ and $t > \max(r,s,h) := T_0$.[4] Each model is defined in terms of a vector of parameters $\boldsymbol{\theta}_m$. Set $\boldsymbol{\theta}^*_m$ the pseudo-true parameter and $\widehat{\boldsymbol{\theta}}_m$ is its estimator. In addition, set $\widehat{Y}_{m,t+h|t} := \widehat{Y}_{m(\widehat{\boldsymbol{\theta}}_m),t+h|t}$ to be the forecast of Y_{t+h} produced by a given model $m \in \mathcal{M}_h$.

The loss function $\mathcal{L}_{m,t}$ associated to the mth model is written as

$$\mathcal{L}_{m(\widehat{\theta}),t+h} := \mathcal{L}(Y_{t+h}, \widehat{Y}_{m,t+h|t}), \quad t = R,\ldots,T-h. \tag{4.9}$$

The equality of forecasts produced by two different models, say m_1 and m_2, can be compared, for instance, by one of the following two hypotheses:

$$\text{(population)} \quad \mathcal{H}_0 : \mathbb{E}\left[\mathcal{L}_{m_1(\theta_1^*),t+h}\right] = \mathbb{E}\left[\mathcal{L}_{m_2(\theta_2^*),t+h}\right] \tag{4.10}$$

$$\text{(sample)} \quad \mathcal{H}_0 : \mathbb{E}\left[\mathcal{L}_{m_1(\widehat{\theta}_1),t+h}\right] = \mathbb{E}\left[\mathcal{L}_{m_2(\widehat{\theta}_2),t+h}\right]. \tag{4.11}$$

Although the difference between the hypotheses (4.10) and (4.11) seem minor, there are important consequences about their properties in practice. Note that the first one is testing equality of forecasts at population values of the parameters and the main goal of testing such null hypotheses is to validate population models. On the other hand, testing the second null hypothesis is the same as taken the forecasts as given (model-free) and just test the equality of the forecasts with respect to some expected loss function.

Testing (4.10) is way more complicated than testing (4.11). A nice discussion about the differences can be found in West (2006), Diebold (2015), and Patton (2015). In this chapter we focus on testing (4.11).

[4] Again we are assuming that $t > \max(r, s, h)$ just to simplify notation.

4.3.1 The Diebold-Mariano Test

The Diebold-Mariano (DM) approach, proposed in Diebold and Mariano (1995), considers the forecast errors as primitives and makes assumptions directly on those errors. Therefore, the goal is to test the null (4.11).

As before, consider two models, m_1 and m_2, which yield two sequence of forecasts $\{\widehat{Y}_{m_1,t+h|t}\}_{t=R}^{T-h}$ and $\{\widehat{Y}_{m_1,t+h|t}\}_{t=R}^{T-h}$ and the respective sequence of forecast errors: $\{\widehat{Z}_{m_1,t+h|t}\}_{t=R}^{T-h}$ and $\{\widehat{Z}_{m_1,t+h|t}\}_{t=R}^{T-h}$.

Let $d_{(12),h,t} := \mathcal{L}_{m_1,t+h} - \mathcal{L}_{m_2,t+h}$, $t = R, \ldots, T-h$ be the loss differential between models m_1 and m_2. The DM statistic is given by

$$DM = \frac{\bar{d}_{(12),h}}{\widehat{\sigma}(\bar{d}_{(12),h})}, \tag{4.12}$$

where $\bar{d}_{(12),h} = \frac{1}{P_h}\sum_{t=R}^{T-h} d_{(12),h,t}$ and $\widehat{\sigma}(\bar{d}_{(12),h})$ is an estimator of the variance of the sample average of $d_{h,t}$. Note that the loss differential will be a sequence of dependent random variables and this must be taken into account.

McCracken (2020) showed that if $d_{(12),h,t}$ is a covariance-stationary random variable with autocovariances decaying to zero,

$$DM \xrightarrow{d} \mathsf{N}(0,1),$$

as $P_h \longrightarrow \infty$.[5] The above result is valid, for instance, when the forecasts are computed in a rolling window framework with fixed length.

The DM statistic can be trivially computed by the t-statistic of a regression of the loss differential on an intercept. Furthermore, the DM test can be extended by controlling for additional variables in the regression that may explain the loss differential, thereby moving from an unconditional to a conditional expected loss perspective; see Giacomini and White (2006) for a discussion.

Finally, there is also evidence that the use of Bartlett kernel should be avoided when computing the standard error in the DM statistic and a rectangular kernel should be used instead. Furthermore, it is advisable to use the small-sample adjustment of Harvey, Leybourne and Newbold (1997):

$$MDM = \sqrt{\frac{P_h + 1 - 2h + h(h-1)/P_h}{p_h}} \times DM.$$

For a deeper discussion on the Diebold-Mariano test, see Clark and McCracken (2013) or Diebold (2015).

---

[5] In the original paper, Diebold and Mariano (1995) imposed only covariance-stationarity of $d_{h,t}$. McCracken (2020) gave a counterexample where the asymptotic normality does not hold even in the case where $d_{h,t}$ is covariance-stationary but the autocovariances do not converge to zero.

4.3.2 Li-Liao-Quaedvlieg Test

The unconditional version of the DM test is based on the unconditional average performance of the competing forecasts. Therefore, it 'integrates out' potential heterogeneity across subsample periods. Recently, Li, Liao and Quaedvlieg (2021) proposed a conditional test for superior predictive ability (SPA) where the null hypothesis states that the conditional expected loss of a benchmark model is no larger than those of the competing alternatives, uniformly across all conditioning states. Such conditioning states are determined by a conditioning variable chosen ex-ante by the practitioner. As a consequence, the conditional SPA (CSPA) null hypothesis proposed by the authors asserts that the benchmark method is a uniformly weakly dominating method among all predictive models under consideration.

Set $d_{(1m),h,t}$ to be the loss differential between model $m \in \mathcal{M}_h$ and model m_1 (benchmark). Given a user-specified conditioning variable C_t, define:

$$h_{m,h}(c) := \mathbb{E}(d_{(1m),h,t}|C_t = c).$$

Note that $h_{m,h}(c) \geq 0$ indicates that the benchmark method is expected to (weakly) outperform the competitor conditional on $C_t = c$. The null hypothesis of the CSPA test is written as:

$$\mathcal{H}_0 : h_{m,h}(c) \geq 0, \forall c \in C, \text{and} \, m \in \mathcal{M}_h, \tag{4.13}$$

where C is the support of c.

Under $\mathcal{H}_0$, the benchmark (model m_1) outperforms all m models uniformly across all conditioning states. Evidently, by the law of iterated expectations, this also implies that the unconditional expected loss of the benchmark is smaller than those of the alternative methods. However, the CSPA null hypothesis is generally much more stringent than its unconditional counterpart. As such, the (uniform) conditional dominance criterion may help the researcher differentiate competing forecasting methods that may appear unconditionally similar.

The practical implementation of the test poses one key difficulty which is the estimation of the unknown conditional expectation function $h_{m,h}(c)$. One way of overcoming such problem is to nonparametrically estimate $h_{m,h}(c)$ by a sieve method as described earlier in this chapter. Hence, define $\boldsymbol{P}(c) = [p_1(c), \ldots, p_J(c)]'$, where $p_i(c), i = 1, \ldots, J$ is an approximating basis function. Therefore,

$$\widehat{h}_{m,h}(c) = \boldsymbol{P}(c)'\widehat{\boldsymbol{b}}_m, \tag{4.14}$$

where

$$\widehat{\boldsymbol{b}}_m = \widehat{\boldsymbol{Q}}^{-1} \left[\frac{1}{P_h} \sum_{t=R+1}^{T-h} \boldsymbol{P}(c_t) d_{(1m),h,t} \right]$$

and

$$\widehat{\boldsymbol{Q}} = \frac{1}{P_h} \sum_{t=R+1}^{T-h} \boldsymbol{P}(c_t)\boldsymbol{P}(c_t)'.$$

Let $\widehat{Z}_{m,h,t} = d_{(1m),h,t} - \widehat{h}_{m,h,t}$, where $\widehat{h}_{m,h,t} := h_{m,h}(c_t)$ and

$$\widehat{\boldsymbol{\Omega}} := \left(\boldsymbol{I}_M \otimes \widehat{\boldsymbol{Q}}\right)^{-1} \widehat{\boldsymbol{S}} \left(\boldsymbol{I}_M \otimes \widehat{\boldsymbol{Q}}\right)^{-1},$$

where $\boldsymbol{I}_M$ is a $(M \times M)$ identity matrix and $\widehat{\boldsymbol{S}}$ is HAC estimator of the long-run covariance matrix of $\widehat{\boldsymbol{Z}}_{h,t} \otimes \boldsymbol{P}(c_t)$. $\widehat{\boldsymbol{Z}}_{h,t}$ is the vector stacking $\widehat{Z}_{m,h,t}$ for different models.

The standard error of $\widehat{h}_{m,h,t}$ is given by

$$\widehat{\sigma}_m(c) := \left[\boldsymbol{P}(c)'\widehat{\boldsymbol{\Omega}}_{(m,m)}\boldsymbol{P}(c)\right]^{1/2},$$

where $\widehat{\boldsymbol{\Omega}}_{(m,m)}$ is the $M \times M$ block of $\widehat{\boldsymbol{\Omega}}$ corresponding to model m.

For a given significance level α, the rejection decision of the CSPA test is determined in the following steps.

1. Simulate a M-dimensional zero-mean Gaussian random vector $\boldsymbol{\xi}^*$ with covariance matrix $\widehat{\boldsymbol{\Omega}}$. Set $t_m^*(c) := \boldsymbol{P}(c)'\xi_m^*/\widehat{\sigma}_m(c)$.
2. Repeat step one many times. For some constant $z > 0$, define $\widehat{q}$ as the $1 - z/\log(P_h)$-quantile of $\max_{1 \leq m \leq M} \sup_{c \in C} \widehat{t}_m^*$ in the simulated sample and set

$$\widehat{\mathcal{V}} := \left\{(m,c) : \widehat{h}_{m,h}(c) \leq \min_{1 \leq m \leq M} \inf_{c \in C} \left[\widehat{h}_{m,h}(c) + P_h^{-1/2}\widehat{q}\widehat{\sigma}_m(c)\right] + 2P_h^{-1/2}\widehat{q}\widehat{\sigma}_m(c)\right\}. \tag{4.15}$$

   The value of z suggested by the authors is 0.1.
3. Set $\widehat{q}_{1-\alpha}$ as the $(1-\alpha)$-quantile of $\sup_{(m,c) \in \widehat{\mathcal{V}}} \widehat{t}_m^*(c)$. Reject the null hypothesis if and only if

$$\widehat{\eta}_{1-\alpha} := \min_{1 \leq m \leq M} \inf_{c \in C} \left[\widehat{h}_{m,h}(c) + P_h^{-1/2}\widehat{q}_m(c)\right] < 0. \tag{4.16}$$

The set $\mathcal{V}$ defined in (4.15) defines an adaptive inequality selection such that, with probability tending to 1, $\mathcal{V}$ contains all (m,c)'s that minimize $h_{m,h}(c)$. By inspection of the null hypothesis (4.13), it is clear that whether the null hypothesis holds or not is uniquely determined by the functions' values at these extreme points. The selection step focuses the test on the relevant conditioning region.

4.3.3 Model Confidence Sets

The MCS method, proposed by Hansen, Lunde and Nason (2011), consists of a sequence of statistic tests which yields the construct of a set of superior models, where the null hypothesis of equal predictive ability (EPA) is not rejected at a certain confidence level. The EPA statistic tests is calculated for an arbitrary loss function that satisfies general weak stationarity conditions

The MCS procedure starts from an initial set of models $\mathcal{M}^0$ of dimension M encompassing all the model specifications available to the user, and delivers, for a given confidence level $1-\alpha$, a smaller set $\mathcal{M}^*_{1-\alpha}$ of dimension $M^* \leq M$. $\mathcal{M}^*_{1-\alpha}$ consists as the set of the superior models. The best scenario is when the final set consists of a single model, i.e., $M^* = 1$.

Formally, let $d_{mn,t+h}$ denotes the loss differential between models m and n: $d_{mn,t+h} = \mathcal{L}(m,t+h) - \mathcal{L}(n,t+h)$, $m,n = 1,\ldots,M$ and $t = R,\ldots,T-h$. Let

$$d_{m,\cdot,t+h} = \frac{1}{M-1}\sum_{n\in\mathcal{M}} d_{m,n,t+h}, \quad m = 1,\ldots,M, \tag{4.17}$$

be the simple loss of model m relative to any other model n at time $t+h$.

The EPA hypothesis for a given set of models $\mathcal{M}$ can be formulated in two alternative ways:

$$\mathcal{H}_{0,\mathcal{M}} : c_{mn} = 0, \quad \forall m,n = 1,\ldots,M, \tag{4.18}$$
$$\mathcal{H}_{A,\mathcal{M}} : c_{mn} \neq 0, \quad \text{for some}\, m,n = 1,\ldots,M, \tag{4.19}$$

or

$$\mathcal{H}_{0,\mathcal{M}} : c_{m\cdot} = 0, \quad \forall m = 1,\ldots,M, \tag{4.20}$$
$$\mathcal{H}_{A,\mathcal{M}} : c_{m\cdot} \neq 0, \quad \text{for some}\, m = 1,\ldots,M, \tag{4.21}$$

where $c_{mn} = \mathbb{E}(d_{mn,t+h})$ and $c_{m\cdot} = \mathbb{E}(d_{m\cdot t+h})$

In order to test the two hypothesis above, the following two statistics are constructed:

$$t_{mn} = \frac{\bar{d}_{mn}}{\widehat{\sigma}(\bar{d}_{mn})} \quad \text{and} \quad t_{m\cdot} = \frac{\bar{d}_{m\cdot}}{\widehat{\sigma}(\bar{d}_{m\cdot})}, \tag{4.22}$$

where

$$\bar{d}_{mn} := \frac{1}{P_h}\sum_{t=R}^{T-h} d_{mn,t+h}, \quad \text{and} \quad \bar{d}_{m\cdot} := \frac{1}{M-1}\sum_{n\in\mathcal{M}} \bar{d}_{mn},$$

and $\widehat{\sigma}(\bar{d}_{mn})$ and $\widehat{\sigma}(\bar{d}_{m\cdot})$ are estimates of the standard deviation of $\bar{d}_{mn}$ and $\bar{d}_{m\cdot}$, respectively. The standard deviations are estimated by block bootstrap.

The null hypotheses of interest map naturally into the following two test statistics:

$$\mathcal{T}_{R,\mathcal{M}} = \max_{m,n\in\mathcal{M}} |t_{mn}| \tag{4.23}$$

and

$$\mathcal{T}_{\max,\mathcal{M}} = \max_{m\in\mathcal{M}} t_{m\cdot\cdot} \tag{4.24}$$

The MCS procedure consists on a sequential testing procedure which eliminates at each step the worst model, until the EPA hypothesis is not rejected for all the models belonging to the set. The choice of the worst model to be eliminated has been made using an elimination rule as follows:

$$e_{R,\mathcal{M}} = \arg\max_{m}\left(\sup_{n\in\mathcal{M}} t_{mn}\right) \qquad \text{and} \qquad e_{\max,\mathcal{M}} = \arg\max_{m\in\mathcal{M}} t_{m\cdot\cdot} \tag{4.25}$$

Therefore, the MCS consists of the following steps:

Algorithm (MCS Procedure) 1. Set $\mathcal{M} = \mathcal{M}^0$;
2. test for EPA–hypothesis: if EPA is not rejected terminate the algorithm and set $\mathcal{M}^*_{1-\alpha} = \mathcal{M}$. Otherwise, use the elimination rules defined in equations (9) to determine the worst model;
3. remove the worst model, and go to step 2.

4.4 Linear Models

We start by reviewing some ML methods based on the assumption that the target function $f_h(\boldsymbol{X}_t)$ is linear. We focus on methods that have not been previously discussed in the first chapter of this book.

Under linearity, the pseudo-true model is given as

$$f_h^*(\boldsymbol{x}_t) := \mathbb{E}(Y_{t+h}|\boldsymbol{X}_t = \boldsymbol{x}_t) = \boldsymbol{x}_t'\boldsymbol{\theta}.$$

Therefore, the class of approximating functions is also linear. We consider three different approaches: Factor-based regression, the combination of factors and penalized regression, and ensemble methods.

4.4.1 Factor Regression

The core idea of factor-based regression is to replace the large dimensional set of potential predictors, $\boldsymbol{W}_t \in \mathbb{R}^d$, by a low dimensional set of latent factors $\boldsymbol{F}_t$, which take values on $\mathbb{R}^k$, $k << d$, and that are estimated as rotations of the original dataset. More specifically, let

$$\boldsymbol{W}_t = \boldsymbol{\Lambda}\boldsymbol{F}_t + \boldsymbol{V}_t, \quad t = 1,\dots,T, \tag{4.26}$$

such that the regressors $\boldsymbol{X}_t$ can be re-defined as

$$\tilde{\boldsymbol{X}}_t = \left(Y_t,\dots,Y_{t-r+1},\boldsymbol{F}_t',\dots,\boldsymbol{F}_{t-s+1}'\right)'.$$

Note that the dimension of $\tilde{\boldsymbol{X}}_t$ is $\tilde{p} = r + ks$, such that $\tilde{p} << p$ as $k << d$.

Set $\boldsymbol{\beta} := (\boldsymbol{\beta}_{AR}, \boldsymbol{\beta}_{1,W}, \ldots, \boldsymbol{\beta}_{s,W})'$ such that:

$$Y_{t+h} = \boldsymbol{Y}_t'\boldsymbol{\beta}_{AR} + \boldsymbol{W}_t'\boldsymbol{\beta}_{1,W} + \ldots + \boldsymbol{W}_{t-s+1}'\boldsymbol{\beta}_{s,W} + U_{t+h},$$

where $\boldsymbol{Y}_t = (Y_t, \ldots, Y_{t-r+1})'$. Hence, the forecasting model becomes

$$\begin{aligned} Y_{t+h} &= \boldsymbol{Y}_t'\boldsymbol{\beta}_{AR} + \boldsymbol{F}_t'\boldsymbol{\Lambda}'\boldsymbol{\beta}_{1,W} + \ldots + \boldsymbol{F}_{t-s+1}'\boldsymbol{\Lambda}'\boldsymbol{\beta}_{s,W} \\ &\quad + \boldsymbol{V}_t'\boldsymbol{\beta}_{1,W} + \ldots + \boldsymbol{V}_{t-s+1}'\boldsymbol{\beta}_{s,W} + U_{t+h} \\ &= \tilde{\boldsymbol{X}}_t'\boldsymbol{\theta} + \epsilon_{t+h}, \end{aligned} \tag{4.27}$$

where

$$\epsilon_{t+h} = \boldsymbol{V}_t'\boldsymbol{\beta}_{1,W} + \ldots + \boldsymbol{V}_{t-s+1}'\boldsymbol{\beta}_{s,W} + U_{t+h},$$

and $\boldsymbol{\theta} = \left(\boldsymbol{\beta}_{AR}', \boldsymbol{\Lambda}'\boldsymbol{\beta}_{1,W}, \cdots, \boldsymbol{\Lambda}'\boldsymbol{\beta}_{s,W}\right)'$.

In order to estimate $\boldsymbol{\Lambda}$ and $\boldsymbol{F}_t$ we make the following assumption.

Assumption *(Factor Model)* Assume:

(a) $\mathbb{E}(\boldsymbol{F}_t) = \boldsymbol{0}$, $\mathbb{E}\left(\boldsymbol{F}_t\boldsymbol{F}_t'\right) = \boldsymbol{I}_r$ and $\boldsymbol{\Lambda}'\boldsymbol{\Lambda}$ is a diagonal matrix;
(b) All eigenvalues of $\boldsymbol{\Lambda}'\boldsymbol{\Lambda}/n$ are bounded away from zero and infinity as $n \to \infty$;
(c) $\|\boldsymbol{\Sigma} - \boldsymbol{\Lambda}\boldsymbol{\Lambda}'\| = O(1)$; and
(d) $\|\boldsymbol{\Lambda}\|_{\max} \leq C$. □

Under Assumption 3, the factors can be estimated as the k most important principal components of the sample covariance matrix of $\boldsymbol{W}$ and the parameter $\boldsymbol{\theta}$ can be estimated by OLS.

The principal components should be estimated in steps as follows:

1. For each $t = \max(r, s, h) + 1, \ldots, T - h$, construct the $(R \times d)$ matrix

$$\boldsymbol{W} = \begin{pmatrix} W_{1,t-R-h+1} & \cdots & W_{d,t-R-h+1} \\ \vdots & \ddots & \vdots \\ W_{1,t-h} & \cdots & W_{d,t-R-h+1} \end{pmatrix}.$$

2. Construct the standardized version $\widetilde{\boldsymbol{W}}$ of $\boldsymbol{W}$ such that each column of $\widetilde{\boldsymbol{W}}$ has zero mean and $\frac{1}{R}\widetilde{\boldsymbol{W}}'\widetilde{\boldsymbol{W}} = \boldsymbol{I}$, where $\boldsymbol{I}$ is a $(d \times d)$ identity matrix.
3. Compute the ordered (in ascending order) eigenvalues and the respective eigenvectors of $\widetilde{\boldsymbol{W}}$. The ith factor is given by $\boldsymbol{F}_i = \widetilde{\boldsymbol{W}}\boldsymbol{\gamma}_i$, $i = 1, \ldots, \min(d, R)$, where $\boldsymbol{\gamma}_i$ is the eigenvector associated with the ith eigenvalue.
4. Select the number of factors by one of the methods described below.

The optimal number of factors can be selected in different ways. For example, Ahn and Horenstein (2013) suggest selecting the factors as[6]

[6] See also Onatski (2010).

$$\widehat{k} = \arg \max_{k \in \{1,\ldots,\max(d,R)-1\}} \frac{\lambda_i}{\lambda_{i+1}},$$

where λ_i is the ith larger eigenvalue of $\widetilde{\boldsymbol{W}}$.

Bai and Ng (2002) suggest the following estimator for the number of factors:

$$\widehat{k} = \arg \max_{k \in \{1,\ldots,\min(d,R)-1\}} IC(k),$$

where IC is a information criterion and possible options are:

$$\begin{aligned}
IC_1(k) &= \log[S(k)] + k\frac{d+R}{dR}\log\left(\frac{dR}{d+R}\right),\\
IC_2(k) &= \log[S(k)] + k\frac{n+T}{dR}\log C_{dR}^2,\\
IC_3(k) &= \log[S(k)] + k\frac{\log C_{dR}^2}{C_{dR}^2}, \text{ or}\\
IC_4(k) &= \log[S(k)] + k\frac{(d+R-k)\log(dR)}{dR},
\end{aligned}$$

where $S(k) = \frac{1}{dR}\|\widetilde{\boldsymbol{W}} - \widehat{\boldsymbol{\Lambda}}_k \widehat{\boldsymbol{F}}_k\|_2^2$ and $C_{dR} := \sqrt{\min(d,R)}$.

The factor-based regression model described here assumes that all regressors are important but the relevant information can be summarized by a small number of factors. On the other hand, several models presented in Chapter 1 take an opposite direction by assuming a sparsity hypothesis, i.e., only a small number of regressors are in fact relevant.

Although, factor and sparse models are two widely used methods to impose a low-dimensional structure in high-dimension, they are seemingly mutually exclusive and there is a large debate in the literature on the relevance of each of them; see, for example, Giannone, Lenza and Primiceri (2021) and Fava and Lopes (2020).

Recently, Fan, Masini and Medeiros (2021) propose a lifting method that combines the merits of these two models in a supervised learning methodology that allows to efficiently explore all the information in high-dimensional datasets. We describe the method in the next section.

4.4.2 Bridging Sparse and Dense Models

The method proposed by Fan et al. (2021) is based on a flexible model for high-dimensional panel data, called factor-augmented regression (`FarmPredict`) model with both observable or latent common factors, as well as idiosyncratic components. This model not only includes both principal component (factor) regression and sparse regression as specific models, but also significantly weakens the cross-sectional dependence and hence, facilitates model selection and interpretability. The methodology consists of three steps. Although their methodology is more general, it can be adapted to forecasting problems.

The method can be implemented in the following steps.

1. Run principal component analysis on the vector $(Y_t, \boldsymbol{W}_t')'$ and select the number of factors by an appropriate method.
2. Compute the idiosyncratic errors $\widehat{\boldsymbol{V}}_t = (Y_t, \boldsymbol{W}_t')' - \widehat{\boldsymbol{\Lambda}}\widehat{\boldsymbol{F}}_t$, where $\widehat{\boldsymbol{\Lambda}}$ is the matrix of estimated loadings.
3. Run a LASSO regression of $\widehat{V}_{1t}$ on $\widehat{\boldsymbol{V}}_{-1t}$, where $\widehat{\boldsymbol{V}}_{-1t}$ is the vector with all the estimated idiosyncratic residuals with the exception of the first one. Call the fitted value $\widetilde{V}_{1t}$.
4. Estimate the forecasting model by OLS:

$$\begin{aligned}\widehat{Y}_{t+h|t} =& Y_t\beta_{1,AR} + \ldots + Y_{t-r+1}\beta_{r,AR} \\ &+ \widehat{\boldsymbol{F}}_t'\widehat{\boldsymbol{\beta}}_{1,F} + \ldots + \widehat{\boldsymbol{F}}_{t-s+1}'\widehat{\boldsymbol{\beta}}_{s,F} + \widetilde{V}_t\boldsymbol{\pi}_1 + \cdots + \widetilde{V}_{t-p+1}\boldsymbol{\pi}_p.\end{aligned} \tag{4.28}$$

Another variation of the method is as follows.

1. Define a benchmark model for forecasting Y_{t+h}. For example, an autoregressive model. Call the forecasts $\widehat{Y}_{t+h}^{\text{bench}}$. Set $\widehat{R}_t = Y_{t+h} - \widehat{Y}_{t+h}^{\text{bench}}$.
2. Run principal component analysis for the vector $\left(\widehat{R}_t, \boldsymbol{W}_t'\right)'$ and compute the idiosyncratic residuals: $\widehat{\boldsymbol{V}}_t = \left(\widehat{R}_t, \boldsymbol{W}_t'\right)' - \widehat{\boldsymbol{\Lambda}}\widehat{\boldsymbol{F}}_t$.
3. Set $\widehat{Y}_{t+h}^{\text{PCR}} = \widehat{Y}_{t+h}^{\text{bench}} + \widehat{\boldsymbol{\lambda}}_1{}'\widehat{\boldsymbol{F}}_t$. $\widehat{\boldsymbol{\lambda}}_1$ is the OLS estimate of the coefficient of the regression of $\widehat{R}_{t+h}$ on $\widehat{\boldsymbol{F}}_t$. Run a LASSO regression of $\widehat{V}_{1t}$ on $\widehat{\boldsymbol{V}}_{-1t}$. Call the fitted value $\widetilde{V}_{1t}$.
4. Finally, construct the forecast as:

$$Y_{t+h} = \widehat{Y}_{t+h}^{\text{PCR}} + \widetilde{V}_{1t}. \tag{4.29}$$

4.4.3 Ensemble Methods

4.4.3.1 Bagging

The term *bagging* means *Bootstrap Aggregating* and was proposed by Breiman (1996) to reduce the variance of unstable predictors[7]. For example, imagine a linear predictive regression model, where the variables included in the final model depends on the result of a standard t-test. Furthermore, suppose that some of the regressors are relevant but the 'true' parameters, although being non-zero, are quite small as compared to the size of the sample used to estimate the model. In this case, the t-statistic can yield the rejection or not of the null hypothesis of a zero coefficient due

[7] An *unstable predictor* has large variance. Intuitively, small changes in the data yield large changes in the predictive model.

to sample variability and the regressor can be erroneously removed from the model, generating poorer forecasts. The idea of Baggging is to alleviate this problem by estimating the model, running the test, and computing forecasts, in several different samples and taking the average of the forecasts over the samples. Each sample is constructed by bootstrap, which is a well-known resampling technique.

Bagging was popularized in the time series literature by Inoue and Kilian (2008), who to construct forecasts from multiple regression models with local-to-zero regression parameters and errors subject to possible serial correlation or conditional heteroscedasticity. Bagging is designed for situations in which the number of predictors is moderately large relative to the sample size.

The bagging algorithm in time-series settings have to take into account the time dependence dimension when constructing the bootstrap samples.

Algorithm (Bagging for Time-Series Models) The Bagging algorithm is defined as follows:

1. For each t, arrange the data as

$$\left(Y_{\tau+h}, \boldsymbol{X}_{\tau}'\right), \tau = t - R - h + 1, \ldots, t - h,$$

in the form of a matrix $\boldsymbol{V}$ of dimension $R \times p$.

2. Construct (block) bootstrap samples with R observations of the form

$$\left\{\left(Y_{(i)1}^{*}, \boldsymbol{X}_{(i)1}^{\prime *}\right), \ldots, \left(Y_{(i)R}^{*}, \boldsymbol{X}_{(i)R}^{\prime *}\right)\right\}, i = 1, \ldots, B,$$

by drawing blocks of M rows of $\boldsymbol{V}$ with replacement.

3. Compute the ith bootstrap forecast as

$$\widehat{Y}_{(i)t+h|t}^{*} = \begin{cases} 0 & \text{if } |t_j^*| < c \,\forall j, \\ \widehat{\boldsymbol{\lambda}}_{(i)}^{*} \widetilde{\boldsymbol{X}}_{(i)t}^{*} & \text{otherwise}, \end{cases} \tag{4.30}$$

where $\widetilde{\boldsymbol{X}}_{(i)t}^{*} := \boldsymbol{S}_{(i)t}^{*} \boldsymbol{Z}_{(i)t}^{*}$ and $\boldsymbol{S}_t$ is a diagonal selection matrix with jth diagonal element given by

$$\mathbb{I}_{\{|t_j| > c\}} = \begin{cases} 1 & \text{if } |t_j| > c, \\ 0 & \text{otherwise}, \end{cases}$$

c is a pre-specified critical value of the test. $\widehat{\boldsymbol{\lambda}}_{(i)}^{*}$ is the OLS estimator at each bootstrap repetition.

4. Compute the average forecasts over the bootstrap samples:

$$\tilde{Y}_{t+h|t} = \frac{1}{B} \sum_{i=1}^{B} \widehat{Y}_{(i)t|t-1}^{*}.$$

In algorithm 2, above, one requires that it is possible to estimate and conduct inference in the linear model. This is certainly infeasible if the number of predictors is larger than the sample size ($p > T$), which requires the algorithm to be modified.

Garcia, Medeiros and Vasconcelos (2017) and Medeiros, Vasconcelos, Veiga and Zilberman (2021) adopt the following changes of the algorithm:

Algorithm (Bagging for Time-Series Models and Many Regressors) The Bagging algorithm is defined as follows.

0. Run p univariate regressions of Y_{t+h} on each covariate in $\boldsymbol{X}_t$. Compute t-statistics and keep only the ones that turn out to be significant at a given pre-specified level. Call this new set of regressors as $\check{\boldsymbol{X}}_t$.

1–4. Same as before but with $\boldsymbol{X}_t$ replaced by $\check{\boldsymbol{X}}_t$.

4.4.3.2 Complete Subset Regression

Complete Subset Regression (CSR) is a method for combining forecasts developed by Elliott et al. (2013, 2015). The motivation was that selecting the optimal subset of $\boldsymbol{X}_t$ to predict Y_{t+h} by testing all possible combinations of regressors is computationally very demanding and, in most cases, unfeasible. For a given set of potential predictor variables, the idea is to combine forecasts by averaging[8] all possible linear regression models with fixed number of predictors. For example, with p possible predictors, there are p unique univariate models and

$$p_{q,n} = \frac{p!}{(p-q)!q!}$$

different q-variate models for $q \leq Q$. The set of models for a fixed value of q as is known as the complete subset.

When the set of regressors is large the number of models to be estimated increases rapidly. Moreover, it is likely that many potential predictors are irrelevant. In these cases it was suggested that one should include only a small, $\tilde{q}$, fixed set of predictors, such as five or ten. Nevertheless, the number of models still very large, for example, with $p = 30$ and $q = 8$, there are $5,852,925$ regressions to be estimated. An alternative solution is to follow Garcia et al. (2017) and Medeiros et al. (2021) and adopt a similar strategy as in the case of Bagging high-dimensional models. The idea is to start fitting a regression of Y_{t+h} on each of the candidate variables and save the t-statistics of each variable. The t-statistics are ranked by absolute value, and we select the $\tilde{p}$ variables that are more relevant in the ranking. The CSR forecast is calculated on these variables for different values of q. Another possibility is to pre-select the variables by elastic-net or other selection method; see, Chapter 1 for details.

[8] It is possible to combine forecasts using any weighting scheme. However, it is difficult to beat uniform weighting (Genre, Kenny, Meyler & Timmermann, 2013).

4.5 Nonlinear Models

4.5.1 Feedforward Neural Networks

4.5.1.1 Shallow Neural Networks

Neural Network (NN) is one of the most traditional nonlinear sieve methods. NN can be classified into *shallow* (single hidden layer) or *deep* networks (multiple hidden layers). We start describing the shallow version. The most common shallow NN is the feedforward neural network.

Definition (Feedforward NN model) In the single hidden layer feedforward NN (sieve) model, the approximating function $f_{h,D}(\boldsymbol{X}_t)$ is defined as

$$\begin{aligned} f_{h,D}(\boldsymbol{X}_t) := f_{h,D}(\boldsymbol{X}_t;\boldsymbol{\theta}) &= \beta_0 + \sum_{j=1}^{J} \beta_j S(\boldsymbol{\gamma}_j' \boldsymbol{X}_t + \gamma_{0,j}), \\ &= \beta_0 + \sum_{j=1}^{J} \beta_j S(\tilde{\boldsymbol{\gamma}}_j' \tilde{\boldsymbol{X}}_t), \end{aligned} \tag{4.31}$$

In (4.31), $\tilde{\boldsymbol{X}}_t = (1, \boldsymbol{X}_t')'$, $S_j(\cdot)$ is a basis function and the parameter vector to be estimated is given by $\boldsymbol{\theta} = (\beta_0, \ldots, \beta_K, \boldsymbol{\gamma}_1', \ldots, \boldsymbol{\gamma}_J', \gamma_{0,1}, \ldots, \gamma_{0,J})'$, where $\tilde{\boldsymbol{\gamma}}_j = (\gamma_{0,j}, \boldsymbol{\gamma}_j')'$. □

NN models form a very popular class of nonlinear sieves where the function $g_{h,j}(\boldsymbol{X}_t)$ in (4.5) is given by $S(\tilde{\boldsymbol{\gamma}}_j' \tilde{\boldsymbol{X}}_t)$. Such kind of model has been used in many applications of forecasting for many decades. Usually, the basis functions $S(\cdot)$ are called *activation functions* and the parameters are called *weights*. The terms in the sum are called hidden-neurons as an unfortunate analogy to the human brain. Specification (4.31) is also known as a single hidden layer NN model as is usually represented in the graphical form as in Figure 4.2. The green circles in the figure represent the input layer which consists of the regressors of the model $(\boldsymbol{X}_t)$. In the figure there are four input variables. The blue and red circles indicate the hidden and output layers, respectively. In the example, there are five elements (called neurons in the NN jargon) in the hidden layer. The arrows from the green to the blue circles represent the linear combination of inputs: $\boldsymbol{\gamma}_j' \boldsymbol{X}_t + \gamma_{0,j}$, $j = 1, \ldots, 5$. Finally, the arrows from the blue to the red circles represent the linear combination of outputs from the hidden layer: $\beta_0 + \sum_{j=1}^{5} \beta_j S(\boldsymbol{\gamma}_j' \boldsymbol{X}_t + \gamma_{0,j})$.

There are several possible choices for the activation functions. In the early days, $S(\cdot)$ was chosen among the class of squashing functions as per the definition below.

Definition (Squashing (sigmoid) function) A function $S : \mathbb{R} \longrightarrow [a, b]$, $a < b$, is a squashing (sigmoid) function if it is non-decreasing, $\lim_{x \longrightarrow \infty} S(x) = b$ and $\lim_{x \longrightarrow -\infty} S(x) = a$. □

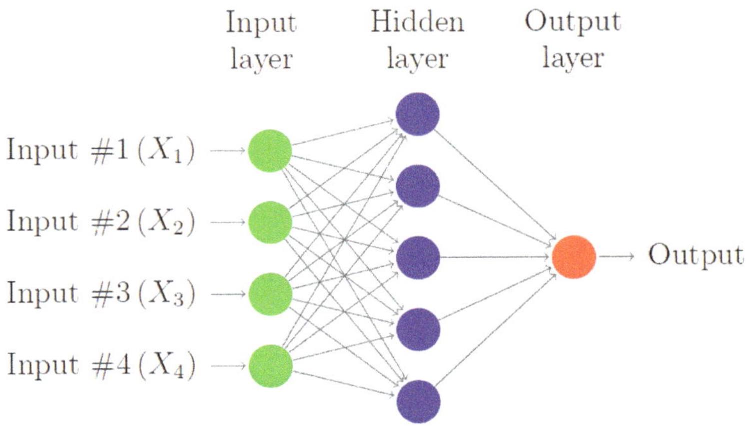

Fig. 4.2: Graphical representation of a single hidden layer neural network

Historically, the most popular choices are the logistic and hyperbolic tangent functions such that:

$$\text{Logistic: } S(x) = \frac{1}{1+\exp(-x)}$$

$$\text{Hyperbolic tangent: } S(x) = \frac{\exp(x)-\exp(-x)}{\exp(x)+\exp(-x)}.$$

The popularity of such functions was partially due to theoretical results on function approximation. Funahashi (1989) establishes that NN models as in (4.31) with generic squashing functions are capable of approximating any continuous functions from one finite dimensional space to another to any desired degree of accuracy, provided that J_T is sufficiently large. Cybenko (1989) and Hornik, Stinchombe and White (1989) simultaneously proved approximation capabilities of NN models to any Borel measurable function and Hornik et al. (1989) extended the previous results and showed the NN models are also capable to approximate the derivatives of the unknown function. Barron (1993) relate previous results to the number of terms in the model.

Stinchcombe and White (1989) and Park and Sandberg (1991) derived the same results of Cybenko (1989) and Hornik et al. (1989) but without requiring the activation function to be sigmoid. While the former considered a very general class of functions, the later focused on radial-basis functions (RBF) defined as:

$$\text{Radial Basis: } S(x) = \exp(-x^2).$$

More recently, Yarotsky (2017) showed that the rectified linear units (ReLU) as

$$\text{Rectified Linear Unit: } S(x) = \max(0, x),$$

are also universal approximators and ReLU activation function is one of the most popular choices among practitioners due to the following advantages:

1. Estimating NN models with ReLU functions is more efficient computationally as compared to other typical choices, such as logistic or hyperbolic tangent. One reason behind such improvement in performance is that the output of the ReLU function is zero whenever the inputs are negative. Thus, fewer units (neurons) are activated, leading to network sparsity.
2. In terms of mathematical operations, the ReLU function involves simpler operations than the hyperbolic tangent and logistic functions, which also improves computational efficiency.
3. Activation functions like the hyperbolic tangent and the logistic functions may suffer from the *vanishing gradient problem*, where gradients shrink drastically during optimization, such that the estimates are no longer improved. ReLU avoids this by preserving the gradient since it is an unbounded function.

However, ReLU functions suffer from the *dying activation problem*: many ReLU units yield output values of zero, which happens when the ReLU inputs are negative. While this characteristic gives ReLU its strengths (through network sparsity), it becomes a problem when most of the inputs to these ReLU units are in the negative range. The worst-case scenario is when the entire network dies, meaning that it becomes just a constant function. A solution to this problem is to use some modified versions of the ReLU function, such as the Leaky ReLU (LeReLU):

$$\text{Leaky ReLU: } S(x) = \max(\alpha x, x),$$

where $0 < \alpha < 1$.

For each estimation window, $t = \max(r, s, h) + 1, \ldots, T - h$, model (4.31) can be written in matrix notation. Let $\boldsymbol{\Gamma} = (\tilde{\boldsymbol{\gamma}}_1, \ldots, \tilde{\boldsymbol{\gamma}}_J)$ be a $(p+1) \times J$ matrix,

$$
\boldsymbol{X} = \underbrace{\begin{pmatrix} 1 & X_{1,t-R-h+1} & \cdots & X_{p,t-R-h+1} \\ 1 & X_{1,t-R-h+2} & \cdots & X_{p,t-R-h+2} \\ \vdots & \ddots & \vdots & \\ 1 & X_{1,t-h} & \cdots & X_{p,t-h} \end{pmatrix}}_{R\times(p+1)}, \text{ and}
$$

$$
\boldsymbol{O}(\boldsymbol{X\Gamma}) = \underbrace{\begin{pmatrix} 1 & S(\tilde{\boldsymbol{\gamma}}'_1\tilde{\boldsymbol{X}}_{t-R-h+1}) & \cdots & S(\tilde{\boldsymbol{\gamma}}'_J\tilde{\boldsymbol{X}}_{t-R-h+1}) \\ 1 & S(\tilde{\boldsymbol{\gamma}}'_1\tilde{\boldsymbol{X}}_{t-R-h+2}) & \cdots & S(\tilde{\boldsymbol{\gamma}}'_J\tilde{\boldsymbol{X}}_{t-R-h+2}) \\ \vdots & \vdots & \ddots & \vdots \\ 1 & S(\tilde{\boldsymbol{\gamma}}'_1\tilde{\boldsymbol{X}}_{t-h}) & \cdots & S(\tilde{\boldsymbol{\gamma}}'_J\tilde{\boldsymbol{X}}_{t-h}) \end{pmatrix}}_{R\times(J+1)}.
$$

Therefore, by defining $\boldsymbol{\beta} = (\beta_0, \beta_1, \ldots, \beta_K)'$, the output of a feedforward NN is given by:

$$
\begin{aligned}
\boldsymbol{f}_D(\boldsymbol{X},\boldsymbol{\theta}) &= [f_D(\boldsymbol{X}_{t-R-h+1};\boldsymbol{\theta}), \ldots, f_D(\boldsymbol{X}_{t-h};\boldsymbol{\theta})]' \\
&= \begin{bmatrix} \beta_0 + \sum_{j=1}^{J} \beta_j S(\boldsymbol{\gamma}'_j \boldsymbol{X}_{t-R-h+1} + \gamma_{0,j}) \\ \vdots \\ \beta_0 + \sum_{j=1}^{J} \beta_j S(\boldsymbol{\gamma}'_j \boldsymbol{X}_{t-h} + \gamma_{0,j}) \end{bmatrix} \\
&= \boldsymbol{O}(\boldsymbol{X\Gamma})\boldsymbol{\beta}.
\end{aligned} \tag{4.32}
$$

The number of hidden units (neurons), J, and the choice of activation functions are known as the architecture of the NN model. Once the architecture is defined, the dimension of the parameter vector $\boldsymbol{\theta} = [\boldsymbol{\Gamma}', \boldsymbol{\beta}']'$ is $D = (p+1) \times J + (J+1)$ and can easily get very large such that the unrestricted estimation problem defined as

$$
\widehat{\boldsymbol{\theta}} = \arg\min_{\boldsymbol{\theta}\in\mathbb{R}^D} \|\boldsymbol{Y} - \boldsymbol{O}(\boldsymbol{X\Gamma})\boldsymbol{\beta}\|_2^2
$$

is unfeasible. A solution is to use regularization as in the case of linear models and consider the minimization of the following function:

$$
Q(\boldsymbol{\theta}) = \|\boldsymbol{Y} - \boldsymbol{O}(\boldsymbol{X\Gamma})\boldsymbol{\beta}\|_2^2 + p(\boldsymbol{\theta}), \tag{4.33}
$$

where usually $p(\boldsymbol{\theta}) = \lambda\boldsymbol{\theta}'\boldsymbol{\theta}$. Traditionally, the most common approach to minimize (4.33) is to use Bayesian methods as in MacKay (1992a), (1992b), and Foresee and Hagan (1997). See also Chapter 2 for more examples of regularization with nonlinear models.

A more modern approach is to use a technique known as *Dropout* (Srivastava, Hinton, Krizhevsky, Sutskever & Salakhutdinov, 2014). The key idea is to randomly drop neurons (along with their connections) from the neural network during estimation. A NN with J neurons in the hidden layer can generate 2^J possible *thinned* NN by just removing some neurons. Dropout samples from this 2^J different thinned NN and train the sampled NN. To predict the target variable, we use a single unthinned network that has weights adjusted by the probability law induced by the random drop. This procedure significantly reduces overfitting and gives major improvements over other regularization methods.

We modify equation (4.31) by

$$f_D^-(\boldsymbol{X}_t) = \beta_0 + \sum_{j=1}^{J} s_j \beta_j S(\boldsymbol{\gamma}_j' [\boldsymbol{r} \odot \boldsymbol{X}_t] + v_j \gamma_{0,j}),$$

where s, v, and $\boldsymbol{r} = (r_1, \ldots, r_p)$ are independent Bernoulli random variables each with probability q of being equal to 1. The NN model is thus estimated by using $f_D^-(\boldsymbol{X}_t)$ instead of $f_D(\boldsymbol{X}_t)$ where, for each training example, the values of the entries of s, v, and $\boldsymbol{r}$ are drawn from the Bernoulli distribution. The final estimates for β_j, $\boldsymbol{\gamma}_j$, and $\gamma_{0,j}$ are multiplied by q.

4.5.1.2 Deep Neural Networks

A Deep Neural Network model is a straightforward generalization of specification (4.31), where more hidden layers are included in the model, as represented in Figure 4.3. In the figure, we represent a Deep NN with two hidden layers with the same number of hidden units in each. However, the number of hidden units (neurons) can vary across layers.

As pointed out in Mhaska, Liao and Poggio (2017), while the universal approximation property holds for shallow NNs, deep networks can approximate the class of compositional functions as well as shallow networks but with exponentially lower number of training parameters and sample complexity.

Set J_ℓ as the number of hidden units in layer $\ell \in \{1, \ldots, L\}$. For each hidden layer ℓ define $\boldsymbol{\Gamma}_\ell = (\tilde{\boldsymbol{\gamma}}_{1\ell}, \ldots, \tilde{\boldsymbol{\gamma}}_{k_\ell \ell})$. Then the output $\boldsymbol{O}_\ell$ of layer ℓ is given recursively by

$$\underbrace{\boldsymbol{O}_\ell(\boldsymbol{O}_{\ell-1}(\cdot)\boldsymbol{\Gamma}_\ell)}_{p\times(J_\ell+1)} = \begin{pmatrix} 1 & S(\tilde{\boldsymbol{\gamma}}_{1\ell}' \boldsymbol{O}_{1\ell-1}(\cdot)) & \cdots & S(\tilde{\boldsymbol{\gamma}}_{k_\ell \ell}' \boldsymbol{O}_{1\ell-1}(\cdot)) \\ 1 & S(\tilde{\boldsymbol{\gamma}}_{1\ell}' \boldsymbol{O}_{2\ell-1}(\cdot)) & \cdots & S(\tilde{\boldsymbol{\gamma}}_{k_\ell \ell}' \boldsymbol{O}_{2\ell-1}(\cdot)) \\ \vdots & \vdots & \ddots & \vdots \\ 1 & S(\tilde{\boldsymbol{\gamma}}_{1\ell}' \boldsymbol{O}_{n\ell-1}(\cdot)) & \cdots & S(\tilde{\boldsymbol{\gamma}}_{J_\ell \ell}' \boldsymbol{O}_{n\ell-1}(\cdot)) \end{pmatrix}$$

where $\boldsymbol{O}_0 := \boldsymbol{X}$. Therefore, the output of the Deep NN is the composition

$$\boldsymbol{h}_D(\boldsymbol{X}) = \boldsymbol{O}_L(\cdots \boldsymbol{O}_3(\boldsymbol{O}_2(\boldsymbol{O}_1(\boldsymbol{X}\boldsymbol{\Gamma}_1)\boldsymbol{\Gamma}_2)\boldsymbol{\Gamma}_3)\cdots)\boldsymbol{\beta}.$$

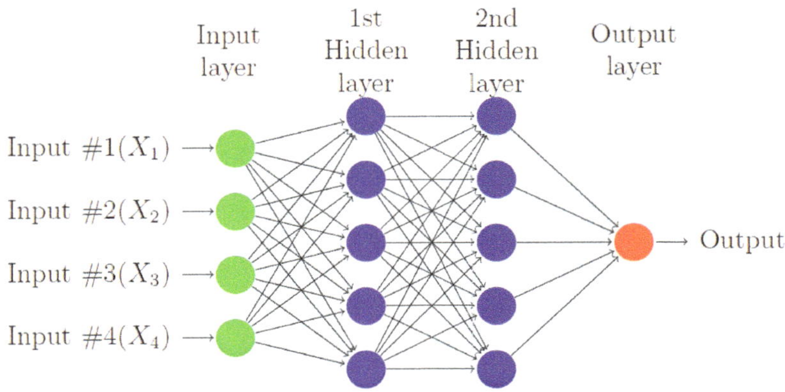

Fig. 4.3: Deep neural network architecture

The estimation of the parameters is usually carried out by stochastic gradient descend methods with dropout to control the complexity of the model.

4.5.2 Long Short Term Memory Networks

Broadly speaking, Recurrent Neural Networks (RNNs) are NNs that allow for feedback among the hidden layers. RNNs can use their internal state (memory) to process sequences of inputs. In the framework considered in this chapter, a generic RNN could be written as

$$\begin{aligned} \boldsymbol{H}_t &= \boldsymbol{f}(\boldsymbol{H}_{t-1}, \boldsymbol{X}_t), \\ \widehat{Y}_{t+h|t} &= g(\boldsymbol{H}_t), \end{aligned}$$

where $\widehat{Y}_{t+h|t}$ is the prediction of Y_{t+h} given observations only up to time t, $\boldsymbol{f}$ and g are functions to be defined and $\boldsymbol{H}_t$ is what we call the k-dimensional (hidden) state. From a time-series perspective, RNNs can be seen as a kind of nonlinear state-space model.

RNNs can remember the order that the inputs appear through its hidden state (memory) and they can also model sequences of data so that each sample can be assumed to be dependent on previous ones, as in time-series models. However, RNNs are hard to be estimated as they suffer from the vanishing/exploding gradient problem. For each estimation window, set the cost function to be

$$Q(\boldsymbol{\theta}) = \sum_{\tau=t-R-h+1}^{t-h} \left(Y_{\tau+h} - \widehat{Y}_{\tau+h|\tau}\right)^2,$$

where $\boldsymbol{\theta}$ is the vector of parameters to be estimated. It is easy to show that the gradient $\frac{\partial Q(\boldsymbol{\theta})}{\partial \boldsymbol{\theta}}$ can be very small or diverge. Fortunately, there is a solution to the problem proposed by Hochreiter and Schmidhuber (1997). A variant of RNN which is called Long-Short-Term Memory (LSTM) network . Figure 4.4 shows the architecture of a typical LSTM layer. A LSTM network can be composed of several layers. In the figure, red circles indicate logistic activation functions, while blue circles represent hyperbolic tangent activation. The symbols 'X' and '+' represent, respectively, the element-wise multiplication and sum operations. The RNN layer is composed of several blocks: the cell state and the forget, input, and ouput gates. The cell state introduces a bit of memory to the LSTM so it can 'remember' the past. LSTM learns to keep only relevant information to make predictions, and forget non relevant data. The forget gate tells which information to throw away from the cell state. The output gate provides the activation to the final output of the LSTM block at time t. Usually, the dimension of the hidden state ($\boldsymbol{H}_t$) is associated with the number of hidden neurons.

Algorithm 4 describes analytically how the LSTM cell works. $\boldsymbol{f}_t$ represents the output of the forget gate. Note that it is a combination of the previous hidden-state ($\boldsymbol{H}_{t-1}$) with the new information ($\boldsymbol{X}_t$). Note that $\boldsymbol{f}_t \in [0, 1]$ and it attenuates the signal coming com $\boldsymbol{c}_{t-1}$. The input and output gates have the same structure. Their function is to filter the 'relevant' information from the previous time period as well as from the new input. $\boldsymbol{p}_t$ scales the combination of inputs and previous information. This signal is then combined with the output of the input gate ($\boldsymbol{i}_t$). The new hidden state is an attenuation of the signal coming from the output gate. Finally, the prediction is a linear combination of hidden states. Figure 4.5 illustrates how the information flows in a LSTM cell.

Algorithm Mathematically, RNNs can be defined by the following algorithm:

1. Initiate with $\boldsymbol{c}_0 = 0$ and $\boldsymbol{H}_0 = 0$.
2. Given the input $\boldsymbol{X}_t$, for $t \in \{1, \ldots, T\}$, do:

$$\begin{aligned}
\boldsymbol{f}_t &= \text{Logistic}(\boldsymbol{W}_f \boldsymbol{X}_t + \boldsymbol{U}_f \boldsymbol{H}_{t-1} + \boldsymbol{b}_f) \\
\boldsymbol{i}_t &= \text{Logistic}(\boldsymbol{W}_i \boldsymbol{X}_t + \boldsymbol{U}_i \boldsymbol{H}_{t-1} + \boldsymbol{b}_i) \\
\boldsymbol{o}_t &= \text{Logistic}(\boldsymbol{W}_o \boldsymbol{X}_t + \boldsymbol{U}_o \boldsymbol{H}_{t-1} + \boldsymbol{b}_o) \\
\boldsymbol{p}_t &= \text{Tanh}(\boldsymbol{W}_c \boldsymbol{X}_t + \boldsymbol{U}_c \boldsymbol{H}_{t-1} + \boldsymbol{b}_c) \\
\boldsymbol{c}_t &= (\boldsymbol{f}_t \odot \boldsymbol{c}_{t-1}) + (\boldsymbol{i}_t \odot \boldsymbol{p}_t) \\
\boldsymbol{H}_t &= \boldsymbol{o}_t \odot \text{Tanh}(\boldsymbol{c}_t) \\
\widehat{Y}_{t+h|t} &= \boldsymbol{W}_y \boldsymbol{H}_t + b_y
\end{aligned}$$

where $\boldsymbol{U}_f$, $\boldsymbol{U}_i$, $\boldsymbol{U}_o$,$\boldsymbol{U}_c$,$\boldsymbol{U}_f$, $\boldsymbol{W}_f$, $\boldsymbol{W}_i$, $\boldsymbol{W}_o$, $\boldsymbol{W}_c$, $\boldsymbol{W}_y$, $\boldsymbol{b}_f$, $\boldsymbol{b}_i$, $\boldsymbol{b}_o$, $\boldsymbol{b}_c$, and b_y are parameters to be estimated.

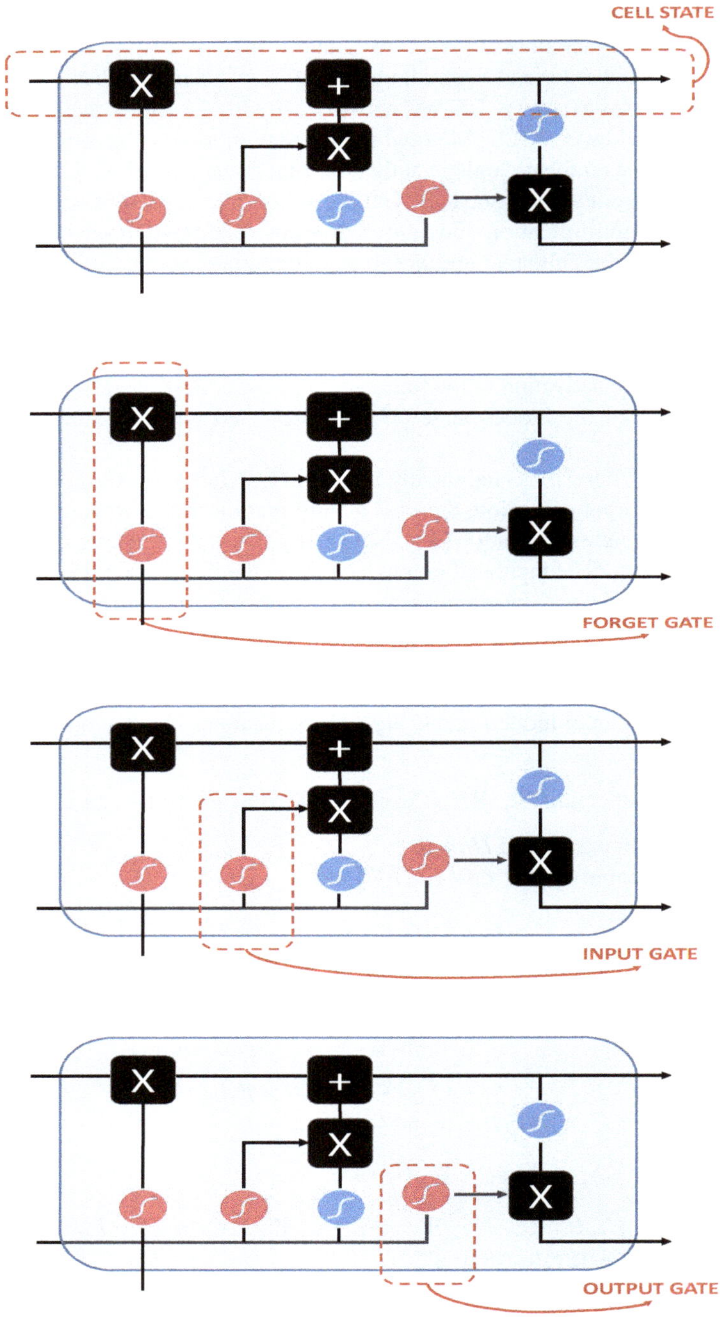

Fig. 4.4: Architecture of the Long-Short-Term Memory Cell (LSTM)

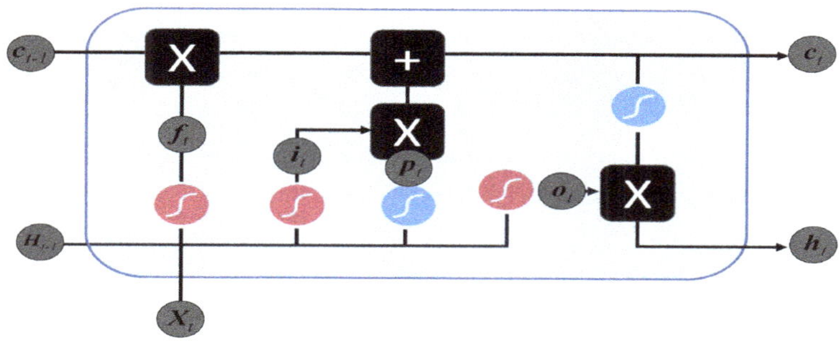

Fig. 4.5: Information flow in a LTSM Cell

4.5.3 Convolution Neural Networks

Convolutional Neural Networks (CNNs) are a class of Neural Network models that have proven to be very successful in areas such as image recognition and classification and are becoming popular for time series forecasting.

Figure 4.6 illustrates graphically a typical CNN. It is easier to understand the architecture of a CNN through an image processing application.

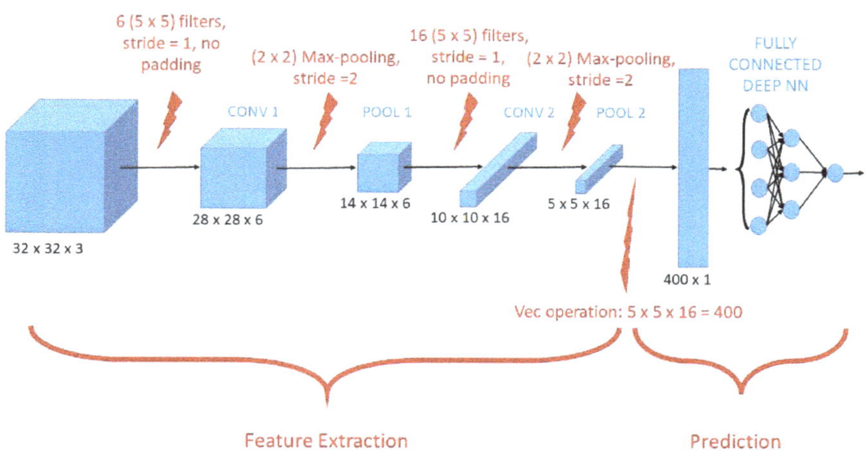

Fig. 4.6: Representation of a Convolution Neural Network

As can be seen in Figure 4.6, the CNN consist of two main blocks: the *feature extraction* and the *prediction* blocks. The prediction block is a feedforward deep NN as previously discussed. The feature extraction block has the following key elements:

one or more convolutional layer; a nonlinear transformation of the data; one or more pooling layers for dimension reduction; and a fully-connected (deep) feed-forward neural network.

The elements above are organized in a sequence of layers:

convolution + nonlinear transformation → pooling → convolution + nonlinear transformation → pooling → ··· → convolution + nonlinear transformation → pooling → Fully-connected (deep) NN

To a computer, an image is matrix of pixels. Each entry of the matrix is the intensity of the pixel: $0-255$. The dimension of the matrix is the resolution of the image. For coloured images, there is a third dimension to represent the colour channels: red, green, and blue. Therefore, the image is a three dimensional matrix (tensor): $\text{Height} \times \text{Width} \times 3$. An image kernel is a small matrix used to apply effects (filters), such as blurring, sharpening, outlining, for example. In CNNs, kernels are used for *feature extraction*, a technique for determining the most important portions of an image. In this context the process is referred to more generally as *convolution*.

The convolution layer is defined as follows. Let $\boldsymbol{X} \in \mathbb{R}^{M \times N}$ be the input data and $\boldsymbol{W} \in \mathbb{R}^{Q \times R}$ the filter kernel.

For $i = 1, \ldots M - Q + 1$, $j = 1, \ldots, N - R + 1$ write:

$$\begin{aligned} O_{ij} &= \sum_{q=1}^{Q} \sum_{r=1}^{R} [\boldsymbol{W} \odot [\boldsymbol{X}]_{i:i+Q-1, j:j+R-1}]_{q,r} \\ &= \boldsymbol{\iota}'_Q \left(\boldsymbol{W} \odot [\boldsymbol{X}]_{i:i+Q-1, j:j+R-1} \right) \boldsymbol{\iota}_R \end{aligned} \tag{4.34}$$

where $\odot$ is the element-by-element multiplication, $\boldsymbol{\iota}_Q \in \mathbb{R}^Q$ and $\boldsymbol{\iota}_R^R$ are vector of ones, $[\boldsymbol{X}]_{i:i+Q-1,j:j+R-1}$ is the block of the matrix X running from row i to row $i+Q-j$ and from column j to column $j+R-1$, and $[\boldsymbol{X}]_{i:i+Q-1,j:j+R-1}]_{q,r}$ is the element of $[\boldsymbol{X}]_{i:i+Q-1,j:j+R-1}$ in position (q,r).

O_{ij} is the *discrete convolution* between $\boldsymbol{W}$ and $[\boldsymbol{X}]_{i:i+Q-1,j:j+R-1}$:

$$O_{ij} = \boldsymbol{W} * [\boldsymbol{X}]_{i:i+Q-1,j:j+R-1}.$$

Figure 4.7 shows an example of the transformation implied by the convolution layer. In this example a (3×3) filter is applied to (6×6) matrix of inputs. The output is a (4×4) matrix. Each element of the output matrix is the sum of the dot product between the entries of the input matrix (shaded red area) and the ones from the weight matrix. Note that the shaded red (3×3) matrix is slided to the right and down by one entry. Sometimes, in order to reduce the dimension of the output, one can apply the *stride* technique, i.e., slide over more than one entry of the input matrix. Figure 4.8 shows an example.

Due to border effects, the output of the convolution layer is of smaller dimension than the input. One solution is to use the technique called *padding*, i.e., filling a border with zeroes. Figure 4.9 illustrates the idea. Note that in the case presented in the figure the input and the ouput matrices have the same dimension.

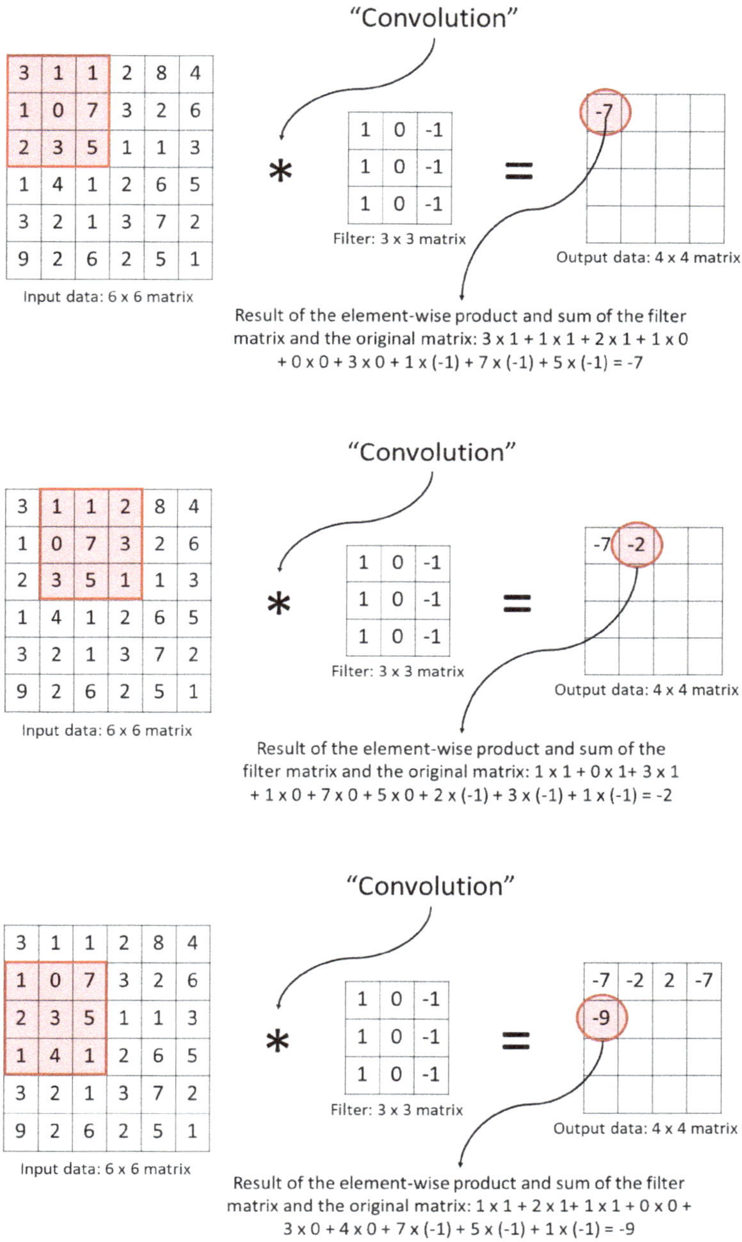

Fig. 4.7: Example of a convolution layer

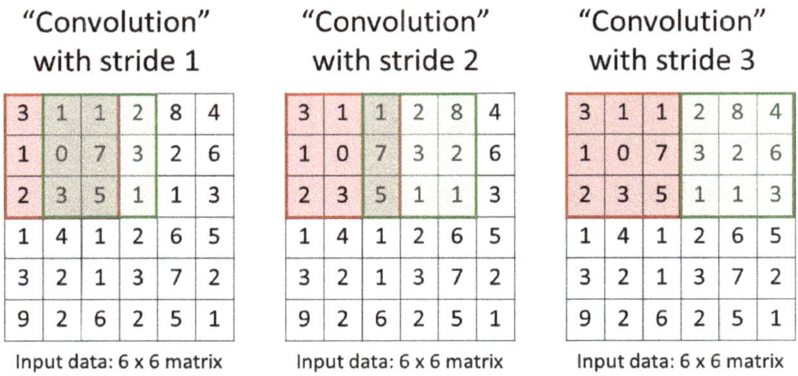

Fig. 4.8: Example of a convolution layer with stride

Each convolution layer may have more than one convolution filter. Figure 4.10 shows a convolution layer where the input is formed by three (6×6) matrices and where there are two filters. The output of the layer is a set of two (4×4) matrices. In this case stride is equal to one and there is no padding.

Usually the outputs of the convolution layer are sent through a nonlinear activation function as, for example, the ReLU. See Figure 4.11 for an illustration.

The final step is the application of a dimension reduction technique called *pooling*. One common polling approach is the *max pooling*, where the final output is the maximum entry in a sub-matrix of the output of the convolution layer. See, for example, the illustration in Figure 4.12

The process describe above is than repeated as many times as the number of convolution layers in the network.

Summarizing, the user has to define the following hyperparameters concerning the architecture of the convolution NN:

1. number of convolution layers (C);
2. number of pooling layers (P);
3. number (K_c) and dimensions (Q_c height, R_c width and S_c depth) of filters in each convolution layer $c = 1, \dots, C$;
4. architecture of the deep neural network.

The parameters to be estimated are

1. Filter weights: $\boldsymbol{W}_{ic} \in \mathbb{R}^{Q_c \times R_c \times S_c}$, $i = 1, \dots, K_c$, $c = 1, \dots, C$;
2. ReLU biases: $\boldsymbol{\gamma}_c \in \mathbb{R}^{K_c}$, $c = 1, \dots, C$;
3. All the parameters of the fully connected deep.

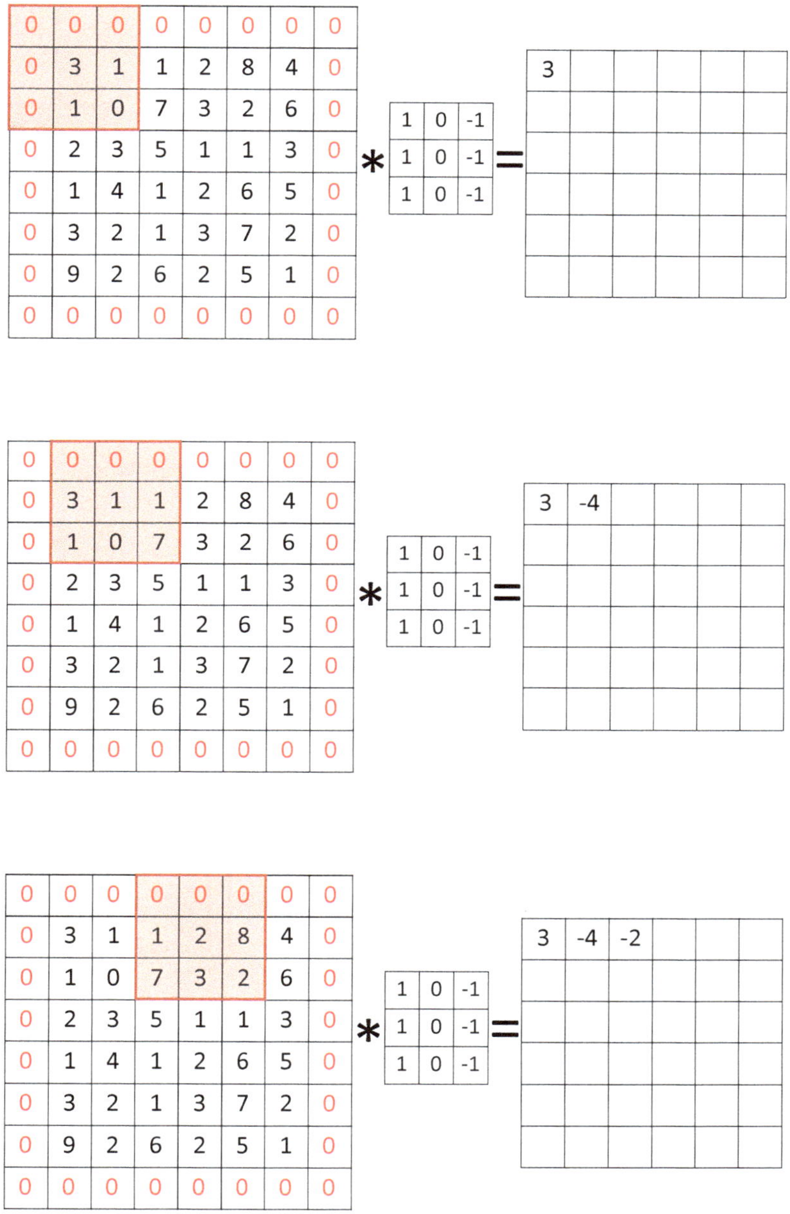

Fig. 4.9: Example of an output of a convolution layer with padding

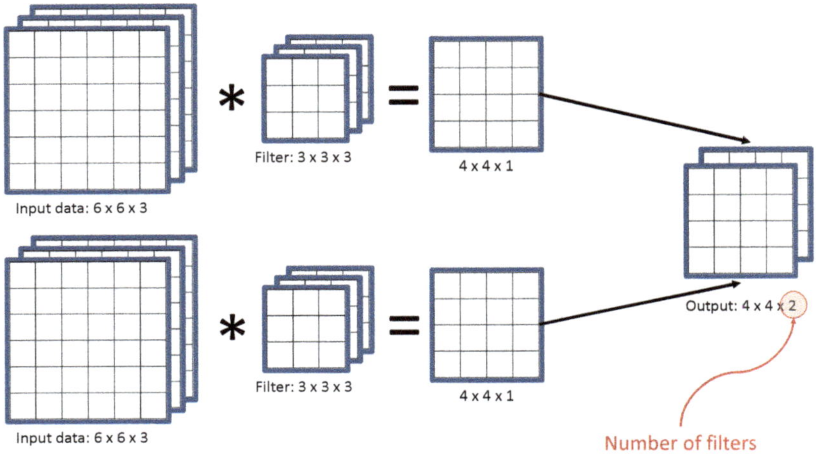

Fig. 4.10: Example of a convolution layer with two convolution filters

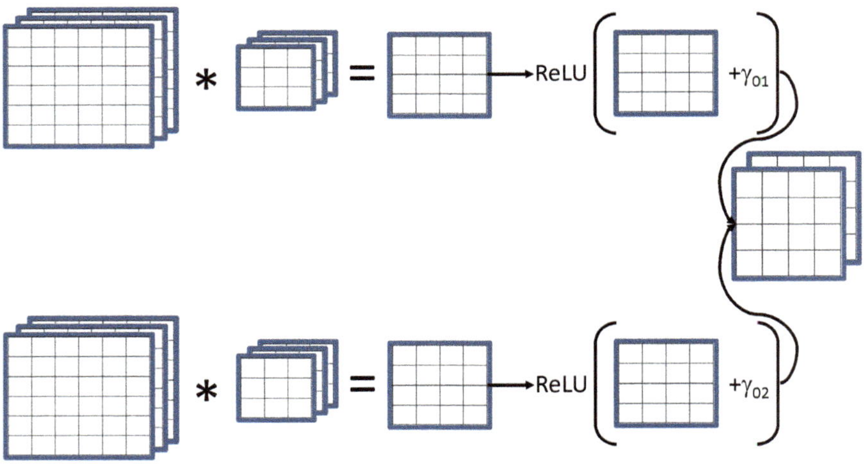

Fig. 4.11: Example of a convolution layer with two convolution filters and nonlinear transformation

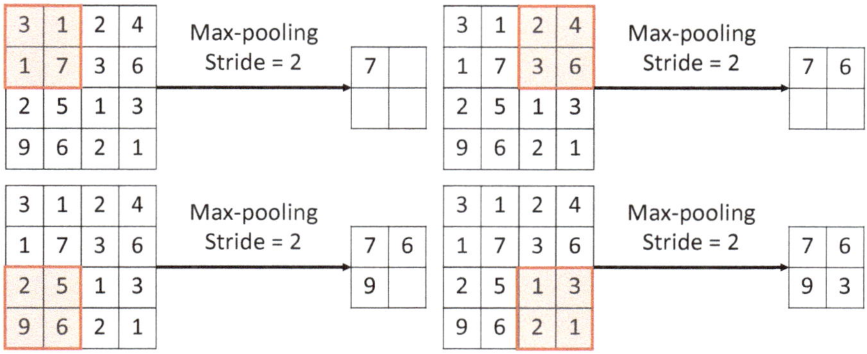

Fig. 4.12: Example of a convolution layer with two convolution filters and nonlinear transformation

4.5.4 Autoenconders: Nonlinear Factor Regression

Autoencoders are the primary model for dimension reduction in the ML literature. They can be interpreted as a nonlinear equivalents to PCA. An autoenconder is a special type of deep neural network in which the outputs attempt to approximate the input variables. The input variables pass through neurons in the hidden layer(s), creating a compressed representation of the input variables. This compressed input is decoded (decompressed) into the output layer. The layer of interest is the hidden layer with the smallest numbers of neurons, since the neurons in this layer represent the latent non-linear factors that we aim to extract. To illustrate the basic structure of an autoencoder, Figure 4.13 illustrates an autoencoder consisting of five inputs and three hidden layers with four, one and four neurons, respectively. The second hidden layer in the diagram represents the latent single factor we wish to extract, O_1^2. The layer preceding it is the encoding layer while the layer that follows it is the decoding layer.

As other deep neural networks, autoencoders can be written using the same recursive formulas as before. The estimated non-linear factors can serve as inputs for linear and nonlinear forecasting models as the ones described in this Chapter or in Chapters 1 and 2.

4.5.5 Hybrid Models

Recently, Medeiros and Mendes (2013) proposed the combination of LASSO-based estimation and NN models. The idea is to construct a feedforward single-hidden layer NN where the parameters of the nonlinear terms (neurons) are randomly generated and the linear parameters are estimated by LASSO (or one of its generalizations).

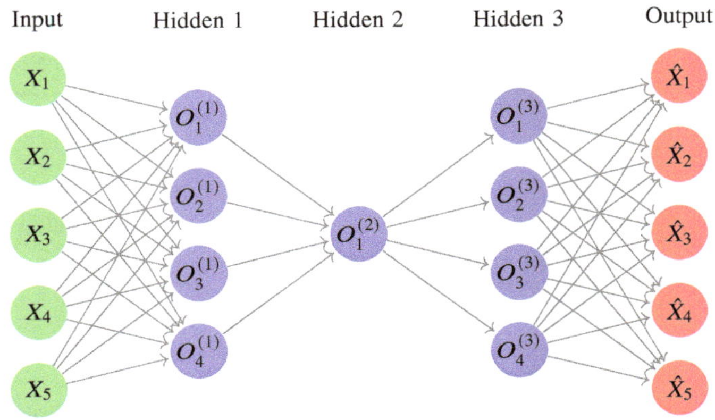

Fig. 4.13: Graphical representation of an Autoecoder

Similar ideas were also considered by Kock and Teräsvirta (2014) and Kock and Teräsvirta (2015).

Trapletti, Leisch and Hornik (2000) and Medeiros, Teräsvirta and Rech (2006) proposed to augment a feedforward shallow NN by a linear term. The motivation is that the nonlinear component should capture only the nonlinear dependence, making the model more interpretable. This is in the same spirit of the semi-parametric models considered in Chen (2007).

Inspired by the above ideas, Medeiros et al. (2021) proposed combining random forests with adaLASSO and OLS. The authors considered two specifications. In the first one, called RF/OLS, the idea is to use the variables selected by a Random Forest in a OLS regression. The second approach, named adaLASSO/RF, works in the opposite direction. First select the variables by adaLASSO and than use them in a Random Forest model. The goal is to disentangle the relative importance of variable selection and nonlinearity to forecast inflation.

4.6 Concluding Remarks

In this chapter we review the most recent advances in using Machine Learning models/methods to forecast time-series data in a high-dimensional setup, where the number of variables used as potential predictors is much larger than the available sample to estimate the forecasting models.

We start the chapter by discussing how to construct and compare forecasts from different models. More specifically, we discuss the Diebold-Mariano test of equal predictive ability and the Li-Liao-Quaedvlieg test of conditional superior predictive ability. Finally, we illustrate how to construct model confidence sets.

In terms of linear ML models, we complement the techniques described in Chapter 1 by focusing of factor-based regression, the combination of factors and penalized regressions and ensemble methods.

After presenting the linear models, we review neural network methods. We discuss both shallow and deep networks, as well as long shot term memory and convolution neural networks.

We end the chapter by discussing some hybrid methods and new proposals in the forecasting literature.

Acknowledgements The author wishes to acknowledge Marcelo Fernandes and Eduardo Mendes as well as the editors, Felix Chan and László Mátyás, for insightful comments and guidance.

References

Ahn, S. & Horenstein, A. (2013). Eigenvalue ratio test for the number of factors. *Econometrica, 81*, 1203–1227.

Bai, J. & Ng, S. (2002). Determining the number of factors in approximate factor models. *Econometrica, 70*, 191–221.

Barron, A. (1993). Universal approximation bounds for superpositions of a sigmoidal function. *IEEE Transactions on Information Theory, 39*, 930–945.

Breiman, L. (1996). Bagging predictors. *Machine Learning, 24*, 123–140.

Chen, X. (2007). Large sample sieve estimation of semi-nonparametric models. In J. Heckman & E. Leamer (Eds.), *Handbook of econometrics.* Elsevier.

Clark, T. & McCracken, M. (2013). Advances in forecast evaluation. In G. Elliott & A. Timmermann (Eds.), *Handbook of economic forecasting* (Vol. 2, p. 1107-1201). Elsevier.

Cybenko, G. (1989). Approximation by superposition of sigmoidal functions. *Mathematics of Control, Signals, and Systems, 2*, 303–314.

Diebold, F. (2015). Comparing predictive accuracy, twenty years later: A personal perspective on the use and abuse of Diebold-Mariano tests. *Journal of Business and Economic Statistics, 33*, 1–9.

Diebold, F. & Mariano, R. (1995). Comparing predictive accuracy. *Journal of Business and Economic Statistics, 13*, 253–263.

Elliott, G., Gargano, A. & Timmermann, A. (2013). Complete subset regressions. *Journal of Econometrics, 177*(2), 357–373.

Elliott, G., Gargano, A. & Timmermann, A. (2015). Complete subset regressions with large-dimensional sets of predictors. *Journal of Economic Dynamics and Control, 54*, 86–110.

Fan, J., Masini, R. & Medeiros, M. (2021). *Bridging factor and sparse models* (Tech. Rep. No. 2102.11341). arxiv.

Fava, B. & Lopes, H. (2020). The illusion of the illusion of sparsity. *Brazilian Journal of Probability and Statistics.* (forthcoming)

Foresee, F. D. & Hagan, M. . T. (1997). Gauss-newton approximation to Bayesian regularization. In *IEEE international conference on neural networks (vol. 3)* (pp. 1930–1935). New York: IEEE.

Funahashi, K. (1989). On the approximate realization of continuous mappings by neural networks. *Neural Networks*, *2*, 183–192.

Garcia, M., Medeiros, M. & Vasconcelos, G. (2017). Real-time inflation forecasting with high-dimensional models: The case of brazil. *International Journal of Forecasting*, *33*(3), 679–693.

Genre, V., Kenny, G., Meyler, A. & Timmermann, A. (2013). Combining expert forecasts: Can anything beat the simple average? *International Journal of Forecasting*, *29*, 108–121.

Giacomini, R. & White, H. (2006). Tests of conditional predictive ability. *Econometrica*, *74*, 1545–1578.

Giannone, D., Lenza, M. & Primiceri, G. (2021). Economic predictions with big data: The illusion of sparsity. *Econometrica*, *89*, 2409–2437.

Grenander, U. (1981). *Abstract inference*. New York, USA: Wiley.

Hansen, P., Lunde, A. & Nason, J. (2011). The model confidence set. *Econometrica*, *79*, 453–497.

Harvey, D., Leybourne, S. & Newbold, P. (1997). Testing the equality of prediction mean squared errors. *International Journal of Forecasting*, *13*, 281–291.

Hochreiter, S. & Schmidhuber, J. (1997). Long short-term memory. *Neural Computation*, *9*, 1735–1780.

Hornik, K., Stinchombe, M. & White, H. (1989). Multi-layer Feedforward networks are universal approximators. *Neural Networks*, *2*, 359–366.

Inoue, A. & Kilian, L. (2008). How useful is bagging in forecasting economic time series? a case study of U.S. consumer price inflation. *Journal of the American Statistical Association*, *103*, 511-522.

Kock, A. & Teräsvirta, T. (2014). Forecasting performance of three automated modelling techniques during the economic crisis 2007-2009. *International Journal of Forecasting*, *30*, 616–631.

Kock, A. & Teräsvirta, T. (2015). Forecasting macroeconomic variables using neural network models and three automated model selection techniques. *Econometric Reviews*, *35*, 1753–1779.

Li, J., Liao, Z. & Quaedvlieg, R. (2021). Conditional superior predictive ability. *Review of Economic Studies*. (forthcoming)

MacKay, D. J. C. (1992a). Bayesian interpolation. *Neural Computation*, *4*, 415–447.

MacKay, D. J. C. (1992b). A practical Bayesian framework for backpropagation networks. *Neural Computation*, *4*, 448–472.

McAleer, M. & Medeiros, M. (2008). A multiple regime smooth transition heterogeneous autoregressive model for long memory and asymmetries. *Journal of Econometrics*, *147*, 104–119.

McCracken, M. (2020). Diverging tests of equal predictive ability. *Econometrica*, *88*, 1753–1754.

Medeiros, M. & Mendes, E. (2013). Penalized estimation of semi-parametric additive time-series models. In N. Haldrup, M. Meitz & P. Saikkonen (Eds.), *Essays in nonlinear time series econometrics*. Oxford University Press.

Medeiros, M., Teräsvirta, T. & Rech, G. (2006). Building neural network models for time series: A statistical approach. *Journal of Forecasting*, *25*, 49–75.

Medeiros, M., Vasconcelos, G., Veiga, A. & Zilberman, E. (2021). Forecasting inflation in a data-rich environment: The benefits of machine learning methods. *Journal of Business and Economic Statistics*, *39*, 98–119.

Mhaska, H., Liao, Q. & Poggio, T. (2017). When and why are deep networks better than shallow ones? In *Proceedings of the thirty-first aaai conference on artificial intelligence (aaai-17)* (pp. 2343–2349).

Onatski, A. (2010). Determining the number of factors from empirical distribution of eigenvalues. *Review of Economics and Statistics*, *92*, 1004–1016.

Park, J. & Sandberg, I. (1991). Universal approximation using radial-basis-function networks. *Neural Computation*, *3*, 246–257.

Patton, A. (2015). Comment. *Journal of Business & Economic Statistics*, *33*, 22-24.

Samuel, A. (1959). Some studies in machine learning using the game of checkers. *IBM Journal of Research and Development*, *3.3*, 210–229.

Srivastava, N., Hinton, G., Krizhevsky, A., Sutskever, I. & Salakhutdinov, R. (2014). Simple way to prevent neural networks from overfitting. *Journal of Machine Learning Research*, *15*, 1929–1958.

Stinchcombe, M. & White, S. (1989). Universal approximation using feedforward neural networks with non-sigmoid hidden layer activation functions. In *Proceedings of the international joint conference on neural networks* (pp. 613–617). Washington: IEEE Press, New York, NY.

Stock, J. & Watson, M. (2002a). Forecasting using principal components from a large number of predictors. *Journal of the American Statistical Association*, *97*, 1167–1179.

Stock, J. & Watson, M. (2002b). Macroeconomic forecasting using diffusion indexes. *Journal of Business & Economic Statistics*, *20*, 147–162.

Suarez-Fariñas, Pedreira, C. & Medeiros, M. (2004). Local-global neural networks: A new approach for nonlinear time series modelling. *Journal of the American Statistical Association*, *99*, 1092–1107.

Teräsvirta, T. (1994). Specification, estimation, and evaluation of smooth transition autoregressive models. *Journal of the American Statistical Association*, *89*, 208–218.

Teräsvirta, T. (2006). Forecasting economic variables with nonlinear models. In G. Elliott, C. Granger & A. Timmermann (Eds.), (Vol. 1, p. 413-457). Elsevier.

Teräsvirta, T., Tjøstheim, D. & Granger, C. (2010). *Modelling nonlinear economic time series*. Oxford, UK: Oxford University Press.

Trapletti, A., Leisch, F. & Hornik, K. (2000). Stationary and integrated autoregressive neural network processes. *Neural Computation*, *12*, 2427–2450.

West, K. (2006). Forecast evaluation. In G. Elliott, C. Granger & A. Timmermann (Eds.), (Vol. 1, pp. 99–134). Elsevier.

Yarotsky, D. (2017). Error bounds for approximations with deep ReLU networks. *Neural Networks*, *94*, 103–114.

Chapter 5
Causal Estimation of Treatment Effects From Observational Health Care Data Using Machine Learning Methods

William Crown

Abstract The econometrics literature has generally approached problems of causal inference from the perspective of obtaining an unbiased estimate of a parameter in a structural equation model. This requires strong assumptions about the functional form of the model and data distributions. As described in Chapter 3, there is a rapidly growing literature that has used machine learning to estimate causal effects. Machine learning models generally require far fewer assumptions. Traditionally, the identification of causal effects in econometric models rests on theoretically justified controls for observed and unobserved confounders. The high dimensionality of many datasets offers the potential for using machine learning to uncover potential instruments and expand the set of observable controls. Health care is an example of high dimensional data where there are many causal inference problems of interest. Epidemiologists have generally approached such problems using propensity score matching or inverse probability treatment weighting within a potential outcomes framework. This approach still focuses on the estimation of a parameter in a structural model. A more recent method, known as doubly robust estimation, uses mean differences in predictions versus their counterfactual that have been updated by exposure probabilities. Targeted maximum likelihood estimators (TMLE) optimize these methods. TMLE methods are not, inherently, machine learning methods. However, because the treatment effect estimator is based on mean differences in individual predictions of outcomes for those treated versus the counterfactual, super learning machine learning approaches have superior performance relative to traditional methods. In this chapter, we begin with the same assumption of selection of observable variables within a potential outcomes framework. We briefly review the estimation of treatment effects using inverse probability treatment weights and doubly robust estimators. These sections provide the building blocks for the discussion of TMLE methods and their estimation using super learner methods. Finally, we consider the extension of the TMLE estimator to include instrumental variables in order to control for bias from unobserved variables correlated with both treatment and outcomes.

William Crown ✉
Brandeis University, Waltham, Massachusetts, USA, e-mail: wcrown@brandeis.edu

F. Chan, L. Mátyás (eds.), *Econometrics with Machine Learning*, Advanced Studies in Theoretical and Applied Econometrics 53, https://doi.org/10.1007/978-3-031-15149-1_5

5.1 Introduction

Several aspects of the changing healthcare data landscape including the rapid growth in the volume of healthcare data, the fact that much of it is unstructured, the ability to link different types of data together (claims, EHR, sociodemographics, genomics), and the speed with which the data are being refreshed create serious challenges for traditional statistical methods from epidemiology and econometrics while, simultaneously, creating opportunities for the use of machine learning methods.

Methods such as logistic regression have long been used to predict whether a patient is at risk of developing a disease or having a health event such as a heart attack—potentially enabling intervention before adverse outcomes occur. Such models are rapidly being updated with machine learning methods such as lasso, random forest, support vector machines, and neural network models to predict such outcomes as hospitalization (Hong, Haimovich & Taylor, 2018; Futoma, Morris & Lucas, 2015; Shickel, Tighe, Bihorac & Rashidi, 2018; Rajkomar et al., 2018) or the onset of disease (Yu et al., 2010). These algorithms that offer the potential to improve the sensitivity and specificity of predictions used in health care operations and to guide clinical care (Obermeyer & Emanuel, 2016). In some areas of medicine, such as radiology, machine learning methods show great promise for improved diagnostic accuracy (Obermeyer & Emanuel, 2016; Ting et al., 2017). However, the applications of machine learning in health care have been almost exclusively about prediction; rarely, are machine learning methods used for causal inference.

5.2 Naïve Estimation of Causal Effects in Outcomes Models with Binary Treatment Variables

Everything else equal, randomized trials are the strongest design for estimating unbiased treatment effects because when subjects are randomized to treatment, the alternative treatment groups are asymptotically ensured to balance on the basis of both observed and unobserved covariates. In the absence of randomization, economists often use quasi-experimental designs that attempt to estimate the average treatment effect (ATE) that one would have attained from a randomized trial on the same patient group to answer the same question. Randomized controlled trials highlight the fact that, from a conceptual standpoint, it is important to consider the estimation problem in the context of both observable and unobservable variables.

Suppose we wish to estimate:

$$Y = B_0 + B_1 T + B_2 X + B_3 U + e,$$

where Y is a health outcome of interest, T is a treatment variable, X is a matrix of observed covariates, U is a matrix of unobserved variables, e is vector of residuals, and B_0, B_1, B_2, and B_3 are parameters, or parameter vectors to be estimated. Theoretically, randomization eliminates the correlation between any covariate and treatment. This is important because any unobserved variable that is correlated both with treatment and with outcomes will introduce a correlation between treatment and the residuals. This, by definition, introduces bias (Wooldridge, 2002). In health economic evaluations, the primary goal is to obtain unbiased and efficient estimates of the treatment effect, B_1. However, the statistical properties of B_1 may be influenced by a variety of factors that may introduce correlation between the treatment variable and the residuals.

Consider the matrix of unobserved variables U. If U has no correlation with T, then its omission from the equation will have no effect on the bias of the estimate of B_1. However, if $cov(T,U)$ is not equal to zero, then

$$E(B_1) = B_1 + B_3 cov(T,U),$$

where $E(B_1)$ is the expected value of B_1. In other words, the estimator for treatment effect will be biased by the amount of $B_3 cov(T,U)$. In brief, a necessary condition for obtaining unbiased estimates in any health economic evaluation using observational data is the inclusion of strong measures on all important variables hypothesized to be correlated with both the outcome and the treatment (i.e., measurement of all important confounders noted earlier).

Leaving aside the issue of unobserved variables for the moment, one difficulty with the standard regression approach is that there is nothing in the estimation method that assures that the groups are, in fact, comparable. In particular, there may be subpopulations of patients in the group receiving the intervention for whom there is no overlap with the covariate distributions of patients in the comparison group and vice versa. This concept is known as positivity or lack of common support. Positivity requires that, for each value of X in the treated group, the probability of observing X in the comparison group is positive. In the absence of positivity or common support, the absence of bias in the treatment effect estimate is possible only if the estimated functional relationship between the outcome Y and the covariate matrix X holds for values outside of the range of common support (Jones & Rice, 2009). Crump, Hotz, Imbens and Mitnik (2009) point out that, even with some evidence of positivity, bias and variance of treatment effect estimates can be very sensitive to the functional form of the regression model.

One approach to this problem is to match patients in the intervention group with patients in the comparison group who have the exact same pattern of observable covariates (e.g., age, gender, race, medical comorbidities). However, in practice this exact match approach requires enormous sample sizes—particularly when the match is conducted on numerous strata. Instead, Rosenbaum and Rubin (1983) proposed matching on the propensity score. Propensity score methods attempt to control for distributional differences in observed variables in the treatment cohorts so that the groups being compared are at least comparable on observed covariates.

5.3 Is Machine Learning Compatible with Causal Inference?

A number of papers have reviewed the role of machine learning for estimating treatment effects (e.g., Athey & Imbens, 2019; Knaus, Lechner & Strittmatter, 2021). Some machine learning approaches use regression-based methods for prediction. For example, Lasso, Ridge, and elastic net methods utilize correction factors to reduce the risk of over fitting (Hastie, Tibshirani & Friedman, 2009; Tibshirani, 1996). However, as noted by Mullainathan and Spiess (2017) and discussed in Chapter 3, causal interpretation should not be given to the parameters of such models. Unfortunately, there is nothing magical about machine learning that protects against the usual challenges encountered in drawing causal inferences in observational data analysis. In particular, just because machine learning methods are operating on high-dimensional data does not protect against bias. Increasing sample size—for example, assembling more and more and more medical claims data for an outcomes study–does not correct the problem of bias if the dataset is lacking in key clinical severity measures such as cancer stage in a model of breast cancer outcomes (Crown, 2015). Moreover, even in samples with very large numbers of observations such as medical claims or electronic medical record datasets, models with very large numbers of variables become sparse in high dimensional space. This can lead to a breakdown in assumptions of positivity for example. Perhaps most importantly, machine learning can provide a statistical method for estimating causal effects but only in the context of an appropriate causal framework. The use of machine learning without a causal framework is fraught.

On the other hand, economists, epidemiologists, and health services researchers have been trained that they must have a theory that they test through model estimation. The major limitation of this approach is that it makes it very difficult to escape from the confines of what we already know (or think we know). Machine learning methods such as Lasso can provide a manner for systematically selecting variables to be included in the model, as well as exploring alternative functional forms for the outcome equation. The features thus identified then can become candidates for inclusion in a causal modeling framework that is estimated using standard econometric methods—preferably using a different set of observations than those used to identify the features. However, as discussed in Chapter 3, parameter estimates from machine learning models cannot be assumed to be unbiased estimates of causal parameters in a structural equation.

Machine learning methods may also help in the estimation of traditional econometric or epidemiologic models using propensity score or inverse probability treatment weights. Machine learning can aid in the estimation of causal models in other ways as well. An early application of machine learning for causal inference in health economics and outcomes research (HEOR) was to fit propensity score (PS) models to create matched samples and estimate average treatment effects (ATEs) by comparing these samples (Westreich, Lessler & Jonsson Funk, 2010; Rosenbaum & Rubin, 1983). While traditionally, logistic regression was used to fit PS models, it has been shown that 'off the shelf' ML methods for prediction and classification (e.g., random forests, classification and regression trees, Lasso) are sometimes more flexible and can lead to lower bias in treatment effect estimates (Setoguchi, Schneeweiss,

Brookhart, Glynn & Cook, 2008). However, these approaches in themselves are imperfect as they are tailored to minimize root mean square error (RMSE) as opposed to targeting the causal parameter. Some extensions of these methods have addressed specific challenges of using the PS for confounding adjustment, by customizing the loss function of the ML algorithms (e.g., instead of minimizing classification error, to maximize balance in the matched samples) (Rosenbaum & Rubin, 1983). However, the issue remains that giving equal importance to many covariates when creating balance may not actually minimize bias (e.g., if many of the covariates are only weak confounders). It is recommended that balance on variables that are thought to be the most prognostic to the outcome should be prioritized; however, this ultimately requires subjective judgement (Ramsahai, Grieve & Sekhon, 2011).

Finally, machine learning methods can be used to estimate causal treatment effects directly. As discussed in Chapter 3, Athey and Imbens (2016), Wager and Athey (2018), and Athey, Tibshirani and Wager (2019) propose the use of tree-based approaches to provide a step-function approximation to the outcome equation. These methods place less emphasis on parametric estimation of treatment effects. Chapter 3 also discusses the estimation of double debiased estimators (Belloni, Chen, Chernozhukov & Hansen, 2012; Belloni, Chernozhukov & Hansen, 2013; Belloni, Chernozhukov & Hansen, 2014a; Belloni, Chernozhukov & Hansen, 2014b; Belloni, Chernozhukov, Fernndez-Val & Hansen, 2017; Chernozhukov et al., 2017; and Chernozhukov et al., 2018).

In this chapter, we discuss the estimation of Targeted Maximum Likelihood Estimation (TMLE) models (Schuler & Rose, 2017; van der Laan & Rose, 2011). TMLE is not, itself, a machine learning method. However, the fact that TMLE uses the potential outcomes framework and bases estimates of ATE upon predictions of outcomes and exposures makes it a natural approach to implement using machine learning techniques—particularly, super learner methods. This has many potential advantages including reducing the necessity of correctly chosing the specification of the outcome model and reducing the need to make strong distributional assumptions about the data.

5.4 The Potential Outcomes Model

Imbens (2020) provides a comprehensive review and comparison of two major causal inference frameworks with application for health economics—(1) directed acyclic graphs (DAGs) (Pearl, 2009) and (2) potential outcomes (Rubin, 2006). Imbens (2020) notes that the potential outcomes framework has been more widely used in economics but that DAGs can be helpful in clarifying the assumptions made in the analysis such as the role of 'back door' and 'front door' criteria in identifying treatment effects. Richardson (2013) introduce Single World Intervention Graphs (SWIGs) as a framework for unifying the DAG and potential outcomes approaches to causality. Dahabreh, Robertson, Tchetgen and Stuart (2019) use SWIGs to examine the conditions under which it is possible to generalize the results of a randomized trial to a target population of trial-eligible individuals.

The potential outcomes framework has mainly focused upon the estimation of average effects of binary treatments. It has made considerable progress not only on questions related to the identification of treatment effects but also problems of study design, estimation, and inference (Imbens, 2020). For these reasons, the remainder of this chapter will focus on the estimation of causal effects using the potential outcomes framework. In particular, we will focus on the building blocks for TMLE which, as mentioned above, is a statistical technique that can be implemented within the potential outcomes framework. Because TMLE estimates ATE as the average difference in predicted outcomes for individuals exposed to an intervention relative to their counterfactual outcome, it lends itself to implementation using machine learning methods.

Potential outcomes. In observational studies we observe outcomes only for the treatment that individuals receive. There is a potential outcome that could be observed if that individual was exposed to an alternative treatment but this potential outcome is not available. A straightforward estimate of treatment effect is the expected difference between the outcome that an individual had for the treatment received versus the outcome that they would have had if exposed to an alternative treatment $E[Y_1 - Y_0]$ The most straightforward estimate of this ATE is the parameter estimate for a binary treatment variable in a regression model. It is also possible to estimate the conditional average treatment effect (CATE) given a vector of patient attributes X. We assume that the observed data are n independent and identically distributed copies of $O = (Z, T, Y, X) \sim P_0$, where Y is a vector of observed outcomes, T indicates the treatment group, X is a matrix of observed control variables, Z is a matrix of instruments for unobserved variables correlated with both outcomes and treatment assignment, and P_0 is the true underlying distribution from which the data are drawn. If there are no unobserved confounders to generate endogeneity bias Z is not needed.

Assumptions needed for causal inference with observational data. Drawing causal inference in observational studies requires several assumptions (Robins, 1986; van der Laan & Rose, 2011; van der Laan & Rubin, 2006; Rubin, 1974). The first of these—the Stable Unit Value Assumption (SUTVA)—is actually a combination of several assumptions. It states that (1) an individual's potential outcome under his or her observed exposure history is the outcome that will actually be observed for that person (also known as consistency) (Cole & Frangakis, 2009), (2) the exposure of any given individual does not affect the potential outcomes of any other individuals (also known as non-interference) and (3) the exposure level is the same for all exposed individuals (Rubin, 1980; Rubin, 1986; Cole & Hernán, 2008).

In addition, causal inference with observational data requires an assumption that there are no unmeasured confounders. That is, all common causes of both the exposure and the outcome have been measured (Greenland & Robins, 1986) and the exposure mechanism and potential outcomes are independent after conditioning on the set of covariates. Unmeasured confounders are a common source of violation of the assumption of exchangeability of treatments (Hernán, 2011). When results from

RCTs and observational studies have been found to differ it is often assumed that this is due to the failure of observational studies to adequately control for unmeasured confounders.

Finally, there is the assumption of positivity which states that, within a given strata of X, every individual has a nonzero probability of receiving either exposure condition; this is formalized as $0 < P(A = 1|X) < 1$ for a binary exposure (Westreich & Cole, 2010; Petersen, Porter, Gruber, Wang & Laan, 2012). If the positivity assumption is violated, causal effects will not be identifiable (Petersen et al., 2012).

5.5 Modeling the Treatment Exposure Mechanism–Propensity Score Matching and Inverse Probability Treatment Weights

One mechanism for testing the positivity assumption is to model treatment exposure as a function of baseline covariates for the treated and comparison groups in the potential outcomes model. Propensity score methods are widely used for this purpose in empirical research. Operationally, propensity score methods begin with the estimation of a model to generate the fitted probability, or propensity, to receive the intervention versus comparison treatment. (The term 'treatment' is very broad and can be anything from a pharmaceutical intervention to alternative models of benefit design or organizing patient care such as accountable care organizations [ACOs]). Observations that have a similar estimated propensity to be in either the study group or comparison group will tend to have similar observed covariate distributions (Rosenbaum & Rubin, 1983). Once the propensity scores have been estimated, it is possible to pair patients receiving the treatment with patients in the comparison group on the basis of having similar propensity scores. This is known as propensity score matching. Thus, propensity score methods can be thought of as a cohort balancing method undertaken prior to the application of traditional multivariate methods (Brookhart et al., 2006; Johnson et al., 2006). More formally, Rosenbaum and Rubin (1983) define the propensity score for subject i as the conditional probability of assignment to a treatment ($T = 1$) versus comparison ($T = 0$) given covariates, X:

$$Pr(T = 1|X).$$

The validity of this approach assumes that there are no variables that influence treatment selection for which we lack measures. Note that the propensity score model of treatment assignment and the model of treatment outcomes share the common assumption of strong ignorability. That is, both methods assume that any missing variables are uncorrelated with both treatment and outcomes and can be safely ignored.

One criticism of propensity score matching is that it is sometimes not possible to identify matches for all the patients in the intervention group. This leads to lose of sample size. Inverse probability treatment weighting (IPTW) retains the full sample by using the propensity score to develop weights that, for each subject, are the inverse of the predicted probability of the treatment that the subject received. This approach gives more weight to subjects who have a lower probability of receiving a particular treatment and less weight to those with a high probability of receiving the treatment.

Hirano, Imbens and Ridder (2003) propose using inverse probability treatment weights to estimate average treatment effects:

$$ATE = \frac{1}{N}\sum_{i=0}^{n}\left[\frac{Y_i T}{PS} - \frac{Y_i(1-T)}{1-PS}\right].$$

IPTW is also useful for estimating more complex causal models such as marginal structural models (Joffe, Have, Feldman & Kimmel, 2004). When the propensity score model is correctly specified, IPTW can lead to efficient estimates of ATE under a variety of data-generating processes. However, IPTW can generate biased estimates of ATE if the propensity score model is mis-specified. This approach is similar to weighting methodologies long used in survey research.

Intuitively, propensity score matching is very appealing because it forces an assessment of the amount of overlap (common support) in the populations in the study and comparison groups. There is a large, and rapidly growing, literature using propensity score methods to estimate treatment effects with respect to safety and health economic outcomes (Johnson, Crown, Martin, Dormuth & Siebert, 2009; Mitra & Indurkhya, 2005). In application, there are a large number of methodologies that can be used to define the criteria for what constitutes a 'match' (Baser, 2006; Sekhon & Grieve, 2012).

5.6 Modeling Outcomes and Exposures: Doubly Robust Methods

Most of the medical outcomes literature using observational data to estimate ATE focuses on the modeling of the causal effect of an intervention on outcomes by balancing the comparison groups with propensity score matching or IPTW. Doubly robust estimation is a combination of propensity score matching and covariate adjustment using regression methods (Bang & Robins, 2005; Scharfstein, Rotnitzky & Robins, 1999). Doubly robust estimators help to protect against bias in treatment effects because the method is consistent when either the propensity score model or the regression model is mis-specified (Robins, Rotnitzky & Zhao, 1994). See Robins et al. (1994), Imbens and Wooldridge (2009) and Abadie and Cattaneo (2018) for surveys.

After estimating the propensity score (PS) in the usual fashion the general expressions for the doubly robust estimates in response to the presence or absence of exposure (DR_1) and (DR_0), respectively are given by Funk et al. (2011):

$$DR_1 = \frac{Y_{x=1}X}{PS} - \frac{\hat{Y}_1(X-PS)}{PS}$$

$$DR_0 = \frac{Y_{x=0}(1-X)}{1-PS} - \frac{\hat{Y}_0(X-PS)}{1-PS}.$$

When $X = 1$, DR_1 and DR_0 simplify to

$$DR_1 = \frac{Y_{x=1}}{PS} - \frac{\hat{Y}_1(1-PS)}{PS}$$

and

$$DR_0 = \hat{Y}_0.$$

Similarly, when $X = 0$, DR_1 and DR_0 simplify to

$$DR_1 = \hat{Y}_1$$

and

$$DR_0 = \frac{Y_{x=0}}{1-PS} + \frac{\hat{Y}_0 PS}{1-PS}.$$

Note that for exposed individuals (where $X = 1$), DR is a function of their observed outcomes under exposure and predicted outcomes under exposure given covariates, weighted by a function of the PS. The estimated value for DR_0 is simply the individuals' predicted response, had they been unexposed based on the parameter estimates from the outcome regression among the unexposed and the exposed individuals' covariate values (Z).

Similarly, for the unexposed ($Y = 0$) DR_0 is calculated as a function of the observed response combined with the predicted response weighted by a function of the PS, while DR_1 is simply the predicted response in the presence of exposure conditional on covariates. Finally, the ATE is estimated as the difference between the means of DR_1 and DR_0 calculated across the entire study population.

With some algebraic manipulation the doubly robust estimator can be shown to be the mean difference in response if everyone was either exposed or unexposed to the intervention plus the product of two bias terms—one from the propensity score model and one from the outcome model. If bias is zero from either the propensity score model or the outcome model it will "zero out" any bias from the other model.

Recently the doubly robust literature has focused on the case with a relatively large number of pretreatment variables (Chernozhukov et al., 2017; Athey, Imbens & Wager, 2018; van der Laan & Rose, 2011; Shi, Blei & Veitch, 2019). Overlap issues in covariate distributions, absent in all the DAG discussions, become prominent among practical problems (Crump et al., 2009; D'Amour, Ding, Feller, Lei & Sekhon, 2021).

Basu, Polsky and Manning (2011) provide Monte Carlo evidence on the finite sample performance of OLS, propensity score estimates, IPTW, and doubly robust estimates. They find that no single estimator can be considered best for estimating treatment effects under all data-generating processes for healthcare costs. IPTW estimators are least likely to be biased across a range of data-generating processes but can be biased if the propensity score model is mis-specified. IPTW estimators are generally less efficient than regression estimators when the latter are unbiased. Doubly robust estimators can be biased as a result of mis-specification of both the propensity score model and the outcome model. The intent is that they offer the opportunity to offset bias from mis-specifying either the propensity score model or the regression model by getting the specification right for at least one; however, this comes at an efficiency cost. As a result, the efficiency of doubly robust estimators tends to fall between that of IPTW and regression estimators. These findings are consistent with several simulation studies that have compared doubly robust estimation to multiple imputation for addressing missing data problems (Kang & Schafer, 2007; Carpenter, Kenward & Vansteelandt, 2006). These studies have found that IPTW by itself, or doubly robust estimation, can be sensitive to the specification of the imputation model—especially when some observations have a small estimated probability of being observed. Using propensity score matching prior to estimating treatment effects (as with doubly robust methods) has obvious appeal as it appears to simulate a randomized trial with observational data. However, it is still possible that unobserved variables may be correlated with both outcomes and the treatment variable, resulting in biased estimates of ATEs. As a result, researchers should routinely test for residual confounding or endogeneity even after propensity score matching. This can be done in a straightforward way by including the residuals from the propensity model as an additional variable in the outcome model (Terza, Basu & Rathouz, 2008; Hausman, 1983; Hausman, 1978). If the coefficient for the residuals variable is statistically significant, this indicates that residual confounding or endogeneity remain.

5.7 Targeted Maximum Likelihood Estimation (TMLE) for Causal Inference

In epidemiologic and econometrics studies, estimation of causal effects using observational data is necessary to evaluate medical treatment and policy interventions. Numerous estimators can be used for estimation of causal effects. In the epidemiologic literature propensity score methods or G-computation have been widely used within the potential outcomes framework. In this chapter, we briefly discuss these other methods as the building blocks for targeted maximum likelihood estimation (TMLE). TMLE is a well-established alternative method with desirable statistical properties but it is not as widely utilized as these other methods such as g-estimation and propensity score techniques. In addition, implementation of TMLE benefits from the use of machine learning super learner methods.

TMLE is related to G-computation and propensity score methods in that TMLE involves estimation of both $E(Y|T,X)$ and $P(T=1|X)$. G-computation is used to estimate the outcome model in TMLE. G-computation (Robins & Hernan, 2009) is an estimation approach that is especially useful for incorporating time-varying covariates. Despite its potential usefulness, however, it is not widely used due to lack of understanding of its theoretical underpinnings and empirical implementation (Naimi, Cole & Kennedy, 2017). TMLE is a doubly robust, maximum-likelihood–based estimation method that includes a secondary "targeting" step that optimizes the bias-variance tradeoff for the parameter of interest. Although TMLE is not specifically a causal modeling method, it has features that make it a particularly attractive method for causal effect estimation in observational data. First, because it is a doubly robust method it will yield unbiased estimates of the parameter of interest if either $E(Y|T,X)$ or $P(T=1|X)$ is consistently estimated. Even if the outcome regression is not consistently estimated, the final ATE estimate will be unbiased as long as the exposure mechanism is also not consistently estimated. Conversely, if the outcome is consistently estimated, the targeting step will preserve this unbiasedness and may remove finite sample bias (van der Laan & Rose, 2011). Additionally, TMLE is an asymptotically efficient estimator when both the outcome and exposure mechanisms are consistently estimated (van der Laan and Rose, 2011). Furthermore, TMLE is a substitution estimator; these estimators are more robust to outliers and sparsity than are nonsubstitution estimators (van der Laan & Rose, 2011). Finally, when estimated using machine learning TMLE has the flexibility to incorporate a variety of algorithms for estimation of the outcome and exposure mechanisms. This can help minimize bias in comparison with use of misspecified regressions for outcomes and exposure mechanisms.

The estimation of ATE using TMLE models is comprised of several steps (Schuler and Rose, 2017). We assume that the observed data are n independent and identically distributed copies of $O=(T,Y,X)\sim P_0$, where P_0 is the true underlying distribution from which the data are drawn.

Step 1. The first step is to generate an initial estimate of $E(Y|T,X)$ using g-estimation. $E(Y|T,X)$ is the conditional expectation of the outcome, given the exposure and the covariates. As noted earlier, g-estimation addresses many common problems in causal modeling such as time-varying covariates while minimizing assumptions about the functional form of the outcome equation and data distribution. We could use any regression model to estimate $E(Y|T,X)$ but the use of super learning allows us to avoid choosing a specific functional form for the model. This model is then used to generate the set of potential outcomes corresponding to T = 1 and T = 0, respectively. That is, the estimated outcome equation is used to generate predicted outcomes for the entire sample assuming that everyone is exposed to the intervention and that no one is exposed to the intervention, respectively. The mean difference in the two sets of predicted outcomes is the g-estimate of the ATE:

$$ATE_{G-comp}=\phi_{G-comp}=\frac{1}{n}\sum E[Y|T=1,X]-E[Y|T=0,X].$$

However, because we used machine learning, the expected outcome estimates have the optimal bias-variance tradeoff for estimating the outcome, not the ATE. As a result, the ATE estimate may be biased. Nor can we compute the standard error of the ATE. (We could bootstrap the standard errors but they would only be correct if the estimand was asymptotically normally distributed.)

Step 2. The second step is to estimate the exposure mechanism $P(T=1|X)$. As with the outcome equation, we use super learner methods to estimate P (T = 1|X) using a variety of machine learning algorithms. For each individual, the predicted probability of exposure to the intervention is given by the propensity score P_1. The individual's predicted probability of exposure to the comparison treatment is $\hat{P}_0$, where $\hat{P}_0 = 1 - \hat{P}_1$. In Step 1 we estimated the expected outcome, conditional on treatment and confounders. As noted earlier these machine learning estimates have an optimal bias-variance trade-off for estimating the outcome (conditional on treatment and confounders), rather than the ATE. In Step 3, the estimates of the exposure mechanism are used to optimize the bias-variance trade-off for the ATE so we can make valid inferences based upon the results.

Step 3. Updating the initial estimate of $E(Y|T,X)$ for each individual. This is done by first calculating $H_t(T=t,X) = \frac{I(T=1)}{\hat{P}_1} - \frac{I(T=0)}{\hat{P}_0}$ based upon the previously calculated values for $\hat{P}_1$ and $\hat{P}_0$ and each patient's actual exposure status. This step is very similar to IPTW but is based upon the canonical gradient (van der Laan & Rose, 2011). We need to use H(X) to estimate the Efficient Influence Function (EIF) of the ATE. Although we do not discuss the details here, in semi-parametric theory an Influence Function is a function that indicates how much an estimate will change if the input changes. If an Efficient Influence Function exists for an estimand (in this case, the ATE), it means the estimand can be estimated efficiently. The existence of the EIF for the ATE is what enables TMLE to use the asymptotic properties of semi-parametric estimators to support reliable statistical inference based upon the estimated ATE.

To estimate the EIF, the outcome variable Y is regressed on H_t specifying a fixed intercept to estimate $logit(E^*(Y|T,X)) = logit(\hat{Y}_t) + \delta H_t$. We also calculate $H_1(T=1,X) = 1/\hat{P}_1$ and $H_0(T=0,X) = -1/\hat{P}_0$. H_1 is interpreted as the inverse probability of exposure to the intervention; H_0 is interpreted as the negative inverse probability of exposure. Finally, we generate updated ("targeted") estimates of the set of potential outcomes using information from the exposure mechanism to reduce bias. Note that in the $logit(\hat{Y}_1^*)$ equation, $logit(\hat{Y}_1)$ is not a constant value B_0. Rather, it is a vector of values. This means that it is a fixed intercept rather than a constant intercept. In the TMLE literature, δ is called the fluctuation parameter, because it provides information about how much to change, or fluctuate, the initial outcome estimates. Similarly, $H(X)$ is referred to as the clever covariate because it 'cleverly' helps us solve for the EIF and then update the estimates.

Step 4. In the final step, the fluctuation parameter and clever covariate are used to update the initial estimates of the expected outcome, conditional on confounders and treatment These estimates have the same interpretation as the original estimates of potential outcomes in step 1 but their values have been updated for potential exposure bias. To update the estimate from step 1 we first need to transform the predicted outcomes to the logit scale. Then, it is a simple matter of adjusting these estimates by $\delta H(X)$:

$$logit(\hat{Y}_1^*) = logit(\hat{Y}_1) + \delta H_1 \text{ and } logit(\hat{Y}_0^*) = logit(\hat{Y}_0) + \delta H_0.$$

After retransforming the updated, estimated values $logit(\hat{Y}_1^*)$ we calculate ATE for the target parameter of interest as the mean difference in predicted values of outcomes for individuals receiving the treatment relative to their predicted counterfactual outcome if they had not received treatment.

$$\widehat{ATE} = \frac{1}{n}\sum_{i=1}^{n}\left[\hat{Y}_1^* - \hat{Y}_0^*\right].$$

The ATE is interpreted as the causal difference in outcomes that would be apparent if all individuals in the population of interest participated in the intervention group versus not participating in the intervention.

To obtain the standard errors for the estimated ATE we need to compute the *influence curve (IC):*

$$IC = (Y - E^*[Y|T, X]) + E^*[Y|T = 1, X] - E^*[Y|T = 0, X] - ATE.$$

Once we have the IC, its standard error is simply

$$SE_{IC} = \frac{1}{n}\sqrt{Var(IC)}.$$

5.8 Empirical Applications of TMLE in Health Outcomes Studies

There is a growing literature of empirical implementations of TMLE, along with simulations comparing TMLE to alternative methods. Kreif et al. (2017) use TMLE to estimate the causal effect of nutritional interventions on clinical outcomes among critically ill children, comparing TMLE causal effect estimates to those from models using g-computation and inverse probability treatment weights. After adjusting for time-dependent confounding they find that three methods generate similar results. Pang et al. (2016) compare the performance of TMLE and IPTW on the marginal causal effect of statin use on the one-year risk of death among patients who had previously suffered a myocardial infarction. Using simulation methods they show that TMLE methods perform better with richer specifications of the outcome model and is less likely to be biased because of its doubly-robust property. On the other hand, IPTW had a better mean square error in a high dimensional setting. And violations of the positivity assumption, which are common in high dimensional data, are an issue for both methods. (Schuler & Rose, 2017) found that TMLE outperformed traditional methods such as g-estimation and inverse probability weighting in simulation.

5.8.1 Use of Machine Learning to Estimate TMLE Models

As just described, TMLE uses doubly robust maximum likelihood estimation to update the initial outcome model using estimated probabilities of exposure (Funk et al., 2011). The average treatment effect (ATE) is then estimated as the average difference in the predicted outcome for treated patients versus their outcome if they had not been treated (their counterfactual). This approach does not require machine learning to accomplish and, as a result, TMLE is not inherently a machine learning estimation technique. However due to complexity of specifying the exposure and outcome mechanisms machine learning methods have a number of advantages for estimating TMLE models. In particular, the Super Learner is an ensembling ML approach that is recommended to be used with TMLE to help overcome bias due to model misspecification (van der Laan & Rose, 2018; Funk et al., 2011; van der Laan & Rubin, 2006). Super Learning can draw upon the full repertoire of ML (even non-parametric neural network models) and traditional econometric/epidemiological methods, and produce estimates that are asymptotically as good as the best performing model—eliminating the need to make strong assumptions about functional form and estimation method up front (van der Laan & Rubin, 2006). Estimation using machine learning methods is facilitated by the fact that the ATE estimated with TMLE is based on the predicted values of outcomes for the treated and counterfactual groups, rather than an estimated parameter value.

5.9 Extending TMLE to Incorporate Instrumental Variables

It is reasonable to expect that health care datasets such as medical claims and electronic medical record data will not contain all variables needed to estimate treatment exposure and treatment outcomes. As a result, TMLE models that utilize only observed variables will be biased. Toth and van der Laan (2016) show that the extension of TMLE to incorporate the effects of unobserved variables correlated with both treatment and outcomes is conceptually straightforward and basically involves the incorporation of an instrument for unobserved variables into the outcome and exposure equations. As above we assume that the observed data are n independent and identically distributed copies of $O = (Z, T, Y, X) \sim P_0$, where P_0 is the true underlying distribution from which the data are drawn. However, now O also includes Z which are instruments for unobserved variables correlated with both treatment exposure and outcomes. Under the IV model

$$E[Y|Z, X] = w_0(X) + m_0(X)\pi_0(Z, X),$$

where $m_0(X)$ is the model of treatment on outcomes, $\pi_0(Z, X)$ is the model of treatment exposure, and $w_0(X) = E[Y - Tm_0(X)|X]$. In other words, $w_0(X)$ returns the expected value of Y conditional on whether $T = 1$ or $T = 0$. Estimation of the TMLE for the IV model begins by obtaining initial estimates of the outcome model $m(Z, X)$, the exposure model $\pi(Z, X)$, and the instrument propensity score $g(X)$. From these, an initial estimate of the potential outcomes for the treated and comparison groups is generated.

The second step of estimating the TMLE for a parameter requires specifying a loss function L(P) where the expectation of the loss function is minimized at the true probability distribution. It is common to use the squared error loss function. The efficient influence function (EIF) can be written as

$$
\begin{aligned}
&D^*(m,g,Q_x)(O) = \\
&= H(X)\{\pi_0(Z,X) - E_0(\pi_0(Z,X)|X\}\{Y - \pi_0(Z,X)m_0(X) - w_0(X)\} - \\
&- H(X)\{\pi_0(Z,X) - E_0(\pi_0(Z,X)|X\}m_0(X)\}(T - \pi_0(Z,X)) + D_X(Q_X),
\end{aligned}
$$

where $H(X)$ is the "clever covariate" that is a function of the inverse probability treatment weights of the exposure variable, along with $\zeta^{-2}(X)$, a term that measures instrument strength. $H(X)$, $\zeta^{-2}(X)$, and $D_X(Q_X)$ are defined as

$$
H(X) = Var(V)^{-1} \begin{bmatrix} E[V^2] - E[V]V \\ V - E[V] \end{bmatrix} \zeta^{-2}(X),
$$

$$
\zeta^{-2}(X) = Var_{Z|X}(\pi(Z,X)|X),
$$

$$
D_X(Q_X) = c\{m_0(X) - m_\psi(V)\}.
$$

Here V is a variable in X for which we wish to estimate the treatment effect. In the targeting step, a linear model for $m_0(X)$ is fitted using only the clever covariate $H(X)$. This model is used to generate potential outcomes for $[Y|T=1]$ and $[Y|T=0]$. Finally, the average treatment effect is estimated as the mean difference in predicted values of outcomes for individuals receiving the treatment relative to their predicted counterfactual outcome if they had not received treatment.

$$
\widehat{ATE} = \frac{1}{n}\sum_{i=1}^{n}\left[\hat{Y}_1^* - \hat{Y}_0^*\right].
$$

5.10 Some Practical Considerations on the Use of IVs

Although estimation of TMLE models with IVs is theoretically straightforward, estimates can be very sensitive to weak instruments. Weak instruments are also likely to be associated with residual correlation of the instrument with the residuals in the outcome equation creating opportunities to introduce bias in the attempt to correct for it. There is an extensive literature on the practical implications of implementing IVs which has direct relevance for their use in TMLE models.

Effective implementation of instrumental variables methods requires finding variables that are correlated with treatment selection but uncorrelated with the outcome variable. This turns out to be extremely difficult to do. The difficulty of finding instrumental variables that are correlated with treatment selection but uncorrelated with outcomes often leads to variables that have weak correlations with treatment. It is important to recognize that an extensive literature has now shown that the use of variables that are only weakly correlated with treatment selection and/or that are even weakly correlated with the residuals in the outcome equation can lead to larger bias than ignoring the endogeneity problem altogether (Bound, Jaeger &

Baker, 1995; Staiger & Stock, 1997; Hahn & Hausman, 2002; Kleibergen & Zivot, 2003; Crown, Henk & Vanness, 2011). Excellent introductions and summaries of the instrumental variable literature are provided in Basu, Navarro and Urzua (2007), Brookhart, Rassen and Schneeweiss (2010), and Murray (2007).

Bound et al. (1995) show that the incremental bias of instrumental variable versus ordinary least squares (OLS) is inversely proportional to strength of the instrument and number of variables that are correlated with treatment but not the outcome variable. Crown et al. (2011) conducted a Monte Carlo simulation analysis of the Bound et al. (1995) results to provide empirical estimates of the magnitude of bias in instrumental variables under alternative assumptions related to the strength of the correlation between the instrumental variable and the variable that it is intended to replace. They also examine how bias in the instrumental variable estimator is related to the strength of the correlation between the instrumental variable and the observed residuals (the contamination of the instrument). Finally, they examine how bias changes in relation to sample size for a range of study sizes likely to be encountered in practice. The results were sobering. For the size of samples used in most studies, the probability that instrumental variable is outperformed by OLS is substantial, even when the asymptotic results indicate bias to be lower for instrumental variable, when the endogeneity problem is serious, and when the instrumental variable has a strong correlation with the treatment variable. This suggests that methods focusing upon observed data, such as propensity score matching or IPTW, will generally be more efficient than those that attempt to control for unobservables, although it is very important to test for whether any residual confounding or endogeneity remains. These results have implications for attempts to include IVs in the estimation of TMLE as well. In particular, more research is needed on the effects of residual correlation of instruments on the bias and efficiency of TMLE.

5.11 Alternative Definitions of Treatment Effects

This chapter has focused upon the use of TMLE for the estimation of ATEs. There are multiple potential definitions for treatment effects, however, and it is important to distinguish among them. The most basic distinctions are between the average treatment effect (ATE), the average treatment effect of the treated (ATT), and the marginal treatment effect (MTE) (Jones & Rice, 2009; Basu, 2011; Basu et al., 2007). These alternative treatment effect estimators are defined as differences in expected values of an outcome variable of interest (Y) conditional on covariates (X) as follows (Heckman & Navarro, 2003):

$$
\begin{aligned}
&ATE : E(Y_1 - Y_0|X),\\
&ATT : E(Y_1 - Y_0|X, T = 1),\\
&MTE : E(Y_1 - Y_0|X, Z, V = 0),
\end{aligned}
$$

where T refers to the treatment, Z is an instrumental variable (or variables) that, conditional on X, is correlated with treatment selection but not outcomes, and V measures the net utility arising from treatment. Basu et al. (2007) show that the MTE is the most general of the treatment effects, since both the ATE and ATT can be derived from the MTE once it has been estimated. ATEs are defined as the expected difference in outcomes between two groups, conditional upon their observed covariates; some patients may not, in fact, have received the treatment at all. ATEs are very common in clinical trials testing the efficacy or safety of a treatment. An example from observational data might be the parameter estimate for a dummy variable comparing diabetes patients enrolled in a disease management program with those not enrolled in the program. Similarly, ATTs are defined as the expected difference in outcomes between one treated group versus another treated group. Researchers often attempt to estimate ATTs for therapeutic areas where multiple treatments exist and make head-to-head comparisons among treatments (e.g., depressed patients treated with selective serotonin reuptake inhibitors vs. tricyclic antidepressants). As with all statistical parameters, ATEs and ATTs can be defined for both populations and samples. For most of the estimators discussed in this chapter, the distinction between ATEs and ATTs has little implication for choice of statistical estimator. Although researchers generally refer to the estimation of ATEs in medical outcomes studies, most such studies are actually estimates of ATT. MTEs, on the other hand, are relatively new to the empirical literature and have important implications for choice of the statistical estimator. In particular, the estimation of MTEs highlights two key issues: (i) the existence of common support among the treatment groups; and (ii) the presence of unobserved essential heterogeneity in treatment selection and outcomes. The first of these characteristics links the estimation of MTEs to propensity score methods while the second links the estimation of MTEs to instrumental variables. Notably, the TMLE estimator described in this chapter, when implemented using IV, is an estimator of MTEs.

By referring to the econometric literature on IV estimation, it is clear that the modeling of IVs in TMLE can be given a utility maximization interpretation. When heterogeneity in treatment response exists and patients select into treatment based upon their expectation of the utility that they will receive from treatment, it becomes necessary to model treatment selection in order to interpret the instrumental variable estimates (Basu et al., 2007). The probability of a patient selecting treatment T can be modeled as a function of the utility, T^*, that a person expects to receive from the treatment. Let Z be an instrumental variable correlated with treatment selection but uncorrelated with Y,

$$Pr(T = 1|X, Z).$$

Note that the probability of treatment includes the instrumental variable Z. This leads to the sample selection model that has the following form:

$$T = 1 \text{ if } Pr(T^* > 0); \quad T = 0 \text{ otherwise.}$$

That is, if the expected utility T^* associated with treatment T is greater than 0 (standard normal scale), the individual will choose treatment T over the alternative.

$$T^* = C_0 + C_1 X_1 + C_c Z + e_T,$$
$$Y = B_0 + B_{1iv} T * + B_2 X + e_Y,$$

where C_0, C_1 and C_2 are parameters to be estimated, B_{1iv} is the instrumental variable estimate of treatment effectiveness and the remaining variables and parameters are as previously defined. There are many extensions of the basic sample selection model to account for different functional forms, multiple outcome equations, etc. (Cameron & Trivedi, 2013; Maddala, 1983).

Vytlacil (2002) points out that the semiparametric sample selection model is equivalent to the method of local instrumental variables (LIV). LIV estimation enables the identification of MTEs, which are defined as the average utility gain to patients who are indifferent to the treatment alternatives given X and Z (Basu et al., 2007; Heckman & Navarro, 2003; Evans & Basu, 2011; Basu, 2011). A particularly attractive feature of MTEs is that all mean treatment effect estimates can be derived from MTEs. For instance, the ATT is derived as a weighted average of the MTEs over the support of the propensity score (conditional on X). Evans and Basu (2011) provide a very clear description of LIV methods, MTEs, and the relationship of MTEs to other mean treatment effect estimates.

5.12 A Final Word on the Importance of Study Design in Mitigating Bias

Most studies comparing average treatment effects (ATEs) from observational studies with randomized controlled trials (RCTs) for the same disease states have found a high degree of agreement (Anglemyer, Horvath & Bero, 2014; Concato, Shah & Horwitz, 2000; Benson & Hartz, 2000). However, other studies have documented considerable disagreement in such results introduced by the heterogeneity of datasets and other factors (Madigan et al., 2013). In some cases, apparent disagreements have been shown to be due to avoidable errors in observational study design which, upon correction, found similar results from the observational studies and RCTs (Dickerman, García-Albéniz, Logan, Denaxas & Hernán, 2019; Hernán et al., 2008).

For any question involving causal inference, it is theoretically possible to design a randomized trial to answer that question. This is known as designing the target trial (Hernán, 2021). When a study is designed to emulate a target trial using observational data some features of the target trial design may be impossible to emulate. Emulating treatment assignment requires data on all features associated with the implementation of the treatment intervention. This is the basis for the extensive use of propensity score matching and inverse probability weighting in the health outcomes literature. It has been estimated that a relatively small percentage of clinical trials can be emulated using observational data (Bartlett, Dhruva, Shah, Ryan & Ross, 2019). However, observational studies can still be designed with a theoretical target trial in mind–specifying a randomized trial to answer the question of interest and then examining where the available data may limit the ability to emulate this trial (Berger & Crown, 2021). Aside from lack of comparability in defining treatment groups, there are a number of other problems frequently encountered in the design of observational health outcomes studies including immortal time bias, adjustment for intermediate variables, and reverse causation. The target trial approach is one method for avoiding such issues.

Numerous observational studies have designed target trials designed to emulate existing RCTs in order to compare the results from RCTs to those of the emulations using observational data (Seeger et al., 2015; Hernán et al., 2008; Franklin et al., 2020; Dickerman et al., 2019). In general, such studies demonstrate higher levels of agreement than comparisons of ATE estimates from observational and RCTS within a disease area that do not attempt to emulate study design characteristics such as inclusion/exclusion criteria, follow-up periods, etc. For example, a paper comparing RCT emulation results for 10 cardiovascular trials found that the hazard ratio estimate from the observational emulations was within the 95% CI from the corresponding RCT in 8 of 10 studies. In 9 of 10, the results had the same sign and statistical significance.

To date, all of the trial emulations have used traditional propensity score or IPTW approaches. None have used doubly-robust methods such as TMLE implemented with Super Learner machine learning methods. In addition to simulation studies, it would be useful to examine the ability of methods like TMLE to estimate similar treatment effects as randomized trials—particularly in cases where traditional methods have failed to do so.

References

Abadie, A. & Cattaneo, D., Matias. (2018). Econometric methods for program evaluation. *Annual Review of Economics*, *10*, 465–503.

Anglemyer, A., Horvath, H. & Bero, L. (2014). Healthcare outcomes assessed with observational study designs compared with those assessed in randomized trials. *The Cochrane Database of Systematic Reviews*, *4*.

Athey, S. & Imbens, G. (2016). Recursive partitioning for heterogeneous causal effects. *Proceedings of the National Academy of Sciences*, *113*, 7353–7360.

Athey, S. & Imbens, G. (2019). Machine learning methods that economists should know about. *Annual Review of Economics*, *11*, 685–725.

Athey, S., Imbens, G. & Wager, S. (2018). Approximate residual balancing: Debiased inference of average treatment effects in high dimensions. *Journal of the Royal Statistical Society, Series B (Methodological)*, *80*, 597–623.

Athey, S., Tibshirani, J. & Wager, S. (2019). Generalized random forests. *Annals of Statistics*, *47*, 399–424.

Bang, H. & Robins, J. (2005). Doubly robust estimation in missing data and causal inference models. *Biometrics*, *61*, 692–972.

Bartlett, V., Dhruva, S., Shah, N., Ryan, P. & Ross, J. (2019). Feasibility of using real-world data to replicate clinical trial evidence. *JAMA Network Open, 2*, e1912869.

Baser, O. (2006). Too much ado about propensity score models? comparing methods of propensity score matching. *Value in Health: The Journal of the International Society for Pharmacoeconomics and Outcomes Research, 9*, 377–385.

Basu, A. (2011). Economics of individualization in comparative effectiveness research and a basis for a patient-centered health care. *Journal of Health Economics, 30*, 549-59.

Basu, A., Navarro, S. & Urzua, S. (2007). Use of instrumental variables in the presence of heterogeneity and self-selection: An application to treatments of breast cancer patients. *Health Economics, 16*, 1133–1157.

Basu, A., Polsky, D. & Manning, W. (2011). Estimating treatment effects on healthcare costs under exogeneity: Is there a 'magic bullet'? *Health Services & Outcomes Research Methodology, 11*, 1-26.

Belloni, A., Chen, D., Chernozhukov, V. & Hansen, C. (2012). Sparse models and methods for optimal instruments with an application to eminent domain. *SSRN Electronic Journal, 80*, 2369–2429.

Belloni, A., Chernozhukov, V., Fernndez-Val, I. & Hansen, C. (2017). Program evaluation and causal inference with high-dimensional data. *Econometrica, 85*, 233–298.

Belloni, A., Chernozhukov, V. & Hansen, C. (2013). Inference for high-dimensional sparse econometric models. *Advances in Economics and Econometrics: Tenth World Congress Volume 3, Econometrics*, 245–295.

Belloni, A., Chernozhukov, V. & Hansen, C. (2014a). High-dimensional methods and inference on structural and treatment effects. *The Journal of Economic Perspectives, 28*, 29–50.

Belloni, A., Chernozhukov, V. & Hansen, C. (2014b). Inference on treatment effects after selection among high-dimensional controls. *The Review of Economic Studies, 81*, 29–50.

Benson, K. & Hartz, A. (2000). A comparison of observational studies and randomized, controlled trials. *The New England Journal of Medicine, 342*, 1878–1886.

Berger, M. & Crown, W. (2021). How can we make more rapid progress in the leveraging of real-world evidence by regulatory decision makers? *Value in Health, 25*, 167–170.

Bound, J., Jaeger, D. & Baker, R. (1995). Problems with instrumental variables estimation when the correlation between the instruments and the endogenous explanatory variable is weak. *Journal of the American Statistical Association, 90*, 443–450.

Brookhart, M., Rassen, J. & Schneeweiss, S. (2010). Instrumental variable methods for comparative effectiveness research. *Pharmacoepidemiology and Drug safety, 19*, 537-554.

Brookhart, M., Schneeweiss, S., Rothman, K., Glynn, R., Avorn, J. & Sturmer, T. (2006). Variable selection for propensity score models. *American Journal of*

Epidemiology, *163*, 1149-1156.
Cameron, A. & Trivedi, P. (2013). *Regression analysis of count data* (2nd ed.). Cambridge University Press.
Carpenter, J., Kenward, M. & Vansteelandt, S. (2006). A comparison of multiple imputation and doubly robust estimation for analyses with missing data. *Journal of the Royal Statistical Society Series A*, *169*, 571–584.
Chernozhukov, V., Chetverikov, D., Demirer, M., Duflo, E., Hansen, C. & Newey, W. (2017). Double/debiased/neyman machine learning of treatment effects. *American Economic Review*, *107*, 261–265.
Chernozhukov, V., Chetverikov, D., Demirer, M., Duflo, E., Hansen, C., Newey, W. & Robins, J. (2018). Double/debiased machine learning for treatment and structural parameters. *The Econometrics Journal*, *21*, 1–C68.
Cole, S. & Frangakis, C. (2009). The consistency statement in causal inference: a definition or an assumption? *Epidemiology*, *20*, 3–5.
Cole, S. & Hernán, M. (2008, 10). Constructing inverse probability weights for marginal structural models. *American Journal of Epidemiology*, *168*, 656–64.
Concato, J., Shah, N. & Horwitz, R. (2000). Randomized, controlled trials, observational studies, and the hierarchy of research designs. *New England Journal of Medicine*, *342*, 1887-1892.
Crown, W. (2015). Potential application of machine learning in health outcomes research and some statistical cautions. *Value in Health*, *18*, 137–140.
Crown, W., Henk, H. & Vanness, D. (2011). Some cautions on the use of instrumental variables estimators in outcomes research: How bias in instrumental variables estimators is affected by instrument strength, instrument contamination, and sample size. *Value in Health*, *14*, 1078–1084.
Crump, R., Hotz, V., Imbens, G. & Mitnik, O. (2009). Dealing with limited overlap in estimation of average treatment effects. *Biometrika*, *96*, 187–199.
Dahabreh, I., Robertson, S., Tchetgen, E. & Stuart, E. (2019). Generalizing causal inferences from randomized trials: Counterfactual and graphical identification. *Biometrics*, *75*, 685–694.
D'Amour, A., Ding, P., Feller, A., Lei, L. & Sekhon, J. (2021). Overlap in observational studies with high-dimensional covariates. *Journal of Econometrics*, *221*, 644–654.
Dickerman, B., García-Albéniz, X., Logan, R., Denaxas, S. & Hernán, M. (2019, 10). Avoidable flaws in observational analyses: an application to statins and cancer. *Nature Medicine*, *25*, 1601–1606.
Evans, H. & Basu, A. (2011). *Exploring comparative effect heterogeneity with instrumental variables: prehospital intubation and mortality* (Health, Econometrics and Data Group (HEDG) Working Papers). HEDG, c/o Department of Economics, University of York.
Franklin, J., Patorno, E., Desai, R., Glynn, R., Martin, D., Quinto, K., . . . Schneeweiss, S. (2020). Emulating randomized clinical trials with nonrandomized real-world evidence studies: First results from the RCT DUPLICATE initiative. *Circulation*, *143*, 1002–1013.
Funk, J., Westreich, D., Wiesen, C., Stürmer, T., Brookhart, M. & Davidian, M.

(2011, 03). Doubly robust estimation of causal effects. *American Journal of Epidemiology*, *173*, 761–767.

Futoma, J., Morris, M. & Lucas, J. (2015). A comparison of models for predicting early hospital readmissions. *Journal of Biomedical Informatics*, *56*, 229–238.

Greenland, S. & Robins, J. (1986). Identifiability, exchangeability, and epidemiological confounding. *International Journal of Epidemiology*, *15*, 413–419.

Hahn, J. & Hausman, J. (2002). A new specification test for the validity of instrumental variables. *Econometrica*, *70*, 163–189.

Hastie, T., Tibshirani, R. & Friedman, J. (2009). *The elements of statistical learning: Data mining, inference and prediction* (2nd ed.). Springer Verlag, New York.

Hausman, J. (1978). Specification tests in econometrics. *Econometrica*, *46*, 1251–1271.

Hausman, J. (1983). Specification and estimation of simultaneous equation models. In *Handbook of econometrics* (pp. 391–448). Elsevier.

Heckman, J. & Navarro, S. (2003). Using matching, instrumental variables and control functions to estimate economic choice models. *Review of Economics and Statistics*, *86*.

Hernán, M. (2011). Beyond exchangeability: The other conditions for causal inference in medical research. *Statistical Methods in Medical Research*, *21*, 3–5.

Hernán, M. (2021). Methods of public health research–strengthening causal inference from observational data. *The New England Journal of Medicine*, *385*, 1345–1348.

Hernán, M., Alonso, A., Logan, R., Grodstein, F., Michels, K., Willett, W., . . . Robins, J. (2008). Observational studies analyzed like randomized experiments an application to postmenopausal hormone therapy and coronary heart disease. *Epidemiology*, *19*, 766–779.

Hirano, K., Imbens, G. & Ridder, G. (2003). Efficient estimation of average treatment effects using the estimated propensity score. *Econometrica*, *71*, 1161–1189.

Hong, W., Haimovich, A. & Taylor, R. (2018). Predicting hospital admission at emergency department triage using machine learning. *PLOS ONE*, *13*, e0201016.

Imbens, G. (2020). Potential outcome and directed acyclic graph approaches to causality: Relevance for empirical practice in economics. *Journal of Economic Literature*, *58*, 1129–1179.

Imbens, G. & Wooldridge, J. (2009). Recent developments in the econometrics of program evaluation. *Journal of Economic Literature*, *47*, 5–86.

Joffe, M., Have, T., Feldman, H. & Kimmel, S. (2004). Model selection, confounder control, and marginal structural models: Review and new applications. *The American Statistician*, *58*, 272–279.

Johnson, M., Bush, R., Collins, T., Lin, P., Canter, D., Henderson, W., . . . Petersen, L. (2006). Propensity score analysis in observational studies: outcomes after abdominal aortic aneurysm repair. *American Journal of Surgery*, *192*, 336–343.

Johnson, M., Crown, W., Martin, B., Dormuth, C. & Siebert, U. (2009). Good research practices for comparative effectiveness research: analytic methods to improve causal inference from nonrandomized studies of treatment effects using secondary data sources: the ispor good research practices for retrospective database analysis task force report–part iii. *Value Health*, *12*, 1062–1073.

Jones, A. & Rice, N. (2009). Econometric evaluation of health policies. In *The Oxford Handbook of Health Economics.*

Kang, J. & Schafer, J. (2007). Demystifying double robustness: A comparison of alternative strategies for estimating a population mean from incomplete data. *Statistical Science*, *22*, 523–539.

Kleibergen, F. & Zivot, E. (2003). Bayesian and classical approaches to instrumental variable regression. *Journal of Econometrics*, 29-72.

Knaus, M., Lechner, M. & Strittmatter, A. (2021). Machine learning estimation of heterogeneous causal effects: Empirical monte carlo evidence. *The Econometrics Journal*, *24*.

Kreif, N., Tran, L., Grieve, R., Stavola, B., Tasker, R. & Petersen, M. (2017). Estimating the comparative effectiveness of feeding interventions in the pediatric intensive care unit: A demonstration of longitudinal targeted maximum likelihood estimation. *American Journal of Epidemiology*, *186*, 1370–1379.

Maddala, G. S. (1983). *Limited-dependent and qualitative variables in econometrics.* Cambridge University Press.

Madigan, D., Ryan, P., Schuemie, M., Stang, P., Overhage, J. M., Hartzema, A., ... Berlin, J. (2013). Evaluating the impact of database heterogeneity on observational study results. *American Journal of Epidemiology*, *178*, 645–651.

Mitra, N. & Indurkhya, A. (2005). A propensity score approach to estimating the cost-effectiveness of medical therapies from observational data. *Health Economics*, *14*, 805—815.

Mullainathan, S. & Spiess, J. (2017). Machine learning: An applied econometric approach. *Journal of Economic Perspectives*, *31*, 87–106.

Murray, M. (2007). Avoiding invalid instruments and coping with weak instruments. *Journal of Economic Perspectives*, *20*, 111–132.

Naimi, A., Cole, S. & Kennedy, E. (2017). An introduction to g methods. *International Journal of Epidemiology*, *46*, 756–762.

Obermeyer, Z. & Emanuel, E. (2016). Predicting the future — big data, machine learning, and clinical medicine. *The New England Journal of Medicine*, *375*, 1216–1219.

Pang, M., Schuster, T., Filion, K., Schnitzer, M., Eberg, M. & Platt, R. (2016). Effect estimation in point-exposure studies with binary outcomes and high-dimensional covariate data - a comparison of targeted maximum likelihood estimation and inverse probability of treatment weighting. *The international Journal of Biostatistics*, *12*.

Pearl, J. (2009). *Causality* (2nd ed.). Cambridge University Press.

Petersen, M., Porter, K., Gruber, S., Wang, Y. & Laan, M. (2012). Diagnosing and responding to violations in the positivity assumption. *Statistical Methods in Medical Research*, *21*, 31–54.

Rajkomar, A., Oren, E., Chen, K., Dai, A., Hajaj, N., Liu, P., ... Dean, J. (2018). Scalable and accurate deep learning for electronic health records. *npj Digital*

Medicine, 18.

Ramsahai, R., Grieve, R. & Sekhon, J. (2011, 12). Extending iterative matching methods: An approach to improving covariate balance that allows prioritisation. *Health Services and Outcomes Research Methodology, 11*, 95–114.

Richardson, T. (2013, April). *Single world intervention graphs (swigs): A unification of the counterfactual and graphical approaches to causality* (Tech. Rep. No. Working Paper Number 128). Center for Statistics and the Social Sciences. University of Washington.

Robins, J. (1986). A new approach to causal inference in mortality studies with sustained exposure periods - application to control of the healthy worker survivor effect. *Computers & Mathematics With Applications, 14*, 923–945.

Robins, J. & Hernan, M. (2009). Estimation of the causal effects of time varying exposures. In *In: Fitzmaurice g, davidian m, verbeke g, and molenberghs g (eds.) advances in longitudinal data analysis* (pp. 553–599). Boca Raton, FL: Chapman & Hall.

Robins, J., Rotnitzky, A. G. & Zhao, L. (1994). Estimation of regression coefficients when some regressors are not always observed. *Journal of The American Statistical Association, 89*, 846–866.

Rosenbaum, P. & Rubin, D. (1983). The central role of the propensity score in observational studies for causal effects. *Biometrika, 70*, 41–55.

Rubin, D. B. (1974). Estimating causal effects if treatment in randomized and nonrandomized studies. *Journal of Educational Psychology, 66*, 688–701.

Rubin, D. B. (1980). Randomization analysis of experimental data: The fisher randomization test. *Journal of the American Statistical Association, 75*(371), 575–582.

Rubin, D. B. (1986). Statistics and causal inference: Comment: Which ifs have causal answers. *Journal of the American Statistical Association, 81*, 961–962.

Rubin, D. B. (2006). *Matched sampling for causal effects.* Cambridge University Press, Cambridge UK.

Scharfstein, D., Rotnitzky, A. G. & Robins, J. (1999). Adjusting for nonignorable drop-out using semiparametric nonresponse models. *JASA. Journal of the American Statistical Association, 94*, 1096–1120. (Rejoinder, 1135–1146).

Schuler, M. & Rose, S. (2017). Targeted maximum likelihood estimation for causal inference in observational studies. *American Journal of Epidemiology, 185*, 65–73.

Seeger, J., Bykov, K., Bartels, D., Huybrechts, K., Zint, K. & Schneeweiss, S. (2015, 10). Safety and effectiveness of dabigatran and warfarin in routine care of patients with atrial fibrillation. *Thrombosis and Haemostasis, 114*, 1277–1289.

Sekhon, J. & Grieve, R. (2012). A matching method for improving covariate balance in cost-effectiveness analyses. *Health Economics, 21*, 695–714.

Setoguchi, S., Schneeweiss, S., Brookhart, M., Glynn, R. & Cook, E. (2008). Evaluating uses of data mining techniques in propensity score estimation: A simulation study. *Pharmacoepidemiology and Drug Safety, 17*, 546–555.

Shi, C., Blei, D. & Veitch, V. (2019). Adapting neural networks for the estimation of treatment effects..

Shickel, B., Tighe, P., Bihorac, A. & Rashidi, P. (2018). Deep ehr: A survey of recent advances on deep learning techniques for electronic health record (ehr) analysis. *Journal of Biomedical and Health Informatics.*, *22*, 1589–1604.

Staiger, D. & Stock, J. H. (1997). Instrumental variables regression with weak instruments. *Econometrica*, *65*, 557–586.

Terza, J., Basu, A. & Rathouz, P. (2008). Two-stage residual inclusion estimation: Addressing endogeneity in health econometric modeling. *Journal of Health Economics*, *27*, 531–543.

Tibshirani, R. (1996). Regression shrinkage and selection via the lasso. *Journal of the Royal Statistical Society: Series B (Methodological)*, *58*, 267–288.

Ting, D., Cheung, C., Lim, G., Tan, G., Nguyen, D. Q., Gan, A., ... Wong, T.-Y. (2017). Development and validation of a deep learning system for diabetic retinopathy and related eye diseases using retinal images from multiethnic populations with diabetes. *JAMA*, *318*, 2211-2223.

Toth, B. & van der Laan, M. J. (2016, June). *TMLE for marginal structural models based on an instrument* (Tech. Rep. No. Working Paper 350). U.C. Berkeley Division of Biostatistics Working Paper Series.

van der Laan, M. & Rose, S. (2011). *Targeted learning: Causal inference for observational and experimental data.* Springer.

van der Laan, M. & Rose, S. (2018). *Targeted learning in data science: Causal inference for complex longitudinal studies.*

van der Laan, M. & Rubin, D. (2006). Targeted maximum likelihood learning. *International Journal of Biostatistics*, *2*, 1043–1043.

Vytlacil, E. (2002). Independence, monotonicity, and latent index models: An equivalence result. *Econometrica*, *70*, 331–341.

Wager, S. & Athey, S. (2018). Estimation and inference of heterogeneous treatment effects using random forests. *Journal of the American Statistical Association*, *113*(523), 1228-1242.

Westreich, D. & Cole, S. (2010, 02). Invited commentary: Positivity in practice. *American Journal of Epidemiology*, *171*, 674–677; discussion 678–681.

Westreich, D., Lessler, J. & Jonsson Funk, M. (2010). Propensity score estimation: Neural networks, support vector machines, decision trees (cart), and meta-classifiers as alternatives to logistic regression. *Journal of Clinical Epidemiology*, *63*, 826–833.

Wooldridge, J. (2002). *Econometric analysis of cross-section and panel data.* MIT Press.

Chapter 6
Econometrics of Networks with Machine Learning

Oliver Kiss and Gyorgy Ruzicska

Abstract Graph structured data, called networks, can represent many economic activities and phenomena. Such representations are not only powerful for developing economic theory but are also helpful in examining their applications in empirical analyses. This has been particularly the case recently as data associated with networks are often readily available. While researchers may have access to real-world network structured data, in many cases, their volume and complexities make analysis using traditional econometric methodology prohibitive. One plausible solution is to embed recent advancements in computer science, especially machine learning algorithms, into the existing econometric methodology that incorporates large networks. This chapter aims to cover a range of examples where existing algorithms in the computer science literature, machine learning tools, and econometric practices can complement each other. The first part of the chapter provides an overview of the challenges associated with high-dimensional, complex network data. It discusses ways to overcome them by using algorithms developed in computer science and econometrics. The second part of this chapter shows the usefulness of some machine learning algorithms in complementing traditional econometric techniques by providing empirical applications in spatial econometrics.

6.1 Introduction

Networks are fundamental components of a multitude of economic interactions. Social relationships, for example, might affect how people form new connections,

Oliver Kiss ✉
Central European University, Budapest, Hungary and Vienna, Austria e-mail: Kiss_Oliver@phd.ceu.edu

Gyorgy Ruzicska
Central European University, Budapest, Hungary and Vienna, Austria, e-mail: Ruzicska_Gyorgy@phd.ceu.edu

F. Chan, L. Mátyás (eds.), *Econometrics with Machine Learning*, Advanced Studies in Theoretical and Applied Econometrics 53, https://doi.org/10.1007/978-3-031-15149-1_6

while ownership and managerial networks could affect how companies interact in a competitive environment. Likewise, geographic networks can influence where nations export to and import from in international trade. Researchers should incorporate the observable network dependencies in their analyses whenever social, geographical, or other types of linkages influence economic outcomes. Such data is also increasingly available due to the rise of digitization and online interactions. In the literature, economic studies with networks have analyzed, among other topics, peer effects (Sacerdote, 2001), social segregation (Lazarsfeld & Merton, 1954), production networks (Acemoglu, Carvalho, Ozdaglar & Tahbaz-Salehi, 2012), and migration networks (Ortega & Peri, 2013).

There is a growing literature on incorporating network structured data into the econometric estimation framework. Besides, there is an increasing number of papers in machine learning that extract information from large-scale network data and perform predictions based on such data sets. On the other hand, there are relatively few network-related topics studied both by econometricians and machine learning experts. This chapter aims to provide an overview of the most used econometric models, machine learning methods, and algorithmic components in their interaction. We further discuss how different approaches can augment or complement one another when incorporated into social and economic analyses.

The chapter proceeds as follows. The following section introduces the terminology used throughout the chapter whenever we refer to a network or its components. This section is not exhaustive and only provides definitions necessary for understanding our discussion. In Section 6.3, we highlight the most significant difficulties that arise when network data is used for econometric estimation. Section 6.4 discusses graph representation learning, a way to reduce graph dimensionality and extract valuable information from usually sparse matrices describing a graph. In Section 6.5, we discuss the problem of sampling networks. Due to the complex interrelations and the often important properties encoded in neighborhoods, random sampling almost always destroys salient information. We discuss methods proposed in the literature aiming to extract better representations of the population. While the techniques mentioned above have received significant attention in the computer science literature, they have – to our knowledge – not been applied in any well-known economic work yet. Therefore, in Section 6.6, we turn our attention to a range of canonical network models that have been used to analyze spatial interactions and discuss how the spatial weight matrix can be estimated using machine learning techniques. Then, we introduce gravity models, which have been the main building blocks of trade models, and provide a rationale for using machine learning techniques instead of standard econometric methods for forecasting. The chapter closes with the geographically weighted regression model and shows an example where econometric and machine learning techniques effectively augment each other.

6.2 Structure, Representation, and Characteristics of Networks

Networks have been studied in various contexts and fields ranging from sociology and economics through traditional graph theory to computer science. Due to this widespread interest, notations and terminology also differ across fields of study. Throughout this chapter, we rely on a unified notation and terminology introduced in the paragraphs below.

A network (or graph) G is given by a pair $(\mathcal{V},\mathcal{E})$ consisting of a set of nodes or vertices $\mathcal{V} = \{1,2,...,n\}$ and a set of edges $\mathcal{E} \subseteq \{(i,j)|i,j \in \mathcal{V}\}$ between them. An edge (i,j) is incident to nodes i and j. Networks can be represented with an n dimensional positive adjacency matrix $\mathbf{A} \in \mathbb{R}_0^{+|\mathcal{V}|\times|\mathcal{V}|}$, where each column and row corresponds to a node in the network. The a_{ij} element of this matrix contains the weight of the directed edge originating at node i targeting node j. Throughout this chapter, the terms *network* and *graph* refer to the same object described above.

The values in the adjacency matrix can be understood as the strength of interactions between the two corresponding nodes. There are various types of interactions that can be quantified with such links. For example, in spatial econometrics, an edge weight can denote the distance (or its inverse) between two separate locations represented by the nodes. In social network analysis, these edge weights may indicate the number of times two individuals interact with each other in a given period of time.

The diagonal elements (a_{ii}) of the adjacency matrix have a special interpretation. They indicate if there exists an edge originating at a node pointing to itself. Such representation is mainly useful in dynamic networks where the actions of an agent can have effects on its future self. In most static applications, however, the diagonal elements of the adjacency matrix are zero.

Directed and undirected networks. A network may be undirected if all of its edges are bidirectional (with identical weights in both directions) or directed if some are one directional (or if the weights differ). In both types of networks, the lack of a link between nodes i and j is represented by the a_{ij} element of the adjacency matrix being zero. In a directed network having a one directional link from node i to node j means that $a_{ij} > 0$ and $a_{ji} = 0$. If the network is undirected, then its adjacency matrix is symmetric, i.e., $a_{ij} = a_{ji}\ \forall i,j \in \mathcal{V}$.

Different economic and social relationships can be represented by different types of networks. In trade networks, edges are usually directed as they describe the flow of goods from one country to another. On the other hand, friendship relationships are generally considered reciprocal and are, therefore, characterized by undirected edges.

Weighted and unweighted networks. A network may be unweighted if all its ties have the same strength. In unweighted networks, the elements of the adjacency matrix are usually binary $(a_{ij} \in \{0,1\}\forall i,j \in \mathcal{V})$, indicating the existence of a link between two nodes. In a weighted network setting, edges can be assigned different weights. These weights usually represent a quantifiable measure of the strength of connection, and they are incorporated into the elements of the adjacency matrix.

In some applications, such as spatial econometrics, networks are usually weighted, as edge weights can denote the spatial distance between locations or the number of times two agents interact in a given time period. In contrast, some settings make it difficult or impossible to quantify the strength of connections (consider, for example, the strength of friendship relations) and are, therefore, more likely to be represented by unweighted networks.

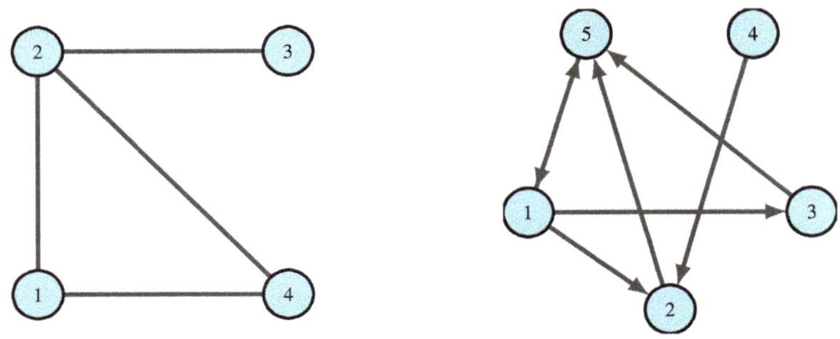

Fig. 6.1: Example for an undirected network (left) and a directed network (right)

Network structured data. Throughout this section, network data or network structured data refers to a data set containing nodes, potentially with their characteristics, and relationships between these nodes described by edges, and possibly edge characteristics. How this data is stored usually depends on the size and type of the network. Small networks can easily be represented by their adjacency matrices which are able to capture weights and edge directions at the same time. In this case, node and edge characteristics are usually given in separate cross-sectional data files. The size of the adjacency matrix (which is usually a sparse matrix) scales quadratically in the number of nodes. Consider, for example, the networks shown in Figure 6.1. Let us denote the adjacency matrix of the undirected network by $\mathbf{A}$ and that of the directed network by $\mathbf{B}$. Then corresponding adjacency matrices are

$$\mathbf{A} = \begin{bmatrix} 0\,1\,0\,1 \\ 1\,0\,1\,1 \\ 0\,1\,0\,0 \\ 1\,1\,0\,0 \end{bmatrix} \quad \text{and} \quad \mathbf{B} = \begin{bmatrix} 0\,1\,1\,0\,1 \\ 0\,0\,0\,0\,1 \\ 0\,0\,0\,0\,1 \\ 0\,1\,0\,0\,0 \\ 1\,0\,0\,0\,0 \end{bmatrix}.$$

In practice, large networks are therefore rather described by edge lists instead of the adjacency matrix. In this representation, we have a data set containing the source and target node identifiers and additional columns for edge characteristics (such as weight). Node characteristics are usually stored in a separate file.

Characteristics of networks. Due to the unique structure of network data, core statistics characterizing a network are also specific and have been designed to capture certain aspects of the underlying relationships. There are two main types of statistics regarding graph structured data. One set (usually related to nodes) aims to describe local relationships by summarizing the neighborhoods of nodes, while others aim to characterize the network as a whole. Although there is a multitude of such measures, this chapter relies predominantly on the following:

- Two nodes are *neighbors* if there is a common edge incident to both of them.
- The *set of neighbors* of node i is $\mathcal{N}(i) = \{j \in \mathcal{V} | (i,j) \in \mathcal{E} \text{ or } (j,i) \in \mathcal{E}\}$.
- The *degree* of node i is the number of its neighbors $d(i) = |\mathcal{N}(i)|$.
- The *degree distribution* of a graph $G(\mathcal{V},\mathcal{E})$ is the distribution over $d(i)|i \in \mathcal{V}$.
- A *path* between the nodes i and j is a set of n edges in $\mathcal{E}$

  $$\{(s_1,t_1),(s_2,t_2),\ldots,(s_{n-1},t_{n-1}),(s_n,t_n)\},$$

  such that $s_1 = i$, $t_n = j$ and $s_k = t_{k-1}\,\forall k \in 2,\ldots,n$.
- Two nodes belong to the same *connected component* if there exists a path between the two nodes.
- A *random walk of length n* from node i is a randomly generated path from node i. The next edge is always chosen uniformly from the set of edges originating in the last visited node.
- The *centrality* of a node is a measure describing its relative importance in the network. There are several ways to measure this. For example, *degree centrality* uses the degree of each node, while *closeness centrality* uses the average length of the shortest path between the node and all other nodes in the graph. More complex centrality measures usually apply a different aggregation of degrees or shortest paths. The *Katz centrality*, for example, uses the number of all nodes that can be connected through a path, while the contributions of distant nodes are penalized. These measures can often be used for efficient stratified sampling of graphs. For example, the *PageRank* of a node (another centrality measure) is used in PageRank node sampling, a method presented in Section 6.5.
- The *local clustering coefficient* of a node measures how sparse or dense the immediate neighborhood of a node is. Given $d(i)$ – the degree of node i – it is straightforward that in a directed network there can be at most $d(i)(d(i)-1)$ edges connecting the neighbors of node i. The local clustering coefficient measures what fraction of this theoretical maximum of edges is present in the network; thus, in a directed network it is given by

  $$C(i) = \frac{|\{e_{jk} \text{ s.t. } j,k \in \mathcal{N}(i) \text{ and } e_{jk} \in \mathcal{E}\}|}{d(i)(d(i)-1)}.$$

  In an undirected setting, the number of edges present must be divided by $d(i)(d(i)-1)/2$.
- The *degree correlation* – or degree assortativity – of a network measures whether similar nodes (in terms of their degree) are more likely to be connected in the network. The phenomenon that high-degree nodes are likely to be connected to other high-degree nodes is called assortative mixing. On the contrary, if high-

degree nodes tend to have low-degree neighbors, we call it disassortative mixing. Details on the calculation of this measure are discussed by Newman (2002).

While the list of network characteristics above is far from being exhaustive, it is sufficient to understand our discussion in the upcoming sections.

Distinction between network structured data and neural networks. This chapter discusses econometric and machine learning methods that utilize network structured data. In Section 6.6, we introduce neural network-based machine learning methods, including deep neural networks, convolutional neural networks, and recurrent neural networks. Importantly, these neural network architectures are not directly related to the network structured data. While networks are often used as inputs to these neural network models, the name "network" in these machine learning models refers to their architectural design. As defined in this section, networks are graphs that represent economic activities, while neural networks are techniques used to identify hidden patterns in the data. Chapter 4 discusses all the neural network architectures described in this chapter.

6.3 The Challenges of Working with Network Data

Network data differs from traditional data structures in many aspects. These differences result in unique challenges requiring specific solutions or the refinement of existing econometric practices. For example, the size of the adjacency matrix can result in computational challenges, while establishing causal relationships is complicated by complex interrelations between agents (represented by nodes) in the networks. This section highlights the most important issues arising from using network structured data in an econometric analysis. These specific aspects must be considered in theoretical modeling and empirical studies that use networks.

Curse of dimensionality. The analysis of large-scale real-world network data (like web graphs or social networks) is often difficult due to the size of the data set. The number of data points in an adjacency matrix increases quadratically with the number of agents in a network. This is in contrast with other structured data sets, where new observations increase the size of the data linearly. Due to its high dimensionality and computational difficulties, the adjacency matrix cannot be directly included in an econometric specification in many cases. The traditional solution in econometric applications has been incorporating network aggregates instead of the whole network into the estimation. Another common alternative is to multiply the adjacency matrix (or a transformation of it) by regressors from the right to control for each node's neighbors' aggregated characteristics when modeling their individual outcomes.

For example, peer effects in networks are modeled by Ballester, Calvo-Armengol and Zenou (2006) using such aggregates. In their model, each agent i chooses the intensity of action y_i to maximize:

$$u_i(y_1, ... y_n) = \alpha_i y_i - \frac{1}{2}\beta_i y_i^2 + \gamma \sum_{j \neq i} a_{ij} y_i y_j,$$

where the adjacency matrix elements a_{ij} represent the strength of connection between agents i and j. The main coefficient of interest in such a setting is γ – often called the peer effect coefficient – measuring the direct marginal effect of an agent's choice on the outcome of a connected peer with a unit connection strength. In this specification, utility is directly influenced by an agent's own action along with all its neighbors' actions. An agent's optimal action is also indirectly affected by the actions of all agents belonging to the same connected component in the network. In fact, the optimal action of nodes in such a setting is directly related to their Katz-Bonancich centrality.

The canonical characterization of outcomes being determined by neighbors' actions and characteristics is attributed to Manski (1993):

$$y_i = \alpha + \beta \sum_{j=1}^{N} a_{ij} y_j + \mu x_i + \gamma \sum_{j=1}^{N} a_{ij} x_j + \epsilon_i.$$

where a_{ij} is defined as in the previous example and x_i, x_j are node-level characteristics of agent i and j, respectively. This specification is discussed further in Section 6.6.1.

While such approaches are easy to interpret, recent advances in computer science provide more efficient machine learning algorithms, which can capture more complex characteristics in a reduced information space. These methods decrease the dimensionality of a network by embedding it into a lower-dimensional space. This lower-dimensional representation can be used to control for node or edge level information present in the network in downstream tasks, such as regressions. Some of the most widely used dimensionality reduction techniques are discussed in Section 6.4.

Sampling. Sampling is crucial when it is impossible to observe the whole population (all nodes and edges) or when the size of the population results in computational challenges. In a network setting, however, random sampling destroys information that might be relevant to a researcher since local network patterns carry useful information. Chandrasekhar and Lewis (2016) provide an early study on the econometrics of sampled networks. They show for two specific random node sampling approaches[1] that sampling leads to a non-classical measurement error, which results in biased regression coefficient estimates. We discuss how applying different sampling algorithms might help preserve important information and discuss common node and edge sampling approaches in Section 6.5. Most of these approaches have been designed to preserve a specific network attribute assuming a particular network type. While their theoretical properties are known, their applicability to real-life data has only been studied in a handful of papers (Rozemberczki, Kiss & Sarkar, 2020).

[1] Both techniques (random node sampling with edge induction and random node-neighbor sampling) are discussed in detail in Section 6.5.

Identification, reverse causality, and omitted variables. Inference based on network data is complicated by the network structure being related to agents' characteristics and actions (observed and unobserved), making it endogenous in most specifications. That is, networks determine outcomes that, in turn, affect the network structure. Therefore, when econometricians try to identify chains of causality, they are often faced with inherently complex problems that may also incorporate a circle of causality. To avoid this issue, researchers must control for all the characteristics that could affect behavior and drive the observed network structure. When some covariates cannot be controlled for – either because of unobservability or unavailable data – the estimation could be biased due to the omitted variables problem.

To illustrate these problems, assume that we would like to document peer effects in an educational setting empirically. In particular, educational outcome (GPA) is regressed on friends' GPA and characteristics:

$$Y_i = \alpha X_i + \beta \bar{X}_{-i} + \gamma \bar{Y}_{-i} + \epsilon_i,$$

where X_i are observed individual characteristics, and ϵ_i incorporates the unobserved individual characteristics and a random error term. $\bar{X}_{-i}$ and $\bar{Y}_{-i}$ measure the average of neighbors' characteristics and GPA, respectively.

This model specifies two types of peer effects. β measures the 'contextual effect' reflecting how peer characteristics affect individual action. On the other hand, γ is the 'endogenous effect' of neighbors' actions. In education, these effects manifest in higher GPA if the agent's peers work hard (endogenous effect) or because peers are intelligent (contextual effect).

We may encounter three problems when identifying coefficients in this regression model. First, there are presumably omitted variables that we cannot control for. For example, if peers in a group are exposed to the same shock, ϵ_i is correlated with ϵ_{-i} and hence $\bar{Y}_{-i}$. Second, if agents select similar peers, the estimation suffers from selection. In such a case, ϵ_i is correlated with both $\bar{X}_{-i}$ and $\bar{Y}_{-i}$. Third, this specification is an example of the reflection problem documented by Manski (1993). The reflection problem occurs when agents' actions are determined jointly in equilibrium. Hence ϵ_i is correlated with $\bar{Y}_{-i}$.

To avoid these problems with identification and disentangle causal effects, econometricians have analyzed experimental and quasi-experimental setups where network connections could be controlled for. For example, Sacerdote (2001) used randomization in college roommate allocation to identify peer effects in educational outcomes. Alternatively, researchers may use instrumental variables to overcome the endogeneity of the social structure in their estimations. Jackson (2010) discusses endogeneity, identification, and instrumental variables in further detail and highlights other problems that arise with identification in social network analysis, including nonlinearities in social interactions and the issue of timing.

In the machine learning literature, spatiotemporal signal processing has been applied to model the co-evolution of networks and outcomes. A spatiotemporal deep learning model combines graph representation learning (extracting structurally representative information from networks) with temporal deep learning techniques.

These models rely on a temporal graph sequence (a graph of agents observed through different time periods). They utilize a graph neural network block to perform message passing at each temporal unit. Then, a temporal deep learning block incorporates the new information into the model. This combination of techniques can be used to model both temporal and spatial autocorrelation across the spatial units and agents (Rozemberczki, Scherer, He et al., 2021).

In general, machine learning methods aimed at establishing causal relationships in networks are rare. The literature primarily focuses on using models for forecasting and prediction. While these methods benefit data-driven decision-making and policy analysis, algorithms focusing on causality remain an important future research domain.

6.4 Graph Dimensionality Reduction

With the rise of digitalization, data became available on an unprecedented scale in many research domains. This abundance of data results in new research problems. One such problem is the availability of too many potentially valuable right-hand side variables. Regularization techniques such as LASSO[2] have a proven track record of separating valuable variables (i.e., those with high explanatory power) from less valuable ones. Another traditional technique in econometrics is dimensionality reduction. This process aims to represent the data in a lower-dimensional space while retaining useful information. Some of these techniques (mainly principal component analysis, which uses a linear mapping to maximize variance in the lower-dimensional representation) have an established history of being used in applied works. However, there is also a growing number of methodologies primarily applied in machine learning to achieve the same goal. The analysis of network data is a prime example of where these algorithms gained importance in the last decade.

Algorithms applied in this domain are collectively called embedding techniques. These techniques can be used to represent the original high dimensional data in a significantly lower-dimensional space where the distance of the embedded objects is traditionally associated with a measure of similarity. Algorithms differ in terms of what is embedded in a lower-dimensional space (nodes, edges, or the whole graph), what graph property is used to measure similarity, and how this similarity is used to encode information in the lower-dimensional space. The representations obtained through these algorithms can then be used to control for the structural information present in the network. This section provides an overview of methodologies prevalent in the machine learning literature that can be useful for applied econometricians working with graph data.

[2] See, e.g., Chapters 1. and 2.

6.4.1 Types of Embeddings

Node embedding. Probably the most common application of embedding techniques is node embedding. In this case, given a graph $G(\mathcal{V},\mathcal{E})$ and the dimensionality of the embedding $d << |\mathcal{V}|$ the goal is to assign a d-dimensional vector to each node in the graph. The pairwise distances between nodes in this lower-dimensional – usually Euclidean – space correspond to a similarity measure of choice. That is, nodes similar to each other in some sense are to be closer in this lower-dimensional space. There is widespread research on how this similarity measure should be defined. There are three major approaches in the literature in this aspect. *Neighborhood preserving* node embeddings (for example, Diff2Vec (Rozemberczki & Sarkar, 2018) or Node2Vec (Grover & Leskovec, 2016)) aim to preserve the distance between nodes in their lower-dimensional representations meaning that neighbors will be closer to each other. This is usually achieved by the decomposition of proximity matrices, a methodology described in Section 6.4.2. Another major set of techniques, called *structural embeddings*, aims to preserve the structural role of nodes in the graph. That is, nodes with similar structural properties (e.g., degree, clustering coefficient, centrality) are to be closer in the embedding space. The third set of techniques exploits additional information encoded in node attributes. In this case, the pairwise similarities of nodes are constructed by considering a set of additional node-level characteristics. In a social network setting, this could, for example, mean a set of variables describing the socioeconomic status of each node. These methods are collectively called *attributed node embeddings*.

Edge embedding. While relationships between edges might seem somewhat less complex than those between nodes, they may contain just as much useful information. In some applications focusing on connections between agents (e.g., friendships or trade), edges are of primary interest. Predicting which of their friends people trust the most or understanding how information spreads in a network are examples where edges have a crucial role. Structural information on edges can be encoded in an adjacency matrix by transforming the original graph into a so-called *line graph*. Given a graph $G(\mathcal{V},\mathcal{E})$, its line graph is another graph in which each edge is represented by a node, and each node is represented by an edge. The nodes in the line graph are connected if the corresponding edges in the original graph $G(\mathcal{V},\mathcal{E})$ have a node that is incident to both of them. Notice that a line graph can be described by an $|\mathcal{E}| \times |\mathcal{E}|$ adjacency matrix. The problem of high dimensionality thus persists when one aims to focus on edges without losing potentially valuable network information. In such cases, edge embeddings might be useful. The core idea is identical to that of the node embeddings. Given a graph $G(\mathcal{V},\mathcal{E})$ and the dimensionality of the embedding $d << |\mathcal{E}|$ the goal is to assign a d-dimensional vector to each edge in the graph.

Whole graph embedding. In some cases, the unit of observation is a graph. For example, one might study ego-nets, where each agent has a corresponding network that consists of the agent and its neighbors. Another example is analyzing discussions in a forum where each graph represents a thread. The users participating in the discussion are the nodes, and the replies to each other are the edges. In such cases, the goal is to embed whole graphs, that is, represent each graph in a set of graphs $\{G_1, G_2, \ldots, G_n\}$ with a d-dimensional vector.

6.4.2 Algorithmic Foundations of Embeddings

Matrix factorization. A major set of embedding algorithms relies on the factorization of matrices encoding graph properties. For example, one might consider a matrix containing pairwise edge or node similarity measures. These matrices are then factorized either directly or – more frequently – using graph Laplacian eigenmaps. Let us denote a pairwise similarity matrix by $\mathbf{W}$. In the case of pairwise node similarities $\mathbf{W} \in \mathbb{R}^{|\mathcal{V}|\times|\mathcal{V}|}$, and $\mathbf{W}_{ij}$ measures the similarity between nodes i and j. Likewise in edge similarities $\mathbf{W} \in \mathbb{R}^{|\mathcal{E}|\times|\mathcal{E}|}$, and $\mathbf{W}_{uv}$ measures the similarity between edges u and v. Given a dimensionality (d), the goal is to find an embedding $\mathbf{Y} \in \mathbb{R}^{|\mathcal{V}|\times d}$ for nodes or $\mathbf{Y} \in \mathbb{R}^{|\mathcal{E}|\times d}$ for edges that minimizes a pre-defined loss function. A traditional loss function used by a variety of algorithms is a classical quadratic loss utilizing the Euclidean (ℓ^2) norm as given in Equation (6.1):

$$\mathbf{Y}^* = \arg\min \sum_{i\neq j} \mathbf{W}_{ij} \cdot ||\mathbf{Y}_i - \mathbf{Y}_j||_2^2. \tag{6.1}$$

$\mathbf{Y}_i$ in Equation (6.1) is the d dimensional representation of node or edge i in the embedding space. Let us define a diagonal matrix $\mathbf{D}$, where

$$\mathbf{D}_{ii} = \sum_{i,j\in\mathcal{V}} \mathbf{W}_{ij}.$$

Then the *Laplacian* of the similarity matrix is given by

$$\mathbf{L} = \mathbf{D} - \mathbf{W}.$$

Notice that using the graph Laplacian Equation (6.1) reduces to

$$\mathbf{Y}^* = \arg\min tr(\mathbf{Y}'\mathbf{L}\mathbf{Y}). \tag{6.2}$$

Belkin and Niyogi (2002) show that one needs to introduce an additional orthogonality constraint in order to obtain a unique solution. This is, in most cases, solved by enforcing

$$\mathbf{Y}'\mathbf{D}\mathbf{Y} = \mathbf{I}.$$

The problem is then given by Equation (6.3):

$$\mathbf{Y}^* = \arg\min_{\mathbf{Y}'\mathbf{D}\mathbf{Y}=\mathbf{I}} tr(\mathbf{Y}'\mathbf{L}\mathbf{Y}). \tag{6.3}$$

By using the Lagrange method it can be seen that the optimal solution in this case consists of eigenvectors that solve the generalized eigenvalue problem given in

Equation (6.4) (Torres, Chan & Eliassi-Rad, 2020; Belkin & Niyogi, 2002):

$$\mathbf{L}\mathbf{Y}_i^* = \lambda_i \mathbf{D}\mathbf{Y}_i^*. \tag{6.4}$$

To be more precise, the eigenvectors corresponding to the d lowest eigenvalues of the normalized Laplacian $\mathbf{D}^{-1}\mathbf{L}$ are the columns of $\mathbf{Y}^*$; thus, the embedding vector of node j consists of the jth elements of these d eigenvectors. The eigenvalues in such a setting also turned out to be useful. The whole graph embedding method SF (de Lara & Edouard, 2018) applies the same approach, representing each whole graph in a set of graphs with the d lowest eigenvalues of the graph in question.

A core question is how one defines pairwise similarities. Some neighborhood preserving methods, such as the seminal Laplacian Eigenmap (Belkin & Niyogi, 2002) or GLEE (Torres et al., 2020), use the adjacency matrix directly. Others rely on different similarity metrics. For example, Isomap (Balasubramanian & Schwartz, 2002) uses the sum of edge weights along the shortest path between node pairs. Cai, Zheng and Chang (2018) provide a comprehensive survey on Laplacian-based embeddings with different similarity measures and alternative objective functions.

Sequence-based methods. A wide set of algorithms relies on observing the neighborhood of each node (either through direct sampling or through multiple random walks originating at each node) and then learning a representation by maximizing the likelihood of predicting the correct neighborhood for each node using its lower-dimensional representation. Given a dimensionality d the goal is to find an embedding $\mathbf{Y} \in \mathbb{R}^{|\mathcal{V}| \times d}$ for the nodes in a graph $G(\mathcal{V}, \mathcal{E})$. A prominent method in this category is node2vec (Grover & Leskovec, 2016) presented below. Other algorithms such as DeepWalk (Perozzi, Al-Rfou & Skiena, 2014), LINE (Tang et al., 2015), or diff2vec (Rozemberczki & Sarkar, 2018) use similar approaches with modifications in terms of the sampling strategy or the exact likelihood function.

With node2vec, for every node $i \in \mathcal{V}$, we define a set of neighboring nodes $\mathcal{N}(i) \subset \mathcal{V}$. The goal is to maximize a log-likelihood function given in Equation (6.5)

$$\max_{\mathbf{Y}} \sum_{i \in \mathcal{V}} \log Pr(\mathcal{N}(i) | \mathbf{Y}_i), \tag{6.5}$$

where $\mathbf{Y}_i$ is the lower-dimensional representation of node i. Grover and Leskovec (2016) make two important assumptions to make the above problem tractable. By assuming that the probability of observing one node in the neighborhood of node i is independent of observing any other node in the neighborhood given node i's representation, they factorize the joint probability of observing a neighborhood to observing individual nodes. This results in the property given in Equation (6.6).

$$Pr(\mathcal{N}(i) | \mathbf{Y}_i) = \prod_{j \in \mathcal{N}(i)} Pr(j | \mathbf{Y}_i). \tag{6.6}$$

An important question is how to model the conditional probability in Equation (6.6). Grover and Leskovec (2016) propose a simple logit structure using the scalar products of the embedding vectors given in Equation (6.7)

$$Pr(j|\mathbf{Y}_i) = \frac{exp(\mathbf{Y}_i\mathbf{Y}_j')}{\sum_{k\in\mathcal{V}} exp(\mathbf{Y}_i\mathbf{Y}_k')}. \tag{6.7}$$

With the assumptions above, Equation (6.5) reduces to

$$\max_{\mathbf{Y}} \sum_{k\in\mathcal{V}} \left(-\log Z_k + \sum_{j\in\mathcal{N}(i)} \mathbf{Y}_i\mathbf{Y}_j' \right),$$

where

$$\log Z_k = \sum_{k\in\mathcal{V}} exp(\mathbf{Y}_i\mathbf{Y}_k').$$

It is worth noting that Z_k is computationally expensive to calculate throughout the optimization process. Therefore, practical implementations usually approximate it with the so-called negative sampling method (Mikolov, Sutskever, Chen, Corrado & Dean, 2013).

6.5 Sampling Networks

Real-world graphs are often too large to be analyzed directly. Consider, for example, social networks with millions of users and billions of edges. In these networks, even deriving basic descriptive statistics might be computationally challenging, if not impossible, with resources available to most researchers. Sampling might seem to be a straightforward solution to this problem. However, applying the proper sampling technique is a major challenge in analyzing large, web-scale graphs. Choosing the appropriate sampling technique, in general, has two components. First, large networks often cause computational issues (Kang, Tsourakakis & Faloutsos, 2009; Gonzalez, Low, Gu, Bickson & Guestrin, 2012) rendering some traditional sampling algorithms unusable. Second, networks describe complex underlying relations between agents, and therefore choosing an inappropriate sampling technique might destroy valuable information. Sampling severely affects a multitude of core descriptive network characteristics such as degree distribution, diameter, clustering coefficient, or transitivity (Easley & Kleinberg, 2010). The choice of the sampling technique is, therefore, less obvious. It has to be reflective of the downstream task. That is, one has to choose the sampling technique that retains the valuable information needed for the task at hand. In general, there are two practical points where graph sampling can happen during the analysis of graph structured data:

1. At data collection. Consider, for example, field experiments collecting data on the social network of individuals (see e.g., Crépon, Devoto, Duflo & Parienté, 2015 or Ferrali, Grossman, Platas & Rodden, 2020). In such cases, collecting the whole network would practically imply collecting data on all 7.75 billion individuals in the world. By construction, these experiments rely on samples (for example, the network of people living in the same village or students in the same

school). Analysis relying on such samples should consider the consequences of such sampling approaches on network features. Another example where sampling might be necessary at the point of collecting the data is web graphs. Web graphs are large-scale graph structured data sets from the web. For example, the graph of users on social network websites (Twitter, Facebook, LinkedIn, etc.) and their connections. Another web graph is the world wide web itself, where websites are the nodes and hyperlinks pointing to each other are the links. In these cases collecting the whole graph might be impossible or impractical.
2. At data analysis. Consider having access to a whole web graph. The analysis of these graphs is impossible on commercial hardware. In business environments, hardware constraints might be less of an issue. It is, however, still unnecessarily costly in many cases to analyze the whole network where performing the analysis on a proper sample is possible.

There is a wide range of algorithms proposed (Hu & Lau, 2013) and implemented (Rozemberczki et al., 2020) for different objectives. This section overviews the most important approaches and the potential applications.

6.5.1 Node Sampling Approaches

One approach is to sample the network by selecting a subset of vertices $\mathcal{N}_s \subseteq \mathcal{N}$. There are four widely discussed methods in the literature for sampling nodes.

Random node sampling. This sampling technique has a single parameter p corresponding to the probability of a node being included in the sample. That is, each node is picked randomly with probability p to be included in the sampled network. An edge will only be part of the sampled network if the nodes connected by it are both included in the sampled network. While this sampling technique is computationally simple and works fast for large networks, it heavily distorts the degree distribution of the sampled graph unless the original graph has a positive or negative binomial degree distribution (the average degree is different even in these cases) (Stumpf, Wiuf & May, 2005). It is worth noting that Chandrasekhar and Lewis (2016) offer a method to correct for this kind of bias. The results of Rozemberczki et al. (2020) show that a potential application domain of this sampling method is the estimation of network assortativity through the calculation of degree correlation. Degree correlations estimated from samples utilizing this method are closer to the ground truth than those using other types of node sampling for multiple real-life social networks.

Random node sampling with neighborhood. We start by selecting a random subset of nodes $V_s \subseteq V$ using random node sampling. Then the algorithm induces the neighboring nodes as the node set and the edges between all of the nodes. This practically results in observing a set of nodes and their immediate neighborhood. This technique is useful when we are interested in the occurrence of small, local patterns such as three-node cliques.

Degree-based node sampling. In many real-life networks, the number of connections of a node encodes essential information. For example, nodes with a higher degree in social networks represent people with more friends, connections, or followers. Highly connected nodes often have important roles in network stability, information transmission, etc. It is, therefore, a meaningful idea to select nodes in the sample with probabilities weighted according to their degree if we want to include potentially more influential agents with a higher probability (Adamic, Lukose, Puniyani & Huberman, 2001). This approach is, by construction, biased towards the inclusion of high-degree nodes, and therefore graphs sampled using this technique have very different distributional properties than their originals (Leskovec & Faloutsos, 2006).

PageRank node sampling. Similarly to the degree-based node sampling, this approach relies on selecting nodes into the sample using a pre-defined distribution over the set of nodes. The core idea is to calculate the PageRank (Page, Brin, Motwani & Winograd, 1999) of each node in the node set. PageRank is a widely used algorithm to rank nodes in a network. The PageRank of a node $i \in \mathcal{N}$ is given by

$$PR(i) = \frac{1-d}{|\mathcal{N}|} + d \sum_{j \in \mathcal{M}_i} \frac{PR(j)}{D_j^+},$$

where d is the so called damping factor, $\mathcal{M}_i$ is the set of nodes that have an edge pointing to node i , D_j^+ is the out-degree of j, and $|\mathcal{N}|$ is the total number of nodes. The PageRank of a node shows how likely a random walk is to go through the given node. Roughly speaking, 'easily reachable' nodes in the network have a higher PageRank.

6.5.2 Edge Sampling Approaches

An alternative to node sampling is edge sampling. These techniques are predominantly helpful when the underlying research question focuses on the connections rather than the agents connected. In these cases, we are analyzing edge attributes. For example, one might be interested in the differences in friendship characteristics (number of interactions, length of friendship, etc.) between agents from the same and different cities. In this case, we need to focus on the differences between two distinct sets of edges. Next, we outline the most prevalent methodologies for edge sampling.

Random edge sampling. Just like with nodes, it is possible to select edges into the sample with a uniform probability (p). Nodes connected by the edges chosen will also be part of the sample. Unless the goal is to approximate edge-level average characteristics, this sampling method has limited use. The resulting sampled graph is usually (depending on the value of p) very sparse, but a multitude of salient network features are destroyed.

Random node-edge sampling. This method slightly differs from the one mentioned above. The traditional random edge sampling method is biased towards selecting high degree nodes into the sample by construction. Random node-edge sampling is used to solve this issue. First, one selects nodes with a uniform probability, p, and then for each selected node, we pick a single edge incident to the node into the sample at random.

Induced edge sampling methods. Random edge sampling with edge set induction first randomly samples edges with a fixed probability. Nodes incident to these edges will be included in the sample. Finally, edges between nodes already in the sample are retained with an induction step. This method is computationally expensive as it requires iterating through the edge set exactly twice. A similar approach was proposed by Ahmed, Neville and Kompella (2013), performing only a partial induction step. Their algorithm (PIES) is a stream sampling algorithm that produces a sample from an edge stream. This has the advantage of low computational costs as – alongside the already sampled part of the graph – only a single edge is kept in the memory at the time. The algorithm iterates through a set of edges and selects each edge into the sample (along with the nodes incident to it) with a fixed probability. Edges connecting two nodes already in the sample are also added to the sample, resulting in a partial induction (edges are only induced if both the nodes they connect have been added to the sample previously). This method (along with the full induction version) is known to have a bias towards high degree nodes due to the random edge selection step. High degree nodes have more incident edges and thus are more likely to be selected into the sample.

Hybrid Approaches and the Importance of the Problem

Many edge and node selection algorithms can be combined relatively easily to fit the underlying research question better. For example, Krishnamurthy et al. (2005) present an approach where at each sampling step, with probability w, they select an edge into the sample using random edge sampling. With probability $1 - w$, they use random node-edge sampling. They claim that these methods are biased in opposite directions for multiple metrics in their study. This, however, does not hold for every case: Leskovec and Faloutsos (2006) and Rozemberczki et al. (2020) evaluate a variety of sampling algorithms using different networks and different network metrics. They find that metrics calculated from samples generated using a simple random edge sampling algorithm are usually closer to the ground truth metric than those calculated from the hybrid samples in their applications. This is yet another case showing that the choice of the sampling method should always correspond to the downstream estimation task.

6.5.3 Traversal-Based Sampling Approaches

Traversal or exploration-based sampling methods extract samples from a graph by probing the neighborhood on one or more seed nodes. The majority of modern sampling algorithms belong to this category. Next, we outline the main types of these algorithms and the best-known versions in each group.

6.5.3.1 Search Based Techniques

Search-based techniques sample a network by starting from seed node (s) and looking for specific neighbors until well-defined stopping criteria are reached.

Breadth first search. With breadth first search (Doerr & Blenn, 2013) we need to define a starting node n_0 first and add it to a *queue* denoted by Q. We also define the – so far – empty set of sampled nodes V_s. The algorithm proceeds by processing the elements of the queue. It always removes the first element of the queue, adds it to the set of sampled nodes, and then adds all those neighbors of the node to the queue that are neither sampled nor already in the queue. The search process ends once a *budget* is exhausted or if there are no more elements in the *queue*. The easiest example of such a budget is sampling a fixed number of nodes. In this case, the initial budget is the number of nodes we want to sample, and at each sampled node, it decreases by one. The sampling stops once the budget reaches zero. It is also possible to make certain sampled nodes more expensive depending on their characteristics. The final sample is the graph induced by the selected nodes.

Depth first search. This algorithm is identical to breadth first search except for how it selects the next node from the queue into the sample. With depth first search (Doerr & Blenn, 2013), we always choose the last node in the queue into the sample. The names of these search methods are reasonably intuitive. Breadth first search will result in a broad tree of sampled nodes as it iterates through all neighbors of a few nodes, while depth first search prefers newly added neighbors and, thus, results in a deep, narrow tree.

Random first search. As its name suggests, this method randomly selects the next node into the sample from the queue.

Snowball sampling. Snowball sampling (Goodman, 1961) is a restricted version of the breadth first search approach, where we only add a limited number (k) of neighbors to the *queue* for each sampled node.

Forest fire. Forest fire can be thought of as a stochastic version of snowball sampling, where the restriction on the number of neighbors visited only holds in expectation (Leskovec, Kleinberg & Faloutsos, 2005). Instead of selecting k neighbors at each iteration, forest fire selects $K \sim Geometric(p)$ of them at each step. Notice that if $p = \frac{1}{k}$ then it holds that $E[K] = k$.

Pseudo Code for Search-Based Sampling Algorithms.

Algorithm 1 presents the pseudo code for the basic search-based algorithms described in this section.

Algorithm 1: Search based sampling algorithms

Data: Graph G, initial node s, budget B and cost b
Result: G_s: Sampled graph
$V_s \leftarrow \emptyset$
$Q \leftarrow List(s)$
while $|Q| > 0$ *and* $B > 0$ **do**
    n = Q.RemoveFirst (BFS) / n = Q.RemoveLast (DFS) / n = Q.RemoveRandom (RFS) $V_s \leftarrow V_s \cup n$ $B \leftarrow B - b$ **for** *v in neighbors(n)* **do**
        if $v \notin Q$ *and* $v \notin V_s$ **then**
            Q.Append(v);
        end
    end
end
$G_s \leftarrow$ G.Induce(V_s)

Community structure expansion. Starting from a randomly selected node, this algorithm adds new nodes to the sample based on their expansion factor (Maiya & Berger-Wolf, 2010). Let us denote the set of sampled nodes by V_s. At each iteration, we calculate $|N(\{v\}) - (N(V_s) \cup V_s|$, the expansion factor for each node $v \in N(V_s)$, where $N(v)$ is the neighbor set of node v and $N(V_s)$ is the union of the neighbors of all the nodes already in the sample. Then, we select the node with the largest expansion factor into the sample. The process goes on until the desired number of sampled nodes is reached. The main intuition behind the algorithm is that at each iteration the sample is extended by one node. The algorithm selects the node which reaches the largest number of nodes that are not in the immediate neighborhood of the sampled nodes yet. This method is known to provide samples better representing different communities in the underlying graph (Maiya & Berger-Wolf, 2010). This is because nodes acting as bridges between different communities will have larger expansion factors and are, thus, more likely to be selected into the sample than members of communities that already have sampled members.

Shortest path sampling. In shortest path sampling (Rezvanian & Meybodi, 2015), one chooses non-incident node pairs and adds a randomly selected shortest path between these nodes to the sample at each iteration. This continues until the desired sample size is reached. Finally, edges between the sampled nodes are induced.

6.5.3.2 Random Walk-Based Techniques

Random walk-based techniques start from a seed node and traverse the graph inducing a sample between visited nodes. The basic random walk approach has many shortcomings addressed by the extended algorithms presented next.

Random walk sampler. The most straightforward random walk sampling technique (Gjoka, Kurant, Butts & Markopoulou, 2010) starts from a single seed node s and adds nodes to the sample by walking through the graph's edges randomly until a pre-defined sample size is reached.

Rejection-constrained Metropolis-Hastings random walk. By construction, higher-degree nodes are more likely to be included in the sample obtained by a basic random walk. This can cause problems in many settings (especially social network analysis) that require a representative sample degree distribution. The Metropolis-Hastings random walk algorithm (Hübler, Kriegel, Borgwardt & Ghahramani, 2008; Stutzbach, Rejaie, Duffield, Sen & Willinger, 2008; R.-H. Li, Yu, Qin, Mao & Jin, 2015) addresses this problem by making the walker more likely to select lower-degree nodes into the sample. The degree to which lower-degree nodes are preferred can be set by choosing a single rejection constraint parameter α. A random neighbor (w) of the most recently added node (v) is selected at each iteration. We generate a random uniform number $\gamma \sim U[0,1]$. If $\gamma < \left(\frac{|N(v)|}{|N(w)|}\right)^{\alpha}$ then node w is added to the sample, otherwise, it is rejected and the iteration is repeated. The process continues until a pre-determined number of nodes is added to the sample. Notice that a lower degree makes it more likely that a node is accepted into the sample. Increasing α increases the probability that high degree nodes are rejected.

Non-backtracking random walk. A further known issue of the traditional random walk sampler is that it is likely to get stuck in densely connected, small communities. There are multiple solutions to overcome this problem. Non-backtracking random walks restrict the traditional random walk approach so that the walker cannot go to the node it came from. More formally, a walker currently at node j, immediately before at node i, will move to the next node $k \in N(j) \setminus \{i\}$ if $d(j) \geq 2$ randomly. If i is the only neighbor of j, the walker is allowed to backtrack. The walk continues until the desired sample size is reached. It has been shown that this approach greatly reduces the bias in the estimated degree distribution compared to traditional random walks (C.-H. Lee, Xu & Eun, 2012).

Random walk with jumps. An alternative to the non-backtracking approach is random walk with jumps, where – with a given probability – the next node might be selected randomly from the whole node set instead of traversing to one of the neighboring nodes. How this probability is determined can be specific to the application. Ribeiro, Wang, Murai and Towsley (2012) suggest a degree-proportional probability $\frac{w}{w+d(v)}$, where w is a parameter of our choice and $d(v)$ is the degree of the last node added to the sample. Ribeiro et al. (2012) also propose a variety of asymptotically unbiased estimators for samples obtained using this sampling method.

Frontier of random walkers. With frontier sampling, one starts by selecting m random walk seed nodes. Then, at each iteration, a traditional random walk step is implemented in one of the m random walks chosen randomly (Ribeiro & Towsley, 2010). The process is repeated until the desired sample size is reached. This algorithm has been shown to outperform a wide selection of other traversal-based methods in estimating the degree correlation on multiple real-life networks (Rozemberczki et al., 2020).

6.6 Applications of Machine Learning in the Econometrics of Networks

This section presents three applications in which machine learning methods complement traditional econometric estimation techniques with network data. Importantly, we assume that the network data is readily available in these applications and focus on estimating economic/econometric models.

First, we discuss spatial models and how machine learning can help researchers learn the spatial weight matrix from the data. Second, we show how specific machine learning methods can achieve higher prediction accuracy in flow prediction, which is a commonly studied topic in spatial econometrics. Third, we present an example where the econometric model of geographically weighted regression has been utilized to improve the performance of machine learning models. These cases show that econometrics and machine learning methods do not replace but complement each other.

6.6.1 Applications of Machine Learning in Spatial Models

Spatial models have been commonly studied in the literature. In international trade, modeling spatial dependencies is utterly important as geographic proximity has a significant impact on economic outcomes (e.g., Dell, 2015, Donaldson, 2018, and Faber, 2014). In epidemiology, spatial connectedness is a major predictor of the spread of an infectious disease in a population (e.g., Rozemberczki, Scherer, Kiss, Sarkar & Ferenci, 2021). Even in social interactions, geographic structure largely determines whom people interact with and how frequently they do so (e.g., Breza & Chandrasekhar, 2019, Feigenberg, Field & Pande, 2013, and Jackson, Rodriguez-Barraquer & Tan, 2012). The following discussion highlights some of the workhorse models of spatial econometrics and presents an application where machine learning can augment econometric estimations methods. Specifically, we discuss how spatial dependencies can be learned using two distinct machine learning techniques.

Representing spatial autocorrelation in econometrics. Understanding spatial autocorrelation is essential for applications that use spatial networks. Spatial autocorrelation is present whenever an economic variable at a given location is correlated with the values of the same variable at other places. Then, there is a spatial interaction between the outcomes at different locations. When spatial autocorrelation is present, standard econometric techniques often fail, and econometricians must account for such dependencies (Anselin, 2003).

Econometricians commonly specify a particular functional form that generates the spatial stochastic process to model spatial dependencies, which relates the values of random variables at different locations. This, in turn, directly determines the spatial covariance structure. Most papers in the literature apply a non-stochastic and exogenous spatial weight matrix, which needs to be specified by the researcher.

This is practically an adjacency matrix of a weighted network used in the spatial domain. Generally, the weight matrix represents the geographic arrangement and distances in spatial econometrics. For example, the adjacency matrix may correspond to the inverse of the distance between locations. Alternatively, the elements may take the value of one when two areas share a common boundary and zero otherwise. Other specifications for the entries of the spatial weight matrix include economic distance, road connections, relative GDP, and trade volume between the two locations. However, researchers should be aware that the weight matrix elements may be outcomes themselves when using these alternative measures. This can be correlated with the final outcome causing endogeneity issues in the estimation.

Spatial dependencies can also be determined by directly modeling the covariance structure using a few parameters or estimating the covariances non-parametrically. As these methods are less frequently used, the discussion of such techniques is out of the scope of this chapter. For further details, see Anselin (2003).

The benchmark model in spatial econometrics. The benchmark spatial autocorrelation model was first presented by Manski (1993). This model does not directly control for any network characteristics but assumes that outcomes are affected by own and neighboring agents' characteristics and actions. It is specified as follows:

$$\begin{aligned} y &= \rho \mathbf{A} y + \mathbf{X}\beta + \mathbf{A}\mathbf{X}\theta + u \\ u &= \lambda \mathbf{A} u + \epsilon, \end{aligned} \tag{6.8}$$

where y is the vector of the outcome variable, $\mathbf{A}$ is the spatial weight matrix, $\mathbf{X}$ is the matrix of exogenous explanatory variables, u is a spatially autocorrelated error term, and ϵ is a random error. The model incorporates three types of interactions:

- an endogenous interaction, where the economic agent's action depends on its neighbors' actions;
- an exogenous interaction, where the agent's action depends on its neighbors' observable characteristics;
- a spatial autocorrelation term, driven by correlated unobservable characteristics.

This model is not identifiable, as shown by Manski (1993). However, if we constrain one of the parameters ρ, θ, or λ to be zero, as proposed by Manski (1993), the remaining ones can be estimated. When $\rho = 0$, the model is called the spatial Durbin error model; when $\theta = 0$, it is a spatial autoregressive confused model, and when both $\rho = 0$ and $\theta = 0$ are assumed, it is referenced as the spatial error model. These are rarely used in the literature. Instead, most researchers assume that $\lambda = 0$, which results in the spatial Durbin model, or $\theta = \lambda = 0$, which is the spatial autoregressive model. As the spatial autoregressive model is the most frequently used in the literature, our discussion follows this specification.

While this section focuses on geographical connectedness, it is important to note that the benchmark model is also applicable to problems other than spatial interactions. The adjacency matrices do not necessarily have to reflect geographical relations. For example, the model used for peer effects – described in Section 6.3 – takes the form of the spatial Durbin model.

The spatial autoregressive model (SAR). The SAR model can be specified as follows:

$$y = \rho \mathbf{A} y + \mathbf{X}\beta + \epsilon, \tag{6.9}$$

where ρ is the spatial autoregressive parameter and y, $\mathbf{A}$, $\mathbf{X}$, and ϵ are as described in Equation (6.8). The weights of the spatial weight matrix are typically row standardized such that for all i, $\sum_j w_{ij} = 1$. To simplify the reduced form, let us define

$$\mathbf{S} \equiv \mathbf{I_N} - \rho \mathbf{A},$$

then Equation (6.9) can be expressed as

$$y = \mathbf{S}^{-1}\mathbf{X}\beta + \mathbf{S}^{-1}\epsilon.$$

Notice that the right-hand side only contains the exogenous node characteristics, the adjacency matrix, and the error term. The estimation of this model has been widely studied in the literature, including the maximum likelihood estimation (Ord, 1975), the instrumental variable method (Anselin, 1980), and the generalized method of moments (L.-F. Lee, 2007, X. Lin & Lee, 2010 and Liu, Lee & Bollinger, 2010).

In some applications, a significant drawback of estimating the SAR model with the standard econometric techniques is that they assume a non-stochastic adjacency matrix, $\mathbf{A}$, determined by the researcher. When spatial weights involve endogenous socioeconomic variables, the elements of the adjacency matrix are likely to be correlated with the outcome variable. To illustrate this phenomenon, consider the regression equation where the outcome variable is GDP, and the elements of the adjacency matrix are trade weights as a share of total trade. Then, the unobservables that affect the outcome may also be correlated with the weights. As implied by the gravity model, trade flows can be affected by the unobservable multilateral resistance, which can be correlated with unobservables in the regression equation (Qu, fei Lee & Yang, 2021). In such cases, the adjacency matrix is not exogenous, and estimators that assume the opposite lead to biased estimates (Qu et al., 2021). As Pinkse and Slade (2010) point out, the endogeneity of spatial weights is a challenging problem, and there is research to be done in this direction of spatial econometrics.

There have been some efforts in the econometrics literature to make these dependencies endogenous to the estimation. Using a control function approach, Qu and fei Lee (2015) developed an estimation method for the case when the entries of the adjacency matrix are functions of unilateral economic variables, $a_{ij} = h(z_i, z_j)$. In their model, the source of endogeneity is the correlation between the error term in the regression equation for entries of the spatial weight matrix and the error term in the SAR model. Qu et al. (2021) extend this model with a_{ij} being determined by bilateral variables, such as trade flows between countries.

Even when the assumption of the adjacency matrix being non-stochastic is valid, and we can define simple connections for the spatial weight matrix, in some applications, it is more difficult to measure the strength of connectivity accurately. For example, spatial connectedness does not necessarily imply that neighboring locations interact when using geographic distance in the adjacency matrix. Furthermore, the interaction strength may not be commensurate with the distance between two zones. A related issue is when there are multiple types of interactions, such as geographic distance or number of road connections, and the choice of the specification for the spatial weight matrix is not apparent. Furthermore, defining the strength of connections becomes very difficult when the size of the data is large. Determining the spatial weight structure of dozens of observations is challenging, if not impossible.

In the literature, Bhattacharjee and Jensen-Butler (2013) propose an econometric approach to identify the spatial weight matrix from the data. The authors show that the spatial weight matrix is fully identified under the structural constraint of symmetric spatial weights in the spatial error model. The authors propose a method to estimate the elements of the spatial weight matrix under symmetry and extend their approach to the SAR model. Ahrens and Bhattacharjee (2015) propose a two-step LASSO estimator for the spatial weight matrix in the SAR model, which relies on the identifying assumptions that the weight matrix is sparse. Lam and Souza (2020) estimate the optimal spatial weight matrix by obtaining the best linear combination of different linkages and a sparse adjustment matrix, incorporating errors of misspecification. The authors use the adaptive LASSO selection method to select which specified spatial weight matrices to include in the linear combination. When no spatial weight matrix is specified, the method reduces to estimating a sparse spatial weight matrix.

Learning spatial dependencies with machine learning. The machine learning literature has focused on developing models for spatiotemporal forecasting, which can learn spatial dependencies directly from the data. As argued by Rozemberczki, Scherer, He et al. (2021), neural network-based models have been able to best capture spatial interactions. There have been various neural network infrastructures proposed that can capture spatial autocorrelation. While these methods do not provide a straightforward solution to the endogeneity issue, they can certainly be used for estimating spatial dependencies. In this chapter, we discuss two of them.

To use the SAR model for spatiotemporal forecasting, machine learning researchers have incorporated a temporal lag in Equation (6.9). Furthermore, a regularization term must be added to the estimation equation so that the spatial autoregressive parameter, ρ, and the spatial weight matrix, $\mathbf{A}$, can be identified simultaneously. This yields the following model:

$$\mathbf{y}_{t+1} = \rho\mathbf{A}\mathbf{y}_t + \mathbf{X}_{t+1}\beta + \gamma|\mathbf{A}| + u, \tag{6.10}$$

where $|\mathbf{A}|$ is a l_1 regularizing term and γ is a tuning parameter set by the researcher. In this model, controlling for $|\mathbf{A}|$ makes the spatial weight matrix more sparse and also helps identify ρ separately from $\mathbf{A}$. To the best of the authors' knowledge, there are no econometric papers in the literature that discuss this specification.

Learning the spatial weight matrix with recurrent neural networks. Ziat, Delasalles, Denoyer and Gallinari (2017) formalize a recurrent neural network (RNN) architecture for forecasting time series of spatial processes. The model denoted spatiotemporal neural network (STNN) learns spatial dependencies through a structured latent dynamical component. Then, a decoder predicts the actual values from the latent representations. The main idea behind recurrent neural networks is discussed further in Chapter 4.

A brief overview of the model is as follows: Assume there are n temporal series with length T, stacked in $\mathbf{X} \in \mathbb{R}^{T \times n}$. First, we assume that a spatial weight matrix, $\mathbf{A} \in \mathbb{R}^{n \times n}$, is provided. Later, this assumption is relaxed. The model predicts the series τ time-steps ahead based on the input variables and the adjacency matrix. The first component of the model captures the process's dynamic and is expressed in a latent space. Assume that each series has a latent space representation in each time period. Then, the matrix of the latent factors can be denoted by $\mathbf{Z_t} \in \mathbb{R}^{n \times N}$, where N is the dimension of the latent space. The latent representation at time $t+1$, $\mathbf{Z_{t+1}}$, depends on its own latent representation at time t (intra-dependency), and on the latent representation of the neighboring time series at time t (inter-dependency). The dynamical component is expressed as

$$\mathbf{Z_{t+1}} = h(\mathbf{Z_t}\mathbf{\Theta}^{(0)} + \mathbf{A}\mathbf{Z_t}\mathbf{\Theta}^{(1)}), \tag{6.11}$$

where $\mathbf{\Theta}^{(0)} \in \mathbb{R}^{N \times N}$ and $\mathbf{\Theta}^{(1)} \in \mathbb{R}^{N \times N}$ are the parameter matrices to be estimated and $h(.)$ is a non-linear function.

This specification is different from standard RNN models, where the hidden state $\mathbf{Z_t}$ is not only a function of the preceding hidden state $\mathbf{Z_{t-1}}$ but also of the ground truth values $\mathbf{X_{t-1}}$. With this approach, the dynamic of the series is captured entirely in the latent space. Therefore, spatial dependencies can be modeled explicitly in the latent factors.

The second component decodes the latent states into a prediction of the series and is written as $\tilde{X}_t = d(\mathbf{Z_t})$ at time t, where $\tilde{X}_t$ is the prediction computed at time t. Importantly, the latent representations and the parameters of both the dynamic transition function $h(.)$ and the decoder function $d(.)$ can be learned from the data assuming they are differentiable parametric functions. In the paper, the authors use $h(.) = tanh(.)$ and $d(.)$ is a linear function but more complex functions can also be used.

Then, the learning problem, which captures the dynamic latent space component and the decoder component can be expressed as

$$\begin{aligned} d^*, \mathbf{Z}^*, \mathbf{\Theta}^{(0)*}, \mathbf{\Theta}^{(1)*} = & \operatorname*{arg\,min}_{d, \mathbf{Z}, \mathbf{\Theta}^{(0)}, \mathbf{\Theta}^{(1)}} \frac{1}{T} \sum_t \Delta(d(\mathbf{Z_t}), X_t) + \\ & \lambda \frac{1}{T} \sum_{t=1}^{T-1} ||\mathbf{Z_{t+1}} - h(\mathbf{Z_t}\mathbf{\Theta}^{(0)} + \mathbf{A}\mathbf{Z_t}\mathbf{\Theta}^{(1)})||^2, \end{aligned} \tag{6.12}$$

where Δ is a loss function and λ is a hyperparameter set by cross-validation.

The first term measures the proximity of the predictions $d(\mathbf{Z_t})$ and the observed values X_t, while the second term captures the latent space dynamics of the series. The latter term takes its minimum when $\mathbf{Z_{t+1}}$ and $h(\mathbf{Z_t})$ are as close as possible. The learning problem can be solved with a stochastic gradient descent algorithm.

To incorporate the learning of the spatial weight matrix, Equation (6.11) can be modified as follows:

$$\mathbf{Z_{t+1}} = h(\mathbf{Z_t}\mathbf{\Theta}^{(0)} + (\mathbf{A} \odot \mathbf{\Gamma})\mathbf{Z_t}\mathbf{\Theta}^{(1)}), \tag{6.13}$$

where $\mathbf{\Gamma} \in \mathbb{R}^{n \times n}$ is a matrix to be learned, $\mathbf{A}$ is a pre-defined set of observed relations, and $\odot$ is the element-wise multiplication between two matrices. Here, $\mathbf{A}$ can be a simple adjacency matrix where elements may represent connections, proximity, distance, etc. Then, the model learns the optimal weight of mutual influence between the connected sources. In the paper, this model is denoted STNN-R(efining).

Then, the optimization problem over $d, \mathbf{Z}, \mathbf{\Theta}^{(0)}, \mathbf{\Theta}^{(1)}, \mathbf{\Gamma}$ can be expressed as

$$\begin{aligned} d^*, \mathbf{Z}^*, \mathbf{\Theta}^{(0)*}, \mathbf{\Theta}^{(1)*}, \mathbf{\Gamma}^* = \underset{d, \mathbf{Z}, \mathbf{\Theta}^{(0)}, \mathbf{\Theta}^{(1)}, \mathbf{\Gamma}}{\arg\min} \; & \frac{1}{T} \sum_t \Delta(d(\mathbf{Z_t}), X_t) + \gamma |\mathbf{\Gamma}| + \\ & \lambda \frac{1}{T} \sum_{t=1}^{T-1} ||\mathbf{Z_{t+1}} - h(\mathbf{Z_t}\mathbf{\Theta}^{(0)} + (\mathbf{A} \odot \mathbf{\Gamma})\mathbf{Z_t}\mathbf{\Theta}^{(1)})||^2, \end{aligned} \tag{6.14}$$

where $|\mathbf{\Gamma}|$ is a l_1 regularizing term, and γ is a hyper-parameter for tuning the regularization.

If no prior is available, then removing $\mathbf{A}$ from Equation (6.13) gives

$$\mathbf{Z_{t+1}} = h(\mathbf{Z_t}\mathbf{\Theta}^{(0)} + \mathbf{\Gamma}\mathbf{Z_t}\mathbf{\Theta}^{(1)}), \tag{6.15}$$

where $\mathbf{\Gamma}$ represents both the relational structure and the relational weights. This version of the model is named STNN-D(iscovery), and can be estimated by replacing the dynamic transition function, $h(.)$ in Equation (6.14) with Equation (6.15). This specification is the most similar to Equation (6.10) in the latent space. An extension of the model to multiple relations and further specifications for the experiments can be found in Ziat et al. (2017).

The model was evaluated on different forecasting problems, such as wind speed, disease, and car-traffic prediction. The paper shows that the STNN and STNN-R perform superior to the Vector Autoregressive Model (VAR) and various neural network architectures, such as recurrent neural networks and dynamic factor graphs (DFG). However, STNN-D, which does not use any prior information on proximity, performs worse than STNN and STNN-R in all use cases. The authors also describe experiments showing the ability of this approach to extract relevant spatial relations. In particular, when estimating the STTN-D model (when the spatial organization of the series is not provided), the model can still learn the spatial proximity by assigning a strong correlation to neighboring observations. These correlations are reflected in the estimated $\mathbf{\Gamma}$ parameter in Equation (6.15).

Learning the spatial autocorrelation with convolutional neural networks. Spatial autocorrelation may be learned with convolutional layers in a deep neural network architecture. The basics of deep neural networks and convolution neural networks (CNNs) are described in Chapter 4. As pointed out by Dewan, Ganti, Srivatsa and

Stein (2019), the main idea for using convolutional layers comes from the fact that the SAR model, presented in Equation (6.9), can also be formulated as given in Equation (6.16), under the assumption that $||\rho \mathbf{A}|| < 1$.

$$y = \sum_{i=0}^{\infty} \rho^i \mathbf{A}^i (\beta \mathbf{X} + \epsilon). \tag{6.16}$$

This mutatis mutandis reflects the invertibility of AR and MA processes. The expression in Equation (6.16) intuitively means that the explanatory variables and shocks at any location affect all other directly or indirectly connected locations. Furthermore, these spatial effects decrease in magnitude with distance from the location in the network. Using neural networks, this learning can be approximated with convolution filters of different sizes. As described in Chapter 4, convolution filters can extract features of the underlying data, including spatial dependencies between pairs of locations.

Using convolutional layers, Dewan et al. (2019) present a novel Convolutional AutoEncoder (CAE) model, called the NN-SAR, that can learn the spatiotemporal structure for prediction. In the model, convolutional layers capture the spatial dependencies, and the autoencoder retains the most important input variables for prediction. Autoencoders are discussed in detail in Chapter 4.

Their modeling pipeline follows two steps. For convolutional neural networks, it is necessary to represent the input data as images. Geographical data can be transformed into images using geohashing, which creates a rectangular grid of all locations while keeping the spatial locality of observations. Therefore, images representing the input and output variables can be constructed for each time period. Then, these images can be used for time series forecasting using neural networks where the historical images are used as inputs, and the output is a single image in a given time period. Time series forecasting using machine learning is further discussed in Chapter 4.

Second, the authors built a deep learning pipeline for predicting the output image. It consists of an encoder that applies convolutional and max pool layers with Rectifier Linear Unit (ReLU) activations and a decoder that includes convolutional and deconvolutional layers. The encoder aims to obtain a compressed representation of the input variables, while the decoder transforms this representation into the output image. Additionally, the network consists of skip connections to preserve information that might have been lost in the encoding process. Their model is similar to the well-known U-Net architecture by Ronneberger, Fischer and Brox (2015), commonly used for image segmentation problems.

In Dewan et al. (2019), the authors used data over a large spatial region and time range to predict missing spatial and temporal values of sensor data. In particular, they used data on particulate matter, an indicator of air pollution levels, but their method can also be applied to other spatiotemporal variables. Their results indicate that the NN-SAR approach is 20% superior on average to the SAR models in predicting outcomes when the sample size is large.

Dewan et al. (2019) and Ziat et al. (2017) provide examples of how different neural network architectures can be used to learn spatial dependencies from the data.

While understanding the spatial correlation structure from Dewan et al. (2019) is more difficult, Ziat et al. (2017) provide a method for explicitly estimating the spatial weight matrix from the data. As these papers concentrate on forecasting, further research needs to be conducted on their applicability in cross-sectional settings.

6.6.2 Gravity Models for Flow Prediction

In this section, we focus on a specific domain of spatial models, gravity models. Several economic studies use gravity models for flow projections, including the World Economic Outlook by the International Monetary Fund (International Monetary Fund, 2022). While machine learning methods have not gained widespread use in such studies, many papers discuss how machine learning models have achieved higher prediction accuracy. This section provides a rationale for using such methods in economic studies. First, we briefly discuss the econometric formulation of gravity models. Second, we summarize results from the machine learning literature regarding mobility flow prediction using gravity models.

Gravity models have been the workhorse of trade and labor economics as they can quantify the determinants that affect trade or migration flows between geographic areas. In gravity models, the adjacency matrix can be treated as if it contained the reciprocals of distances. A higher weight in the adjacency matrix corresponds to a lower distance in the gravity model. The general formulation of the gravity model is expressed as

$$Y_{ijt} = g(X_{it}, X_{jt}, i, j, t),$$

where Y_{ijt} is the bilateral outcome between country i and country j at time t (response variable), X_{it} and X_{jt} are the sets of possible predictors from both countries, and the set $\{i, j, t\}$ refers to a variety of controls on all three dimensions (Anderson & van Wincoop, 2001). With some simplification, in most specifications, the gravity models assume that the flows between two locations increase with the population of locations but decrease with the distance between them.

Matyas (1997) introduced the most widely used specification of the gravity model in trade economics:

$$lnEXP_{ijt} = \alpha_i + \gamma_j + \lambda_t + \beta_1 lnY_{it} + \beta_2 lnY_{jt} + \beta_3 DIST_{ij} + \ldots + u_{ijt},$$

where EXP_{ijt} is the volume of trade (exports) from country i to country j at time t; Y_{it} is the GDP in country i at time t, and the same for Y_{jt} for country j; $DIST_{ij}$ is the distance between the countries i and j; α_i and γ_j are the origin and target country fixed effects respectively; λ_t is the time (business cycle) effect; and u_{ijt} is a white noise disturbance term.

The model specified can be estimated by OLS when we include a constant term and set the identifying restrictions: $\sum \alpha_i = 1$, $\sum \gamma_j = 1$ and $\sum \lambda_t = 1$. The model can also be estimated with several other techniques, including the penalized regression (discussed in Chapter 1 and empirically tested by H. Lin et al., 2019).

Gravity models have also been studied in the machine learning literature. Machine learning methods can capture non-linear relationships between the explanatory variables and trade/mobility flows, which may better characterize the underlying structure. These models are also more flexible and capable of generating more realistic trade and mobility flows. The literature on the topic has mainly focused on comparing standard econometric methods with neural networks. The concept of neural networks and their mathematical formulation are discussed in Chapter 4.

Fischer and Gopal (1994), Gopal and Fischer (1996), and Fischer (1998) compare how accurately gravity models and neural networks can predict inter-regional telecommunications flows in Austria. To evaluate the models' performance, the authors use two different measures – the average relative variance (ARV) and the coefficient of determination (R^2) – and also perform residual analysis. The papers show that the neural network model approach achieves higher predictive accuracy than the classical regression. In Fischer (1998), the neural network-based approach achieved 14% lower ARV and 10% higher R^2, stable across different trials.

In a different domain, Tillema, Zuilekom and van Maarseveen (2006) compare the performance of neural networks and gravity models in trip distribution modeling. The most commonly used methods in the literature to model trip distributions between origins and destinations have been gravity models – they are the basis for estimating trip distribution in a four-step model of transportation (McNally, 2000). The authors use synthetic and real-world data to perform statistical analyses, which help determine the necessary sample sizes to obtain statistically significant results. The results show that neural networks attain higher prediction accuracy than gravity models with small sample sizes, irrespective of using synthesized or real-world data. Furthermore, the authors establish that the necessary sample size for statistically significant results is forty times lower for neural networks.

Pourebrahim, Sultana, Thill and Mohanty (2018) also study the predictive performance of neural networks and gravity models when forecasting trip distribution. They contribute to the literature by utilizing social media data – the number of tweets posted in origin and destination locations – besides using standard input variables such as employment, population, and distance. Their goal is to predict commuter trip distribution on a small spatial scale – commuting within cities – using geolocated Twitter post data. Gravity models have traditionally used population and distance as key predicting factors. However, mobility patterns within cities may not be predicted reliably without additional predictive factors. The paper compares the performance of neural networks and gravity models, both trying to predict home-work flows using the same data set. The results suggest that social media data can improve the modeling of commuter trip distribution and is a step forward to developing dynamic models besides using static socioeconomic factors. Furthermore, standard gravity models are outperformed by neural networks in terms of R^2. This indicates that neural networks better fit the data and are superior for prediction in this application.

Simini, Barlacchi, Luca and Pappalardo (2020) propose a neural network-based deep gravity model, which predicts mobility flows using geographic data. Specifically, the authors estimate mobility flows between regions with input variables such as areas of different land use classes, length of road networks, and the number of health and education facilities. In brief, the deep gravity model uses these input features to compute the probability $p_{i,j}$ that a trip originated at location l_i has a destination location l_j for all possible locations in the region. The authors use the common part of commuters (CPC) evaluation metric to evaluate the model, which computes the similarity between actual and generated flows. Their results suggest that neural network-based models can predict mobility flows significantly better than traditional gravity models, shallow neural network models, and models that do not use extensive geographic data. In areas with high population density, where prediction is more difficult due to the many locations, the machine learning-based approach outperforms the gravity model by 350% in CPC. Finally, the authors also show that the deep gravity model generalizes well to other locations not used in training. This is achieved by the model being tested to predict flows in a region non-overlapping with the training regions.

The results from the machine learning literature provide strong evidence that neural network-based approaches are better suited for flow prediction than standard econometric techniques. Therefore, such techniques should be part of applied economists' toolbox when incorporating flow projections in economic studies.

6.6.3 The Geographically Weighted Regression Model and ML

This section presents the econometric method of geographically weighted regression (GWR). Our discussion closely follows Fotheringham, Brunsdon and Charlton (2002). Then, we discuss a paper from the machine learning literature that, by incorporating this method, improves the predictive performance of several machine learning techniques.

The geographically weighted regression is an estimation procedure that takes into account the spatial distance of the data points in the sample to estimate local variations of the regression coefficients (Brunsdon, Fotheringham & Charlton, 1996). Contrary to traditional estimation methods like the OLS, GWR applies local regressions on neighboring points for each observation in the sample. Therefore, it allows for estimating spatially varying coefficients and uncovers local features that the global approach cannot measure. If the local coefficients vary by the different estimations in space and move away from their global values, this may indicate that, for example, non-stationarity is present.

In GWR estimation, a separate regression is conducted for all data points. Then, the number of regression equations and estimates equals the number of territorial units. The sample consists of the regression point and the neighboring data points within a defined distance in each regression. Their distance from the regression point weights these data points – observations that are spatially closer to the regression point are assigned a larger weight (Brunsdon et al., 1996).

To estimate a geographically weighted regression model, researchers have to choose two parameters that govern local regressions. First, one must define the weighting scheme that handles how nearby observations are weighted in the local regression. Generally, data points closer to the regression point are assigned larger weights, continuously decreasing by distance. Second, researchers have to choose a kernel that may either be fixed, using a fixed radius set by the econometrician, or adaptive, using a changing radius. Applying an adaptive kernel is particularly useful if there are areas in which data points are more sparsely located. Such a kernel provides more robust estimates in areas with less density as its bandwidth changes with the number of neighboring data points. Notably, the estimated coefficients are particularly sensitive to the choice of kernel shape and bandwidth as they determine which points are included in the local regression.

Weighting scheme options for GWR. The simplest option for local weighting is to include only those observations in the local regression which are within a b radius of the examined point. This is called a uniform kernel with constrained support and can be expressed as

$$w_{ij} = 1 \text{ if } d_{ij} < b; w_{ij} = 0 \text{ otherwise}, \tag{6.17}$$

where i is the examined point in a local regression, d_{ij} is the distance of point j from point i in space, and b is the bandwidth chosen by the researcher.

However, this approach is very sensitive to the choice of bandwidth, and estimated parameters may significantly change if borderline observations are included or excluded. A more robust approach to this issue weights observations based on their Euclidean distance from the analyzed point and is called the Gaussian kernel:

$$w_{ij} = exp(-\frac{1}{2}(d_{ij}/b)^2),$$

where the parameters are as defined in Equation (6.17). Here, the weights are continuously decreasing as distance increases.

A combination of the two methods above is given by

$$w_{ij} = [1 - (d_{ij}/b)^2]^2 \text{ if } d_{ij} < b; w_{ij} = 0 \text{ otherwise}.$$

This formulation assigns decreasing weights until the bandwidth is reached and zero weight beyond.

Adaptive kernels consider the density around the regression point to estimate each local regression with the same sample size. When we order observations by their distance to the regression point, an adaptive kernel may be formulated as

$$w_{ij} = exp(-R_{ij}/b),$$

where R_{ij} is the rank number of point j from point i and b is the bandwidth. Then the weighting scheme disregards the actual distance between data points and is only a function of their ordering relative to each other.

Finally, we may also define the weight by taking into account the N number of nearest neighbors:

$$w_{ij} = [1 - (d_{ij}/b)^2]^2 \text{ if } j \text{ is one of the } N \text{ nearest neighbors of } i,$$
$$\text{and } b \text{ is the distance of the N-th nearest neighbor,}$$
$$w_{ij} = 0 \text{ otherwise.}$$

where N is a parameter and d_{ij} is as defined in Equation (6.17). Therefore, a fixed number of nearest neighbors is included in the regression. The parameter N needs to be defined or calibrated for this weighting scheme.

Finding the optimal bandwidth for GWR. The optimal bandwidth can be calibrated using cross validation (CV). One commonly used method is to minimize the following function with respect to the bandwidth b:

$$CV = \sum_{i=1}^{n} [y_i - \hat{y}_{j \neq i}(b)]^2,$$

where b is the bandwidth and $\hat{y}_{j \neq i}$ is the estimated value of the dependent variable if point i is left out of the calibration. This condition is necessary to avoid zero optimal bandwidth (Fotheringham et al., 2002).

Other approaches to finding the optimal b parameter include minimizing the Akaike information criterion (AIC) or the Schwarz criterion (SC). Such metrics also penalize the model for having many explanatory variables, which helps reduce overfitting. As discussed in Chapters 1 and 2, these criteria can also be used for model selection in Ridge, LASSO, and Elastic Net models.

Estimating the GWR model. Once the sample has been constructed for each local regression, the GWR specification can be estimated. The equation of the geographically weighted regression is expressed as

$$y_i = \beta_0(u_i, v_i) + \sum_k \beta_k(u_i, v_i) x_{ik} + \epsilon_i,$$

where (u_i, v_i) is the geographical coordinates of point i and $\beta_k(u_i, v_i)$ is the estimated value of the continuous function $\beta_k(u, v)$ in point i (Fotheringham et al., 2002).

With the weighted least squares method, the following solution can be obtained:

$$\hat{\beta}(u_i, v_i) = (\mathbf{X}^T \mathbf{W}(u_i, v_i) \mathbf{X})^{-1} \mathbf{X}^T \mathbf{W}(u_i, v_i) y, \tag{6.18}$$

where y is the output vector and $\mathbf{X}$ is the input matrix of all examples. Furthermore, $\mathbf{W}(u_i, v_i)$ is the weight matrix of the target location, (u_i, v_i), whose off-diagonal elements are zero and diagonal elements measure the spatial weights defined by the weighting scheme.

The geographically weighted machine learning model. L. Li (2019) combines the geographically weighted regression model (GWR) with neural networks, XGBoost, and random forest classifiers to improve high-resolution spatiotemporal wind speed predictions in China. Due to the size of the country and the heterogeneity between its regions, modeling meteorological factors at high spatial resolution is a particularly challenging problem. By incorporating the GWR method, which captures local variability, the authors show that prediction accuracy can be improved compared to the performance of the base learners.

The outcome variable in the paper is daily wind speed data collected from several wind speed monitoring stations across the country. The covariates in the estimation included coordinates, elevation, day of the year, and region. Furthermore, the author also utilized coarse spatial resolution reanalysis data (i.e., climate data), which provides reliable estimates of wind speed and planetary boundary layer height (a factor for surface wind gradient and related to wind speed) at a larger scale.

The author developed a two-stage approach. They built a geographically weighted ensemble machine learning model in the first stage. They first trained three base learners, an autoencoder-based deep residual network, XGBoost, and random forest, predicting the outcome with the collected covariates. These models are based on three weakly correlated and fundamentally different algorithms. Therefore, their combination can provide better ensemble predictions in theory. The author combined the projections with the GWR method so that the model also incorporates spatial autocorrelation and heterogeneity for boosting forecasts. In brief, the GWR helps obtain the optimal weights of the three models. Furthermore, it provides spatially varying coefficients for the base learners, which allows for spatial autocorrelation and heterogeneity in the estimation.

In the GWR estimation, there are three explanatory variables (the predictions of the three base learners), and so the geographically weighted regression equation is written as

$$y_i = \beta_0(u_i, v_i) + \sum_{k=1}^{3} \beta_k(u_i, v_i)x_{ik} + \epsilon_i,$$

where (u_i, v_i) are the coordinates of the ith sample, $\beta_k(u_i, v_i)$ is the regression coefficient for the kth base prediction, x_{ik} is the predicted value by the kth base learner, and ϵ_i is random noise ($\epsilon_i \sim N(0, 1)$).

The Gaussian kernel was used to quantify the weight matrix for this study. These weights can be applied for the weighted least squares method in Equation (6.18):

$$w_{ij} = exp(-(d_{ij}/b)^2),$$

where b is the bandwidth and d_{ij} is the distance between locations i and j.

In the second stage, the author used a deep residual network to perform a downscaling that matches the wind speed from coarse resolution meteorological reanalysis data with the average predicted wind speed at high resolution inferred from the first stage. This reduces bias and yields more realistic spatial variation (smoothing) in the predictions. Furthermore, it also ensures that projections are consistent with the observed data at a more coarse resolution. The description of the method is out of the scope of this chapter and therefore left out of this discussion.

The geographically weighted regression ensemble achieved a 12–16% improvement in R^2 and lower Root Mean Square Error (RMSE) compared to individual learners' predictions. Therefore, GWR can effectively capture the local variation of the target variable at high-resolution and account for the spatial autocorrelation present in the data. Overall, the paper provides a clear example of how machine learning and econometrics can be combined to improve performance on challenging prediction problems.

6.7 Concluding Remarks

This chapter has focused on overcoming the challenges of analyzing network structured data. The high dimensionality of such data sets is one of the fundamental problems when working with them. Applied works relying on networks often solve this issue by including simple aggregations of the adjacency matrix in their analyses. In Section 6.4 we discuss an alternative to this approach, namely graph dimensionality reduction. While these algorithms have a proven track record in the computer science literature, they have received limited attention from economists and econometricians. Analyzing whether and how information extracted using these algorithms changes our knowledge about economic and social networks is a promising area for future research.

Next, in Section 6.5, we turned our attention towards questions related to sampling such data sets. Many modern-day networks (e.g., web graphs or social networks) have hundreds of millions of nodes and billions of edges. The size of these data sets renders the corresponding analysis computationally intractable. To deal with this issue, we discuss the best-known sampling methods presented in the literature. They can be used to preserve the salient information encoded in networks. Choosing the proper sampling method should always be driven by the analysis. However, there is no comprehensive study on the choice of the appropriate approach to the best of our knowledge. Theoretical properties of these methods are usually derived by assuming a specific network structure. However, these assumptions are unlikely to hold for real-life networks. Rozemberczki et al. (2020) present a study evaluating many sampling algorithms using empirical data. Their results show that there is no obvious choice, although some algorithms consistently outperform others in terms of preserving population statistics. The way the selection of the sampling method affects estimation outcomes is also a potential future research domain.

In Section 6.6, we present three research areas in spatial econometrics that have been analyzed in both the econometrics and the machine learning literature. We aim to show how machine learning can enhance econometric analysis related to network data. First, learning spatial autocorrelations from the data is essential when the strength of spatial interactions cannot be explicitly measured. As estimates of spatial econometrics models are sensitive to the choice of the spatial weight matrix, researchers may fail to recover the spatial covariance structure when such spatial interactions are not directly learned from the data. In the econometrics literature, estimating the spatial weight matrix relies on identifying assumptions, such as structural constraints or the sparsity of the weight matrix. This chapter shows advances in the machine learning literature that rely on less stringent assumptions to extract such information from the data. Incorporating these algorithms into econometric analyses has vast potential and remains an important avenue for research.

Second, forecasting in the spatial domain has been a widely studied area in economics. There are several studies by international institutions that use econometric methods to predict, for example, future trade patterns, flows of goods, and mobility. On the other hand, machine learning methods have not yet gained widespread recognition among economists who work on forecasting macroeconomic variables. We present papers that show how machine learning methods can achieve higher prediction accuracy. In forecasting problems, where establishing estimators' econometric properties is less important, machine learning algorithms often prove superior to standard econometric approaches. Finally, we have introduced the geographically weighted regression and presented an example where an econometric estimation technique can improve the forecasting performance of machine learning algorithms applied to spatiotemporal data. This empirical application suggests that spatially related economic variables predicted by machine learning models may be modeled with econometric techniques to improve prediction accuracy. The approaches presented in the last section suggest that combining machine learning and traditional econometrics has an untapped potential in forecasting variables relevant to policy makers.

References

Acemoglu, D., Carvalho, V. M., Ozdaglar, A. & Tahbaz-Salehi, A. (2012). The network origins of aggregate fluctuations. *Econometrica*, *80*(5), 1977–2016.

Adamic, L. A., Lukose, R. M., Puniyani, A. R. & Huberman, B. A. (2001). Search in power-law networks. *Phys. Rev. E*, *64*, 046135.

Ahmed, N. K., Neville, J. & Kompella, R. (2013). Network sampling: From static to streaming graphs. *ACM Transactions on Knowledge Discovery from Data (TKDD)*, *8*(2), 1–56.

Ahrens, A. & Bhattacharjee, A. (2015). Two-step lasso estimation of the spatial weights matrix. *Econometrics*, *3*(1), 1–28.

Anderson, J. E. & van Wincoop, E. (2001). *Gravity with gravitas: A solution to the border puzzle* (Working Paper No. 8079). National Bureau of Economic Research.

Anselin, L. (1980). *Estimation methods for spatial autoregressive structures: A study in spatial econometrics*. Program in Urban and Regional Studies, Cornell University.

Anselin, L. (2003). Spatial econometrics. In *A companion to theoretical econometrics* (pp. 310–330). John Wiley & Sons.

Balasubramanian, M. & Schwartz, E. L. (2002). The isomap algorithm and topological stability. *Science*, *295*(5552), 7–7.

Ballester, C., Calvo-Armengol, A. & Zenou, Y. (2006). Who's who in networks. wanted: The key player. *Econometrica*, *74*(5), 1403–1417.

Belkin, M. & Niyogi, P. (2002). Laplacian eigenmaps and spectral techniques for embedding and clustering. In *Advances in neural information processing systems* (pp. 585–591).

Bhattacharjee, A. & Jensen-Butler, C. (2013). Estimation of the spatial weights matrix under structural constraints. *Regional Science and Urban Economics, 43*(4), 617–634.

Breza, E. & Chandrasekhar, A. G. (2019). Social networks, reputation, and commitment: Evidence from a savings monitors experiment. *Econometrica, 87*(1), 175–216.

Brunsdon, C., Fotheringham, A. S. & Charlton, M. E. (1996). Geographically weighted regression: A method for exploring spatial nonstationarity. *Geographical Analysis, 28*(4), 281–298.

Cai, H., Zheng, V. W. & Chang, K. (2018). A comprehensive survey of graph embedding: Problems, techniques, and applications. *IEEE Transactions on Knowledge and Data Engineering, 30*(09), 1616–1637.

Chandrasekhar, A. G. & Lewis, R. (2016). *Econometrics of sampled networks.* (Retrieved: 02.02.2022 from https://stanford.edu/ arungc/CL.pdf)

Crépon, B., Devoto, F., Duflo, E. & Parienté, W. (2015). Estimating the impact of microcredit on those who take it up: Evidence from a randomized experiment in morocco. *American Economic Journal: Applied Economics, 7*(1), 123–50.

de Lara, N. & Edouard, P. (2018). A simple baseline algorithm for graph classification. *arXiv, abs/1810.09155.*

Dell, M. (2015). Trafficking networks and the mexican drug war. *American Economic Review, 105*(6), 1738–1779.

Dewan, P., Ganti, R., Srivatsa, M. & Stein, S. (2019). Nn-sar: A neural network approach for spatial autoregression. In *2019 ieee international conference on pervasive computing and communications workshops (percom workshops)* (pp. 783–789).

Doerr, C. & Blenn, N. (2013). Metric convergence in social network sampling. In *Proceedings of the 5th acm workshop on hotplanet* (p. 45–50). New York, NY, USA: Association for Computing Machinery.

Donaldson, D. (2018). Railroads of the raj: Estimating the impact of transportation infrastructure. *American Economic Review, 108*(4-5), 899–934.

Easley, D. & Kleinberg, J. (2010). *Networks, crowds, and markets* (Vol. 8). Cambridge university press Cambridge.

Faber, B. (2014). Trade integration, market size, and industrialization: Evidence from china's national trunk highway system. *The Review of Economic Studies, 81*(3), 1046–1070.

Feigenberg, B., Field, E. & Pande, R. (2013). The economic returns to social interaction: Experimental evidence from microfinance. *The Review of Economic Studies, 80*(4), 1459–1483.

Ferrali, R., Grossman, G., Platas, M. R. & Rodden, J. (2020). It takes a village: Peer effects and externalities in technology adoption. *American Journal of Political Science, 64*(3), 536–553.

Fischer, M. (1998). Computational neural networks: An attractive class of mathematical models for transportation research. In V. Himanen, P. Nijkamp, A. Reggiani & J. Raitio (Eds.), *Neural networks in transport applications* (pp. 3–20).

Fischer, M. & Gopal, S. (1994). Artificial neural networks: A new approach to modeling interregional telecommunication flows. *Journal of Regional Science*, *34*, 503 – 527.

Fotheringham, A., Brunsdon, C. & Charlton, M. (2002). *Geographically weighted regression: The analysis of spatially varying relationships*. John Wiley & Sons.

Gjoka, M., Kurant, M., Butts, C. T. & Markopoulou, A. (2010). Walking in facebook: A case study of unbiased sampling of osns. In *2010 proceedings ieee infocom* (pp. 1–9).

Gonzalez, J. E., Low, Y., Gu, H., Bickson, D. & Guestrin, C. (2012). Powergraph: Distributed graph-parallel computation on natural graphs. In *Presented as part of the 10th {USENIX} symposium on operating systems design and implementation ({OSDI} 12)* (pp. 17–30).

Goodman, L. A. (1961). Snowball sampling. *The annals of mathematical statistics*, 148–170.

Gopal, S. & Fischer, M. M. (1996). Learning in single hidden-layer feedforward network models: Backpropagation in a spatial interaction modeling context. *Geographical Analysis*, *28*(1), 38–55.

Grover, A. & Leskovec, J. (2016). node2vec: Scalable feature learning for networks. In *Proceedings of the 22nd acm sigkdd international conference on knowledge discovery and data mining* (pp. 855–864).

Hu, P. & Lau, W. C. (2013). A survey and taxonomy of graph sampling. *arXiv, abs/1308.5865*.

Hübler, C., Kriegel, H.-P., Borgwardt, K. & Ghahramani, Z. (2008). Metropolis algorithms for representative subgraph sampling. In *2008 eighth ieee international conference on data mining* (pp. 283–292).

International Monetary Fund. (2022). *World economic outlook, april 2022: War sets back the global recovery*. USA: International Monetary Fund.

Jackson, M. O. (2010). *Social and economic networks*. Princeton University Press.

Jackson, M. O., Rodriguez-Barraquer, T. & Tan, X. (2012). Social capital and social quilts: Network patterns of favor exchange. *American Economic Review*, *102*(5), 1857–97.

Kang, U., Tsourakakis, C. E. & Faloutsos, C. (2009). Pegasus: A peta-scale graph mining system implementation and observations. In *2009 ninth ieee international conference on data mining* (pp. 229–238).

Krishnamurthy, V., Faloutsos, M., Chrobak, M., Lao, L., Cui, J. H. & Percus, A. G. (2005). Reducing large internet topologies for faster simulations. In *Proceedings of the 4th ifip-tc6 international conference on networking technologies, services, and protocols; performance of computer and communication networks; mobile and wireless communication systems* (p. 328–341). Berlin, Heidelberg: Springer-Verlag.

Lam, C. & Souza, P. C. (2020). Estimation and selection of spatial weight matrix in a spatial lag model. *Journal of Business & Economic Statistics*, *38*(3), 693–710.

Lazarsfeld, P. F. & Merton, R. K. (1954). Friendship as a social process: a substantive and method-ological analysis. *Freedom and Control in Modern Society*, 18–66.

Lee, C.-H., Xu, X. & Eun, D. Y. (2012). Beyond random walk and metropolis-hastings samplers: why you should not backtrack for unbiased graph sampling. *ACM SIGMETRICS Performance evaluation review*, *40*(1), 319–330.

Lee, L.-F. (2007). Gmm and 2sls estimation of mixed regressive, spatial autoregressive models. *Journal of Econometrics*, *137*(2), 489–514.

Leskovec, J. & Faloutsos, C. (2006). Sampling from large graphs. In *Proceedings of the 12th acm sigkdd international conference on knowledge discovery and data mining* (pp. 631–636).

Leskovec, J., Kleinberg, J. & Faloutsos, C. (2005). Graphs over time: densification laws, shrinking diameters and possible explanations. In *Proceedings of the eleventh acm sigkdd international conference on knowledge discovery in data mining* (pp. 177–187).

Li, L. (2019). Geographically weighted machine learning and downscaling for high-resolution spatiotemporal estimations of wind speed. *Remote Sensing*, *11*(11).

Li, R.-H., Yu, J. X., Qin, L., Mao, R. & Jin, T. (2015). On random walk based graph sampling. In *2015 ieee 31st international conference on data engineering* (pp. 927–938).

Lin, H., Hong, H. G., Yang, B., Liu, W., Zhang, Y., Fan, G.-Z. & Li, Y. (2019). Nonparametric Time-Varying Coefficient Models for Panel Data. *Statistics in Biosciences*, *11*(3), 548–566.

Lin, X. & Lee, L.-F. (2010). Gmm estimation of spatial autoregressive models with unknown heteroskedasticity. *Journal of Econometrics*, *157*(1), 34–52.

Liu, X., Lee, L.-f. & Bollinger, C. R. (2010). An efficient GMM estimator of spatial autoregressive models. *Journal of Econometrics*, *159*(2), 303–319.

Maiya, A. S. & Berger-Wolf, T. Y. (2010). Sampling community structure. In *Proceedings of the 19th international conference on world wide web* (pp. 701–710).

Manski, C. F. (1993). Identification of endogenous social effects: The reflection problem. *The Review of Economic Studies*, *60*(3), 531–542.

Matyas, L. (1997). Proper econometric specification of the gravity model. *The World Economy*, *20*(3), 363-368.

McNally, M. (2000). The four step model. In D. Hensher & K. Button (Eds.), *Handbook of transport modelling* (Vol. 1, pp. 35–53). Elsevier.

Mikolov, T., Sutskever, I., Chen, K., Corrado, G. & Dean, J. (2013). *Distributed representations of words and phrases and their compositionality.*

Newman, M. E. J. (2002). Assortative mixing in networks. *Phys. Rev. Lett.*, *89*, 208701.

Ord, K. (1975). Estimation methods for models of spatial interaction. *Journal of the American Statistical Association*, *70*(349), 120–126.

Ortega, F. & Peri, G. (2013). The effect of income and immigration policies on international migration. *Migration Studies*, *1*(1), 47–74.

Page, L., Brin, S., Motwani, R. & Winograd, T. (1999). *The pagerank citation ranking: Bringing order to the web.* (Technical Report No. 1999-66). Stanford InfoLab. (Previous number = SIDL-WP-1999-0120)

Perozzi, B., Al-Rfou, R. & Skiena, S. (2014). Deepwalk: Online learning of social representations. In *Proceedings of the 20th acm sigkdd international conference on knowledge discovery and data mining* (pp. 701–710).

Pinkse, J. & Slade, M. E. (2010). The future of spatial econometrics. *Journal of Regional Science*, *50*(1), 103–117.

Pourebrahim, N., Sultana, S., Thill, J.-C. & Mohanty, S. (2018). Enhancing trip distribution prediction with twitter data: Comparison of neural network and gravity models. In *Proceedings of the 2nd acm sigspatial international workshop on ai for geographic knowledge discovery* (p. 5–8). New York, NY, USA: Association for Computing Machinery.

Qu, X. & fei Lee, L. (2015). Estimating a spatial autoregressive model with an endogenous spatial weight matrix. *Journal of Econometrics*, *184*(2), 209–232.

Qu, X., fei Lee, L. & Yang, C. (2021). Estimation of a sar model with endogenous spatial weights constructed by bilateral variables. *Journal of Econometrics*, *221*(1), 180–197.

Rezvanian, A. & Meybodi, M. R. (2015). Sampling social networks using shortest paths. *Physica A: Statistical Mechanics and its Applications*, *424*, 254–268.

Ribeiro, B. & Towsley, D. (2010). Estimating and sampling graphs with multidimensional random walks. In *Proceedings of the 10th acm sigcomm conference on internet measurement* (pp. 390–403).

Ribeiro, B., Wang, P., Murai, F. & Towsley, D. (2012). Sampling directed graphs with random walks. In *2012 proceedings ieee infocom* (p. 1692-1700).

Ronneberger, O., Fischer, P. & Brox, T. (2015). U-net: Convolutional networks for biomedical image segmentation. In N. Navab, J. Hornegger, W. M. Wells & A. F. Frangi (Eds.), *Medical image computing and computer-assisted intervention – miccai 2015* (pp. 234–241). Cham: Springer International Publishing.

Rozemberczki, B., Kiss, O. & Sarkar, R. (2020). Little Ball of Fur: A Python Library for Graph Sampling. In *Proceedings of the 29th acm international conference on information and knowledge management (cikm '20).*

Rozemberczki, B. & Sarkar, R. (2018). Fast sequence-based embedding with diffusion graphs. In *International workshop on complex networks* (pp. 99–107).

Rozemberczki, B., Scherer, P., He, Y., Panagopoulos, G., Riedel, A., Astefanoaei, M., ... Sarkar, R. (2021). Pytorch geometric temporal: Spatiotemporal signal processing with neural machine learning models. In *Proceedings of the 30th acm international conference on information and knowledge management* (p. 4564—4573). New York, NY, USA: Association for Computing Machinery.

Rozemberczki, B., Scherer, P., Kiss, O., Sarkar, R. & Ferenci, T. (2021). Chickenpox cases in hungary: a benchmark dataset for spatiotemporal signal processing with graph neural networks. *arXiv*, *abs/2102.08100.*

Sacerdote, B. (2001). Peer Effects with Random Assignment: Results for Dartmouth Roommates. *The Quarterly Journal of Economics*, *116*(2), 681–704.

Simini, F., Barlacchi, G., Luca, M. & Pappalardo, L. (2020). Deep gravity: enhancing mobility flows generation with deep neural networks and geographic information. *arXiv, abs/2012.00489.*

Stumpf, M. P. H., Wiuf, C. & May, R. M. (2005). Subnets of scale-free networks are not scale-free: Sampling properties of networks. *Proceedings of the National Academy of Sciences, 102*(12), 4221–4224.

Stutzbach, D., Rejaie, R., Duffield, N., Sen, S. & Willinger, W. (2008). On unbiased sampling for unstructured peer-to-peer networks. *IEEE/ACM Transactions on Networking, 17*(2), 377–390.

Tang, J., Qu, M., Wang, M., Zhang, M., Yan, J. & Mei, Q. (2015). Line: Large-scale information network embedding. In *Proceedings of the 24th international conference on world wide web* (pp. 1067–1077).

Tillema, F., Zuilekom, K. M. V. & van Maarseveen, M. (2006). Comparison of neural networks and gravity models in trip distribution. *Comput. Aided Civ. Infrastructure Eng., 21*, 104–119.

Torres, L., Chan, K. S. & Eliassi-Rad, T. (2020). Glee: Geometric laplacian eigenmap embedding. *Journal of Complex Networks, 8*(2).

Ziat, A., Delasalles, E., Denoyer, L. & Gallinari, P. (2017). Spatio-temporal neural networks for space-time series forecasting and relations discovery. *2017 IEEE International Conference on Data Mining (ICDM)*, 705–714.

Chapter 7
Fairness in Machine Learning and Econometrics

Samuele Centorrino, Jean-Pierre Florens and Jean-Michel Loubes

Abstract A supervised machine learning algorithm determines a model from a learning sample that will be used to predict new observations. To this end, it aggregates individual characteristics of the observations of the learning sample. But this information aggregation does not consider any potential selection on unobservables and any status quo biases which may be contained in the training sample. The latter bias has raised concerns around the so-called *fairness* of machine learning algorithms, especially towards disadvantaged groups. In this chapter, we review the issue of fairness in machine learning through the lenses of structural econometrics models in which the unknown index is the solution of a functional equation and issues of endogeneity are explicitly taken into account. We model fairness as a linear operator whose null space contains the set of strictly *fair* indexes. A *fair* solution is obtained by projecting the unconstrained index into the null space of this operator or by directly finding the closest solution of the functional equation into this null space. We also acknowledge that policymakers may incur costs when moving away from the status quo. *Approximate fairness* is thus introduced as an intermediate set-up between the status quo and a fully fair solution via a fairness-specific penalty in the objective function of the learning model.

Samuele Centorrino
Stony Brook University, Stony Brook, NY, USA, e-mail: samuele.centorrino@stonybrook.edu

Jean-Pierre Florens
Toulouse School of Economics, University of Toulouse Capitole, Toulouse, France, e-mail: jean-pierre.florens@tse-fr.eu

Jean-Michel Loubes ✉
Université Paul Sabatier, Institut de Mathématiques de Toulouse, Toulouse, France, e-mail: loubes@math.univ-toulouse.fr

F. Chan, L. Mátyás (eds.), *Econometrics with Machine Learning*, Advanced Studies in Theoretical and Applied Econometrics 53, https://doi.org/10.1007/978-3-031-15149-1_7

7.1 Introduction

Fairness has been a growing field of research in Machine Learning, Statistics, and Economics over recent years. Such work aims to monitor predictions of machine-learning algorithms that rely on one or more so-called *sensitive* variables. That is variables containing information (such as gender or ethnicity) that could create distortions in the algorithm's decision-making process. In many situations, the determination of these sensitive variables is driven by ethical, legal, or regulatory issues. From a moral point of view, penalizing a group of individuals is an unfair decision. From a legal perspective, unfair algorithmic decisions are prohibited for a large number of applications, including access to education, the welfare system, or microfinance.[1] To comply with fairness regulations, institutions may either change the decision-making process to remove biases using affirmative actions or try to base their decision on a fair version of the outcome.

Forecasting accuracy has become the gold standard in evaluating machine-learning models. However, especially for more complex methods, the algorithm is often a black box that provides a prediction without offering insights into the process which led to it. Hence, when bias is present in the learning sample, the algorithm's output can differ for different subgroups of populations. At the same time, regulations may impose that such groups ought to be treated in the same way. For instance, discrimination can occur on the basis of gender or ethnic origin.

A typical example is one of the automatic human resources (HR) decisions that are often influenced by gender. In available databases, men and women may self-select in some job categories due to past or present preferences or cultural customs. Some jobs are considered male-dominant, while other jobs are female-dominant. In such unbalanced datasets, the machine-learning procedure learns that gender matters and thus transforms correlation into causality by using gender as a causal variable in prediction. From a legal point of view, this biased decision leads to punishable gender discrimination. We refer to De-Arteaga et al. (2019) for more insights on this gender gap. Differential treatment for university admissions suffers from the same problems. We point out the example of law school admissions described in McIntyre and Simkovic (2018), which is used as a common benchmark to evaluate the bias of algorithmic decisions.

Imposing fairness is thus about mitigating this unwanted bias and preventing the sensitive variable from influencing decisions. We divide the concept of fairness into two main categories. The first definition of fairness is to impose that the algorithm's output is, on average, the same for all groups. Hence, the sensitive variable does not play any role in the decision. Such equality of treatment is referred to as *statistical parity*.

An alternative fairness condition is to impose that two individuals who are identical except for their value of the sensitive variables are assigned the same prediction. This notion of fairness is known as *equality of odds*. In this case, we wish to ensure

[1] Artificial Intelligence European Act, 2021.

that the algorithm performs equally across all possible subgroups. Equality of odds is violated, for instance, by the predictive justice algorithm described in Angwin, Larson, Mattu and Kirchner (2016). Everything else being equal, this algorithm was predicting a higher likelihood of future crimes for African-American convicts.

Several methods have been developed to mitigate bias in algorithmic decisions. The proposed algorithms are usually divided into three categories. The first method is a post-processing method that removes the bias from the learning sample to learn a fair algorithm. The second way consists of imposing a fairness constraint while learning the algorithm and balancing the desired fairness with the model's accuracy. This method is an in-processing method. Finally, the last method is a post-processing method where the output of a possibly unfair algorithm is processed to achieve the desired level of fairness, modeled using different fairness measures. All three methodologies require a proper definition of fairness and a choice of fairness measures to quantify it. Unfortunately, a universal definition of fairness is not available. Moreover, as pointed out by Friedler, Scheidegger and Venkatasubramanian (2021), complying simultaneously with multiple restrictions has proven impossible. Therefore, different fairness constraints give rise to different fair models.

Achieving *full fairness* consists in removing the effect of the sensitive variables completely. It often involves important changes with respect to the unfair case and comes at the expense of the algorithm's accuracy when the accuracy is measured using the biased distribution of the data set. When the loss of accuracy is considered too critical by the designer of the model, an alternative consists in weakening the fairness constraint. Hence the stakeholder can decide, for instance, to build a model for which the fairness level will be below a certain chosen threshold. The model will thus be called *approximately fair*.

To sum up, fairness with respect to some sensitive variables is about controlling the influence of its distribution and preventing its influence on an estimator. We refer to Barocas and Selbst (2016), Chouldechova (2017), Menon and Williamson (2018), Gordaliza, Del Barrio, Fabrice and Loubes (2019), Oneto and Chiappa (2020) and Risser, Sanz, Vincenot and Loubes (2019) or Besse, del Barrio, Gordaliza, Loubes and Risser (2021) and references therein for deeper insights on the notion of bias mitigation and fairness.

In the following, we present the challenges of fairness constraints in econometrics. Some works have studied the importance of fairness in economics (see, for instance, Rambachan, Kleinberg, Mullainathan and Ludwig (2020), Lee, Floridi and Singh (2021), Hoda, Loi, Gummadi and Krause (2018), Hu and Chen (2020), Kasy and Abebe (2021), and references therein). As seen previously, the term fairness is polysemic and covers various notions. We will focus on the role and on the techniques that can be used to impose fairness in a specific class of econometrics models.

Let us consider the example in which an institution must decide on a group of individuals. For instance, this could be a university admitting new students based on their expected performance in a test; or a company deciding the hiring wage of new employees. This decision is made by an algorithm, which we suppose works in the following way. For a given vector of individual characteristics, denoted by X, this algorithm computes a score $h(X) \in \mathbb{R}$ and makes a decision based on the value of this score, which is determined by a functional $\mathcal{D}$ of h. We are not specific about the exact form of $\mathcal{D}(h)$. For instance, this could be a threshold function in which students are admitted if the score is higher than or equal to some values C, and they are not admitted otherwise. The algorithm is completed by a *learning* model, which is written as follows

$$Y = h(X) + U, \tag{7.1}$$

where Y is the outcome, and U is a statistical error (see equation 2.1).[2] For instance, Y could be the test result from previous applicants. We let $X = (Z, S) \in \mathbb{R}^{p+1}$ and $\mathcal{X} = \mathcal{Z} \times \mathcal{S}$ to be the support of the random vector X. This learning model is used to approximate the score, $h(X)$, which is then used in the decision model.

Let us assume that historical data show that students from private high schools obtain higher test scores than students in public high schools. The concern with fairness in this model is twofold. On the one hand, if the distinction between public and private schools is used as a predictor, students from private schools will always have a higher probability of being admitted to a given university. On the other hand, the choice of school is an endogenous decision that is taken by the individual and may be determined by variables that are unobservable to the econometrician. Hence the bias will be reflected both in the lack of fairness in past decision-making processes and the endogeneity of individual choices in observational data. Predictions and admission decisions may be unfair towards the minority class and bias the decision process, possibly leading to discrimination. To overcome this issue, we consider that decision-makers can embed in their learning model a *fairness* constraint. This fairness constraint limits the relationship between the score $h(X)$ and S. Imposing a fairness constraint directly on h and not on $\mathcal{D}(h)$ is done for technical convenience, as $\mathcal{D}(h)$ is often nonlinear, which complicates the estimation and prediction framework substantially.

More generally, we aim to study the consequences of incorporating a fairness constraint in the estimation procedure when the score, h, solves a linear inverse problem of the type

$$Kh = r, \tag{7.2}$$

where K is a linear operator. Equation 7.2 can be interpreted as an infinite-dimensional linear system of equations, where the operator K is tantamount to an infinite-dimensional matrix, whose properties determine the existence and uniqueness of a solution to 7.2. A leading example of this setting is nonparametric instrumental regressions (Newey & Powell, 2003; Hall & Horowitz, 2005; Darolles, Fan, Florens & Renault, 2011). Nonetheless, many other models, such as linear and nonlinear parametric regressions and additive nonparametric regressions, can fit this general framework (Carrasco, Florens & Renault, 2007).

[2] The *learning* model is defined as the statistical specification fitted to the learning sample to estimate the unknown index h.

Let $\mathbf{P}_X$ be the distribution function of X, and $\mathcal{E}$ be the space of functions of X, which are square-integrable with respect to $\mathbf{P}_X$. That is,

$$\mathcal{E} := \left\{ h : \int h^2(x) d\mathbf{P}_X(x) < \infty \right\}.$$

Similarly, let $\mathcal{G}$ be the space of functions of X, which satisfy a fairness constraint. We model the latter as another linear operator $F : \mathcal{E} \to \mathcal{G}$ such that

$$Fh = 0. \tag{7.3}$$

That is, the null space (or kernel) of the operator F is the space of those functions that satisfy a fairness restriction, $\mathcal{N}(F) = \{g \in \mathcal{E}, Fg = 0\}$. The *full fairness* constraint implies restricting the solutions to the functional problem to the kernel of the operator. To weaken this requirement, we also consider relaxations of the condition and define an *approximate fairness* condition as

$$||Fh|| \leq \rho$$

for a fixed parameter $\rho \geq 0$.

In this work, we consider fairness according to the following definitions.

Definition 7.1 (Statistical Parity) The algorithm h maintains statistical parity if, for every $s \in \mathcal{S}$,

$$E\left[h(Z, s)|S = s\right] = E\left[h(X)\right].$$

Definition 7.2 (Irrelevance in prediction) The algorithm h does not depend on S. That is for all $s \in \mathcal{S}$,

$$\frac{\partial h(x)}{\partial s} = 0.$$

Definition 7.1 states that the function h is fair when individuals are treated the same, on average, irrespective of the value of the sensitive attribute, S. For instance, if S is a binary characteristic of the population, with $S = 1$ being the protected group, Definition 7.1 implies that the average score for group $S = 0$ and the average score for group $S = 1$ are the same. Notice that this definition of fairness does not ensure that two individuals with the same vector of characteristics $Z = z$ but with different values of S are treated in the same way. The latter is instead true for our second definition of fairness. In this case, fairness is defined as the lack of dependence of h on S, which implies the *equality of odds* for individuals with the same vector of characteristics $Z = z$. However, we want to point out that both definitions may fail to deliver fairness if the correlation between Z and S is very strong. In our example above, if students going to private schools have a higher income than students going to public schools, and income positively affects the potential score, then discrimination would still occur based on income.

Other definitions of fairness are possible. In particular, definitions that impose restrictions on the entire distribution of h given S. These constraints are nonlinear and thus more cumbersome to deal with in practice, and we defer their study to future work.

7.2 Examples in Econometrics

We let $\mathcal{F}_1$ and $\mathcal{F}_2$ be the set of square-integrable functions which satisfy definitions 7.1 and 7.2, respectively. We consider below examples in which the function h_F satisfies

$$h_F = \operatorname*{arg\,min}_{f \in \mathcal{F}_j} \mathbb{E}\left[(Y - f(X))^2 \,|W = w\right],$$

with $j = \{1,2\}$, and where W is a vector of exogenous variables.

7.2.1 Linear IV Model

Consider the example of a linear model in which $h(X) = Z'\beta + S'\gamma$, with $Z, \beta \in \mathbb{R}^p$, and $S, \gamma \in \mathbb{R}^q$. We take both Z and S to be potentially endogenous, and we have a vector of instruments $W \in \mathbb{R}^k$, such that $k \geq p+q$ and $E[W'U] = 0$.

We let $X = (Z', S')'$ be the vector of covariates, and $h = (\beta', \gamma')'$ be the vector of unknown coefficients.

For simplicity of exposition, we maintain the assumption that the vector

$$\begin{bmatrix} X \\ W \end{bmatrix} \sim N\left(\mathbf{0}_{p+q+k}, \begin{bmatrix} \Sigma_X & \Sigma'_{XW} \\ \Sigma_{XW} & I_k \end{bmatrix}\right),$$

where $\mathbf{0}_{p+q+k}$ is a vector of zeroes of dimension $p+q+k$, I_k is the identity matrix of dimension k, and

$$\underbrace{\Sigma_X}_{(p+q)\times(p+q)} = \begin{bmatrix} \Sigma_Z & \Sigma'_{ZS} \\ \Sigma_{ZS} & \Sigma_S \end{bmatrix}, \qquad \underbrace{\Sigma_{XW}}_{k\times(p+q)} = \begin{bmatrix} \Sigma_{ZW} \; \Sigma_{SW} \end{bmatrix}.$$

The unconstrained value of h is therefore given by

$$h = \left(\Sigma'_{XW}\Sigma_{XW}\right)^{-1} \Sigma'_{XW} E[WY] = (K^*K)^{-1} K^* r.$$

Because of the assumption of joint normality, we have that $E[Z|S] = \Pi S$, where $\Pi = \Sigma_S^{-1}\Sigma_{ZS}$ is a $p \times q$ matrix. Statistical parity, as defined in 7.1, implies that

$$S'(\Pi\beta + \gamma) = 0,$$

which is true as long as $\Pi\beta + \gamma = \mathbf{0}_q$. In the case of Definition 7.2, the fairness constraint is simply given by $\gamma = 0$.

7.2.2 A Nonlinear IV Model with Binary Sensitive Attribute

Let $Z \in \mathbb{R}^p$ be a continuous variable and $S = \{0, 1\}^q$ a binary random variable. For instance, S can characterize gender, ethnicity, or a dummy for school choice (public vs. private). Because of the binary nature of S

$$h(X) = h_0(Z) + h_1(Z)S.$$

Definition 7.1 implies that we are looking for functions $\{h_0, h_1\}$ such that

$$E\left[h_0(Z)|S = 0\right] = E\left[h_0(Z) + h_1(Z)|S = 1\right].$$

That is

$$E\left[h_1(Z)|S = 1\right] = E\left[h_0(Z)|S = 0\right] - E\left[h_0(Z)|S = 1\right].$$

Definition 7.2 instead simply implies that $h_1 = 0$, almost surely. In particular, under the fairness restriction, $Y = h_0(Z) + U$. We discuss this example in more detail in Section 7.6.

7.2.3 Fairness and Structural Econometrics

More generally, supervised machine learning models are often about predicting a conditional moment or a conditional probability. However, in many leading examples in structural econometrics, the score function h does not correspond directly to a conditional distribution or a conditional moment of the distribution of the learning variable Y. Let Γ be the probability distribution generating the data. Then the function h is the solution to the following equation

$$A(h, \Gamma) = 0.$$

A leading example is one of the Neyman-Fisher-Cox-Rubin *potential outcome* models, in which X represents a treatment and, for $X = \xi$, we can write

$$Y_\xi = h(\xi) + U_\xi.$$

If $E\left[U_\xi|W\right] = 0$, this model leads to the nonparametric instrumental regression model mentioned above, in which the function $A(h, \Gamma) = E\left[Y - h(X)|W\right] = 0$, and the fairness condition is imposed directly on the function h. However, this potential outcome model can lead to other objects of interest. For instance, if we assume for simplicity that $(X, W) \in \mathbb{R}^2$, and under a different set of identification assumptions, it can be proven that

$$A(h,\Gamma) = E\left[\frac{dh(X)}{dX}|W\right] - \frac{\frac{dE[Y|W]}{dW}}{\frac{dE[Z|W]}{dW}} = 0,$$

which is a linear equation in h that combines integral and differential operators (see Florens, Heckman, Meghir & Vytlacil, 2008). In this case, the natural object of interest is the first derivative of $h(x)$, which is the marginal treatment effect. The fairness constraint is therefore naturally imposed on $\frac{dh(x)}{dx}$.

Another class of structural models not explicitly considered in this work is one of nonlinear nonseparable models. In these models, we have that

$$Y = h(X,U), \text{ with } U \perp\!\!\!\perp W \text{ and } U \sim \mathcal{U}[0,1],$$

and $h(\xi,\cdot)$ monotone increasing in its second argument. In this case, h is the solution to the following nonlinear inverse problem

$$\int P(Y \leq h(x,u)|X = x, W = w)\, f_{X|W}(x|w)dx = u.$$

The additional difficulty lies in imposing a *distributional* fairness constraint in this setting. We defer the treatment of this case to future research.

7.3 Fairness for Inverse Problems

Recall that the nonparametric instrumental regression (NPIV) model amounts to solving an inverse problem. Let W be a vector of instrumental variables. The NPIV regression model can be written as

$$E(Y|W = w) = E(h(Z,S)|W = w).$$

We let $X = (Z,S) \in R^{p+q}$ and $\mathcal{X} = \mathcal{Z} \times \mathcal{S}$ to be the support of the random vector X. We further restrict $h \in L^2(X)$, with L^2 being the space of square-integrable functions with respect to some distribution $\mathbf{P}$. We further assume that this distribution $\mathbf{P}$ is absolutely continuous with respect to the Lebesgue measure, and it therefore admits a density, p_{XW}. Using a similar notation, we let p_X and p_W, be the marginal densities of X and W, respectively.

If we let $r = E(Y|W = w)$ and $Kh = E(h(X)|W = w)$, where K is the conditional expectation operator, then the NPIV framework amounts to solving an inverse problem. That is, estimating the true function $h_\dagger \in \mathcal{E}$, defined as the solution of

$$r = Kh_\dagger. \tag{7.4}$$

Equation (7.4) can be interpreted as an infinite-dimensional system of linear equations. In parallel with the finite-dimensional case, the properties of the solution depend on the properties of the operator K. In 1923, Hadamard postulated three requirements for problems of this type in mathematical physics: a solution should exist, the solution should be unique, and the solution should depend continuously on r. That is, $h_\dagger$ is stable to small changes in r. A problem satisfying all three requirements is called *well-posed*. Otherwise, it is called *ill-posed*. The existence of the solution is usually established by restricting r to belong to the range of the operator K. In our case, it is sufficient to let $K : \mathcal{E} \to L^2(W)$, and $r \in L^2(W)$, where $L^2(W)$ is the space of square-integrable functions of W. The uniqueness of a solution to (7.4) is guaranteed by the so-called *completeness condition* (or strong identification, see Florens, Mouchart & Rolin, 1990; Darolles et al., 2011). That is, $Kh = 0$, if and only if, $h = 0$, where equalities are intended almost surely. In particular, Let h_1 and h_2 two solutions to (7.4), then $K(h_1 - h_2) = 0$, which implies that $h_1 = h_2$, by the strong identification assumption.

In the following, for two square-integrable functions of X, $\{h_1, h_2\}$, we let

$$\begin{aligned}\langle h_1, h_2\rangle &= \int h_1(x)h_2(x)d\mathbf{P},\\ \|h_1\|^2 &= \langle h_1, h_1\rangle.\end{aligned}$$

The adjoint operator, K^*, is obtained as

$$\begin{aligned}\langle Kh, r\rangle &= \int_W \left(\int_X h(x)p_{X|W}(x|w)dx\right) r(w)p_W(w)dw\\ &= \int_X \left(\int_W r(w)p_{W|X}(w|x)dw\right) h(x)p_X(x)dx = \langle h, K^*r\rangle.\end{aligned}$$

If the operator K^*K is invertible, the solution of (7.4) is given by

$$h_\dagger = (K^*K)^{-1}K^*r. \tag{7.5}$$

The ill-posedness of the inverse problem in (7.4) comes from the fact that, when the distribution of (X, W) is continuous, the eigenvalues of the operator K^*K have zero an as accumulation point. That is, small changes in r may correspond to large changes in the solution as written in (7.5). The most common solution to deal with the ill-posedness of the inverse problem is to use a regularization technique (see Engl, Hanke & Neubauer, 1996, and references therein). In this chapter, we consider Tikhonov regularization, which imposes an L^2-penalty on the function h (Natterer, 1984). Heuristically, Tikhonov regularization can be considered as the functional extension of Ridge regressions, which are used in linear regression models to obtain an estimator of the parameters when the design matrix is not invertible. In this respect, Tikhonov regularization imposes an L^2-penalty on the functions of interest (see Chapters 1 and 2 of this manuscript for additional details).

A regularized version of h, as presented in (Engl et al., 1996), is defined as the solution of a penalized optimization program

$$h_\alpha = \arg\min_{h \in \mathcal{E}} \|r - Kh\|^2 + \alpha \|h\|^2,$$

which can be explicitly written as

$$h_\alpha = (\alpha \mathrm{Id} + K^* K)^{-1} K^* r = R_\alpha(K) K^* r \tag{7.6}$$

where $R_\alpha(K) = (\alpha \mathrm{Id} + K^* K)^{-1}$ is a Tikhonov regularized operator. The solution depends on the choice of the regularization parameter α, which helps bound the eigenvalues of $K^* K$ away from zero. Therefore, precisely as in Ridge regressions, a positive value of α introduces a bias in estimation but it allows us to bound the variance of the estimator away from infinity. As $\alpha \to 0$, in a suitable way, then $h_\alpha \to h_\dagger$.

We consider the estimation of the function h from the following noisy observational model

$$\hat{r} = K h_\dagger + U_n, \tag{7.7}$$

where U_n is an unknown random function with bounded norm. That is, $\|U_n\|^2 = O(\delta_n)$ for a given sequence δ_n which tends to 0 as n goes to infinity. The operators K and K^* are taken to be known for simplicity. This estimation problem has been widely studied in the econometrics literature, and we provide details on the estimation of the operator in Section 7.5. We refer, for instance, to Darolles et al. (2011) for the asymptotic properties of the NPIV estimator when the operator K is estimated from data.

To summarise, we impose the following regularity conditions on the operator K and the true solution $h_\dagger$.

- **[A1]** $r \in \mathcal{R}(K)$ where $\mathcal{R}(K)$ stands for the range of the operator K.
- **[A2]** The operator $K^* K$ is a one-to-one operator. This condition implies the completeness condition as defined above and ensures the identifiability of $h_\dagger$.
- **[A3]** Source Condition : we assume that there exists $\beta \leq 2$ such that

$$h_\dagger \in \mathcal{R}(K^* K)^{\frac{\beta}{2}}.$$

  This condition relates the smoothness of the solution of equation (7.4) to the decay of the eigenvalues of the SVD decomposition of the operator K. It is commonly used in statistical inverse problems. In particular, it guarantees that the Tikhonov regularized solution h_α converges to the true solution $h_\dagger$ at a rate of convergence given by

$$\|h_\alpha - h_\dagger\|^2 = O(\alpha^\beta).$$

  We refer to Loubes and Rivoirard (2009) for a review of the different smoothness conditions for inverse problems.

7.4 Full Fairness IV Approximation

In this model, full fairness of a function $\psi \in \mathcal{E}$ is achieved when $F\psi = 0$, i.e. when the function belongs to the null space of a fairness operator, F. Hence imposing fairness amounts to considering functions that belong to the null space, $\mathcal{N}(F)$, and that are approximate solutions of the functional equation (7.4). The full fairness condition may be seen as a very restrictive way to impose fairness. Actually, if the functional equation does not have a solution in $\mathcal{N}(F)$, full fairness will induce a loss of accuracy, which we refer to as *price for fairness*. The projection to fairness has been studied in the regression framework in Le Gouic, Loubes and Rigollet (2020), Chzhen, Denis, Hebiri, Oneto and Pontil (2020), and Jiang, Pacchiano, Stepleton, Jiang and Chiappa (2020), for the classification task.
The full fairness solution can be achieved in two different ways: either by looking at the solution of the inverse problem and then imposing a fair condition on the solution; or by directly solving the inverse problem under the restriction that the solution is fair. We prove that the two procedures are not equivalent, leading to different estimators having different properties.

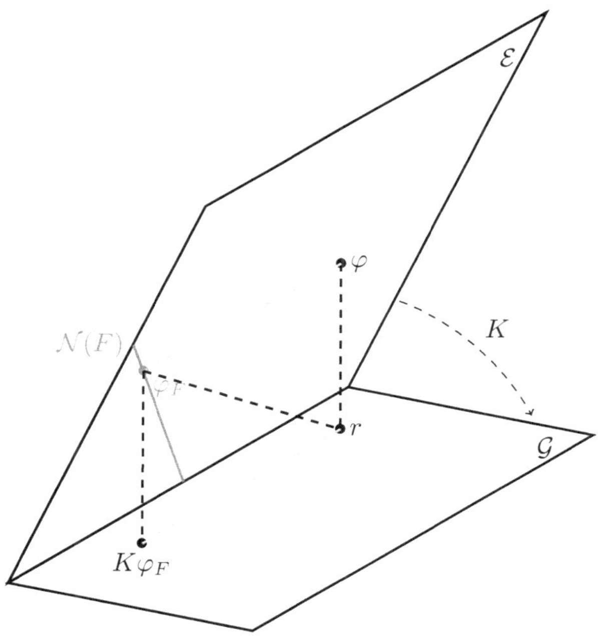

Fig. 7.1: Example of projection onto the space of fair functions.

Figure 7.1 illustrates the situation where either the solution can be solved and then the fairness condition can be imposed. or the solution is directly approximated in the set of fair functions.

7.4.1 Projection onto Fairness

The first way consists in first considering the regularized solution to the inverse problem $\hat{h}_\alpha$ defined as the Tikhonov regularized solution of the inverse problem

$$\hat{h}_\alpha = \arg\min_{h\in\mathcal{E}} \left(\|\hat{r} - Kh\|^2 + \alpha\|h\|^2\right)$$

which can be computed as

$$\hat{h}_\alpha = (\alpha \mathrm{Id} + K^*K)^{-1}K^*\hat{r} = R_\alpha(K)K^*\hat{r}.$$

Then the fair solution is defined as the projection onto the set which models the fairness condition $\mathcal{N}$,

$$\hat{h}_{\alpha,F} = \arg\min_{h\in\mathcal{N}(F)} \|\hat{h}_\alpha - h\|^2$$

In this framework, denote by $P : \mathcal{E} \to \mathcal{N}(F)$ the projection operator onto the kernel of the fairness operator. Hence we have

$$\hat{h}_{\alpha,F} = P\hat{h}_\alpha.$$

Example 7.1 (Linear Model, continued.) Recall that the constraint of statistical parity, as in Definition 7.1, implies that

$$S'(\Pi\beta + \gamma) = 0,$$

which is true as long as $\Pi\beta + \gamma = \mathbf{0}_q$. Thus, we have that

$$\underbrace{F}_{q\times(p+q)} = \left[\Pi \; I_q\right],$$

and

$$P = I_{p+q} - F'(FF')^{-1}F = I_{p+q} - \begin{bmatrix} \Pi'(I_q + \Pi\Pi')^{-1}\Pi & \Pi'(I_q + \Pi\Pi')^{-1} \\ (I_q + \Pi\Pi')^{-1}\Pi & (I_q + \Pi\Pi')^{-1} \end{bmatrix},$$

which immediately gives $FP = \mathbf{0}_q$. Hence, the value of $h_F = Ph$ is the projection of the vector h onto the null space of F.

In the case of definition 7.2, the fairness constraint is simply given by $\gamma = 0$. Let

$$M_{ZW} = I_k - \Sigma_{ZW}\left(\Sigma'_{ZW}\Sigma_{ZW}\right)^{-1}\Sigma'_{ZW},$$

and

$$A_{ZS} = \left(\Sigma'_{ZW}\Sigma_{ZW}\right)^{-1}\Sigma'_{ZW}\Sigma_{SW}.$$

When one wants to project the unconstrained estimator onto the constrained space by the block matrix inversion lemma, we notice that

$$h = \begin{bmatrix} \left(\Sigma'_{ZW}\Sigma_{ZW}\right)^{-1} + A_{ZS}\left(\Sigma'_{SW}M_{ZW}\Sigma_{SW}\right)^{-1}A'_{ZS} & -A_{ZS}\left(\Sigma'_{SW}M_{ZW}\Sigma_{SW}\right)^{-1} \\ -\left(\Sigma'_{SW}M_{ZW}\Sigma_{SW}\right)^{-1}A'_{ZS} & \left(\Sigma'_{SW}M_{ZW}\Sigma_{SW}\right)^{-1} \end{bmatrix} \begin{bmatrix} \Sigma'_{ZW}E[WY] \\ \Sigma'_{SW}E[WY] \end{bmatrix}$$
$$= \begin{bmatrix} \left(\Sigma'_{ZW}\Sigma_{ZW}\right)^{-1}\Sigma'_{ZW}E[WY] - A_{ZS}\left(\Sigma'_{SW}M_{ZW}\Sigma_{SW}\right)^{-1}\left(\Sigma'_{SW}E[WY] - A'_{ZS}\Sigma'_{ZW}E[WY]\right) \\ \left(\Sigma'_{SW}M_{ZW}\Sigma_{SW}\right)^{-1}\left(\Sigma'_{SW}E[WY] - A'_{ZS}\Sigma'_{ZW}E[WY]\right) \end{bmatrix}$$
$$= \begin{bmatrix} \left(\Sigma'_{ZW}\Sigma_{ZW}\right)^{-1}\Sigma'_{ZW}E[WY] - A_{ZS}\gamma \\ \gamma \end{bmatrix}.$$

Therefore, we have that

$$h_F = Ph = \begin{bmatrix} \beta + A_{ZS}\gamma \\ \mathbf{0}_q \end{bmatrix}.$$

The behavior of the projection of the unfair solution onto the space of fair functions is given by the following theorem

Theorem 7.1 *Under Assumptions [A1] to [A3], the fair projection estimator is such that*

$$||\hat{h}_{\alpha,F} - Ph_{\dagger}||^2 = O\left(\frac{1}{\alpha\delta_n} + \alpha^{\beta}\right) \tag{7.8}$$

Proof

$$||\hat{h}_{\alpha,F} - Ph_{\dagger}|| \leq ||P\hat{h}_{\alpha} - Ph_{\dagger}||$$
$$\leq ||\hat{h}_{\alpha} - h_{\dagger}||$$

since P is a projection. The term $||\hat{h}_{\alpha} - h_{\dagger}||$ is the usual estimation term for the structural IV inverse problem. As proved in (Darolles et al., 2011) this term converges at the following rate

$$||\hat{h}_{\alpha} - h_{\dagger}||^2 = O\left(\frac{1}{\alpha\delta_n} + \alpha^{\beta}\right),$$

which proves the result. □

The estimator converges towards the fair part of the function $h_{\dagger}$, i.e. its projection onto the kernel of the fairness operator F. If we consider the difference with respect to the unconstrained solution, we have that

$$||\hat{h}_{\alpha} - h_{\dagger}||^2 = O\left(\frac{1}{\alpha\delta_n} + \alpha^{\beta} + ||h_{\dagger} - Ph_{\dagger}||^2\right).$$

Hence the difference $||h_\dagger - Ph_\dagger||^2$ corresponds to the price to pay for ensuring fairness, which is equal to zero only if the true function satisfies the fairness constraint. This difference between the underlying function $h_\dagger$ and its fair representation is the necessary change of the model that would enable a fair decision process minimizing the quadratic distance between the fair and the unfair functions.

7.4.2 Fair Solution of the Structural IV Equation

A second and alternative solution to impose fairness is to solve the structural IV equation directly on the space of fair functions, $\mathcal{N}(F)$. We denote by K_F the operator K restricted to $\mathcal{N}(F)$, $K_F : \mathcal{N}(F) \mapsto \mathcal{F}$. Since $\mathcal{N}(F)$ is a convex closed space, the projection onto this space is well-defined and unique. We will write P the projection onto $\mathcal{N}(F)$ and $P^\perp$ the projection onto its orthogonal complement in $\mathcal{E}$, $\mathcal{N}(F)^\perp$. With these notations, we get that $K_F = KP$.

Definition 7.3 Define h_{K_F} as the solution of the structural equation $Kh = r$ in the set of fair functions defined as the kernel of the operator F, i.e.

$$h_{K_F} = \arg \min_{h \in \mathcal{N}(F)} \left(||r - Kh||^2 \right).$$

Note that h_{K_F} is the projection of $h_\dagger$ onto $\mathcal{N}(F)$ with the metric defined by K^*K, since

$$h_{K_F} = \arg \min_{h \in \mathcal{N}(F)} \left(||Kh_\dagger - Kh||^2 \right).$$

Note that this approximation depends not only on K but on the properties of the fair kernel $K_F = KP$. Therefore, fairness is quantified here through its effect on the operator K and we denote this solution h_{K_F} to highlight its dependence on the operators K and F.

The following proposition proposes an explicit expression of h_{K_F}.

Proposition 7.1

$$h_{K_F} = (K_F^* K_F)^{-1} K_F^* r.$$

Proof First, h_{K_F} belongs to $\mathcal{N}(F)$. For any function $g \in \mathcal{E}$, $PK^*Kg \in \mathcal{N}(F)$ so the operator $(K_F^* K_F)^{-1} = (PK^*KP)^{-1}$ is defined from $\mathcal{N}(F) \mapsto \mathcal{N}(F)$.
Let $\psi \in \mathcal{N}(F)$ so $P\psi = \psi$. We have that

$$\begin{aligned} 0 &= < r - Kh_{K_F}, K\psi > \\ &= < K^*r - K^*Kh_{K_F}, \psi > \\ &= < K^*r - K^*KPh_{K_F}, P\psi > \\ &= < PK^*r - PK^*KPh_{K_F}, \psi > \end{aligned}$$

which holds for $PK^*r - PK^*KPh_{K_F} = 0$, which leads to $h_{K_F} = (PK^*KP)^{-1}PK^*r$.□

Example 7.2 (Linear model, continued.) For both our definitions of fairness in 7.1 and 7.2, we have that

$$h_{K_F} = \left(P\Sigma'_{XW}\Sigma_{XW}P\right)^{-1} P\Sigma'_{XW}E\left[WY\right],$$

which simply restricts the conditional expectation operators onto the null space of F.

In the case of definition 7.2, the closed-form expression of this estimator is easy to obtain and is equal to

$$h_{K_F} = \begin{pmatrix} \left(\Sigma'_{ZW}\Sigma_{ZW}\right)^{-1}\Sigma'_{ZW}E\left[WY\right] \\ \mathbf{0}_q \end{pmatrix} = \left(P\Sigma'_{XW}\Sigma_{XW}P\right)^{-1} P\Sigma'_{XW}E\left[WY\right],$$

which is equivalent to excluding S from the second stage estimation of the IV model, and where

$$F = \begin{bmatrix} \mathbf{0}_{p\times p} & \mathbf{0}_{p\times q} \\ \mathbf{0}_{q\times p} & I_q \end{bmatrix}, \text{ and } P = I_{p+q} - F.$$

Now consider the fair approximation of the solution of (7.4) as the solution of the following minimization program

$$\hat{h}_{K_F,\alpha} = \arg\min_{h\in\mathcal{N}(F)} \left(\|\hat{r} - Kh\|^2 + \alpha\|h\|^2\right).$$

Proposition 7.2 *The fair solution of the IV structural equation has the following expression*

$$\hat{h}_{K_F,\alpha} = (\alpha\mathrm{Id} + K_F^*K_F)^{-1}K_F^*\hat{r}.$$

It converges to h_{K_F} when α goes to zero as soon as α is chosen such that $\alpha\delta_n \to +\infty$.

Proof As previously, $\hat{h}_{K_F,\alpha}$ minimizes in $\mathcal{N}(F)$, $\|\hat{r} - Kh\|^2 + \alpha\|h\|^2$. Hence the first-order condition is that for all $g \in \mathcal{N}(F)$ we have

$$\begin{aligned} < -Kg, \hat{r} - Kh > + \alpha < g, h > &= 0 \\ < g, K^*Kh - K^*\hat{r} > + \alpha < g, h > &= 0 \\ < g, PK^*Kh - PK^*\hat{r} + \alpha h > &= 0. \end{aligned}$$

Hence using $K_F^* = PK^*$, and since h is in $\mathcal{N}(F)$ and thus $Ph = h$, we obtain the expression of the theorem.

Using this expression, we can compute the estimator as follows :

$$\begin{aligned} &\hat{h}_{K_F,\alpha} - h_{K_F} = \\ &(\alpha\mathrm{Id} + K_F^*K_F)^{-1}K_F^*(\hat{r} - Kh_\dagger) + ((\alpha\mathrm{Id} + K_F^*K_F)^{-1} - (K_F^*K_F)^{-1})K_F^*Kh_\dagger \\ &= (I) \quad + \quad (II). \end{aligned}$$

The first term is a variance term which is such that

$$\|(I)\|^2 = O\left(\frac{1}{\alpha\delta_n}\right).$$

Recall that for two operators

$$A^{-1} - B^{-1} = A^{-1}(B - A)B^{-1}.$$

Hence, the second term can be written as

$$(II) = -\alpha(\alpha\mathrm{Id} + K_F^* K_F)^{-1} h_{K_F}.$$

This tern is the bias of Tikhonov's regularization of the operator $K_F^* K_F = PK^*KP$, which goes to zero when α goes to zero. □

When α decreases to zero, the rate of consistency of the projected fair estimator can be made precise if we assume some Hilbert scale regularity for both the fair part of $h_\dagger$ and the remaining *unfair* part $P^\perp h_\dagger$.

Assume that

- **[E1]** $Ph_\dagger \in \mathcal{R}(PK^*KP)^{\frac{\beta}{2}}$ for $\beta \leq 2$
- **[E2]** $P^\perp h_\dagger \in \mathcal{R}(PK^*KP)^{\frac{\gamma}{2}}$ for $\gamma \leq 2$.

These assumptions are analogous to the source condition in **[A3]** adapted to the fair operator K_F.

Theorem 7.2 *Under Assumptions* **[E1]** *and* **[E2]**, *the estimator* $\hat{h}_{K_F}$ *converges towards* h_{K_F} *at the following rate*

$$\|\hat{h}_{K_F} - h_{K_F}\|^2 = O\left(\frac{1}{\alpha\delta_n} + \alpha^{min(\beta,\gamma)}\right)$$

We recognize the usual rate of convergence of the Tikhonov's regularized estimator. The main change is given here by the fact that the rate is driven by the fair source conditions **[E1]** and **[E2]**. These conditions relate the smoothness of the function with the decay of the SVD of the operator restricted to the kernel of the fairness operator.

Proof The rate of convergence depends on the term (II) previously defined. We decompose it into two terms.

$$\begin{aligned}(II) &= -\alpha(\alpha\mathrm{Id} + K_F^* K_F)^{-1}(K_F^* K_F)^{-1} K_F^*(KPh_\dagger + KP^\perp h_\dagger) \\ &= (A) \quad + \quad (B).\end{aligned}$$

First remark that since $P = P^2$

$$\begin{aligned}(A) &= -\alpha(\alpha\mathrm{Id} + K_F^* K_F)^{-1}(K_F^* K_F)^{-1} K_F^* K_F P h_\dagger \\ &= -\alpha(\alpha\mathrm{Id} + K_F^* K_F)^{-1} P h_\dagger\end{aligned}$$

Assumption **[E1]** provides the rate of decay of this term $||(A)||^2$ and enables to prove that it is of order α^β.
For the second term (B), consider the SVD of the operator $K_F = KP$ denoted by λ_j, ψ_j, e_j for all $j \geq 1$. So we have that

$$\begin{aligned}
||(B)||^2 &= ||\alpha(\alpha \mathrm{Id} + K_F^* K_F)^{-1}(K_F^* K_F)^{-1} K_F^* K P^\perp h_\dagger||^2 \\
&= \alpha^2 \sum_{j\geq 1} \frac{\lambda_j^2}{\lambda_j^4(\alpha+\lambda_j^2)^2} | < KP^\perp h_\dagger, e_j > |^2 \\
&= \alpha^2 \sum_{j\geq 1} \frac{\lambda_j^{2\gamma}}{(\alpha+\lambda_j^2)^2} \frac{| < KP^\perp h_\dagger, e_j > |^2}{\lambda_j^{2(1+\gamma)}} \\
&= O(\alpha^\gamma)
\end{aligned}$$

To ensure that

$$\sum_{j\geq 1} \frac{| < KP^\perp h_\dagger, e_j > |^2}{\lambda_j^{2(1+\gamma)}} < +\infty$$

we assume that

$$\sum_{j\geq 1} \frac{| < P^\perp h_\dagger, \lambda_j \psi_j > |^2}{\lambda_j^{2(1+\gamma)}} = \sum_{j\geq 1} \frac{| < P^\perp h_\dagger, \psi_j > |^2}{\lambda_j^{2\gamma}} < +\infty$$

where $K^* e_j = \lambda_j \psi_j$, which is ensured under Assumption **[E2]**. Finally the two terms are of order $O(\alpha^\beta + \alpha^\gamma)$, which proves the result. □

To summarize, we have defined two fair approximations of the function $h_\dagger$. The first one is its fair projection $h_F = Ph_\dagger$, while the other is the solution of the fair kernel h_{K_F}. The two solutions coincide as soon as

$$h_{K_F} - Ph_\dagger = (K_F^* K_F)^{-1} K_F^* K P^\perp h_\dagger = 0.$$

Under assumption [A2], $K_F^* K_F$ is also one to one. Hence the difference between both approximations is null only if

$$KP^\perp h_\dagger = 0. \tag{7.9}$$

If we consider the case of (IV) regression. This condition is met when

$$E(h(Z,S)|W) - E(E(h(Z,S)|Z)|W) = 0.$$

This is the case when the sensitive variable S is independent w.r.t to the instrument W conditionally to the characteristics Z. Yet, in the general case, both functions are different.

7.4.3 Approximate Fairness

Imposing (7.3) is a way to ensure complete fairness of the solution of (7.4). In many cases, this full fairness leads to bad approximation properties. Hence, we replace it with a constraint on the norm of Fh. Namely, we look for the estimator defined as the solution of the optimization

$$\hat{h}_{\alpha,\rho} = \arg\min_{h\in\mathcal{E}} \left(||\hat{r} - Kh||^2 + \alpha||h||^2 + \rho||Fh||^2 \right) \tag{7.10}$$

This estimator corresponds to the usual Tikhonov regularized estimator with an additional penalty term $\rho||Fh||^2$. The penalty enforces fairness since it enforces $||Fh||$ to be small, which corresponds to a relaxation of the full fairness constraint $Fh = 0$. The parameter ρ provides a trade-off between the level of fairness which is imposed and the closeness to the usual estimator of the NPIV model. We study its asymptotic behavior in the following theorem.

Note first that the solution of (7.10) has a close form and can be written as

$$\hat{h}_{\alpha,\rho} = (\alpha \mathrm{Id} + \rho F^* F + K^* K)^{-1} K^* \hat{r}.$$

The asymptotic behavior of the estimator is provided by the following theorem. It also ensures that the limit solution of (7.10), i.e. when $\rho \longrightarrow +\infty$, is fair in the sense that $\lim_{\rho\rightarrow+\infty} ||Fh_{\alpha,\rho}|| = 0.$. That is, it converges to the structural solution restricted to the set of fair functions h_{K_F}.
We will use the following notations. Consider the collection of operators

$$L_\alpha = (\alpha \mathrm{Id} + K^* K)^{-1} F^* F$$

$$L = (K^* K)^{-1} F^* F.$$

- **[A4]** $\mathcal{R}(F^* F) \subset \mathcal{R}(K^* K)$. This condition guarantees that the operators L and L_α are well-defined operators.

L is an operator $L : \mathcal{E} \longrightarrow \mathcal{E}$, which is not self-adjoint.
Consider also the operator

$$T = (K^* K)^{-1/2} F^* F (K^* K)^{-1/2}$$

which is an self-adjoint operator, and is well-defined as long as

- **[A5]** $\mathcal{R}(F^* F) \subset \mathcal{R}(K^* K)^{1/2}$.

If we assume a source condition on the form

- **[A6]** There exists $\gamma \geq \beta$

$$F^* F P^\perp h_\dagger \in \mathcal{R}(K^* K)^{\frac{\gamma+1}{2}}$$

Theorem 7.3 (Consistency of fair IV estimator) *The approximated fair IV estimator $\hat{h}_{\alpha,\rho}$ is an estimator of the fair projection of the structural function, i.e. h_{K_F}. Its rate of consistency under assumptions [A1] to [A6] is given by*

$$\|\hat{h}_{\alpha,\rho} - h_{K_F}\|^2 = O\left(\alpha^\beta + \frac{1}{\rho^2} + \frac{1}{\alpha\delta_n}\right). \tag{7.11}$$

The rate of convergence is consistent in the following sense. When we increase the level of imposed fairness to the full fairness constraint, i.e when ρ goes to infinity, for appropriate choices of the smoothing parameter α, the estimator converges to a fully fair function. The rate in $\frac{1}{\rho^2}$ corresponds to the fairness part of the rate. If β the Source condition parameter can be chosen large enough such that $\alpha^\beta = \frac{1}{\rho^2}$, hence we recover, for an optimal choice of α_{opt} of order $\delta_n^{-\frac{1}{\beta+1}}$, the usual rate of convergence of the NPIV estimates

$$\|\hat{h}_{\alpha,\rho} - h_{K_F}\|^2 = O\left(\delta_n^{-\frac{\beta}{\beta+1}}\right).$$

Example 7.3 (Linear model, continued.) In the linear IV model, let

$$h_\rho = \left(\rho F'F + \Sigma'_{XW}\Sigma_{XW}\right)^{-1} \Sigma'_{XW} E\left[WY\right],$$

the estimator which imposes the approximate fairness constraint. Notice that

$$\begin{aligned}
&\left(\rho F'F + \Sigma'_{XW}\Sigma_{XW}\right)^{-1} \\
&= \left(\Sigma'_{XW}\Sigma_{XW}\right)^{-1} - \rho\left(\Sigma'_{XW}\Sigma_{XW}\right)^{-1} F'\left(I_q + \rho F\left(\Sigma'_{XW}\Sigma_{XW}\right)^{-1} F'\right)^{-1} F\left(\Sigma'_{XW}\Sigma_{XW}\right)^{-1} \\
&= \left(\Sigma'_{XW}\Sigma_{XW}\right)^{-1} - \left(\Sigma'_{XW}\Sigma_{XW}\right)^{-1} F'\left(\frac{1}{\rho} I_q + F\left(\Sigma'_{XW}\Sigma_{XW}\right)^{-1} F'\right)^{-1} F\left(\Sigma'_{XW}\Sigma_{XW}\right)^{-1}.
\end{aligned}$$

This decomposition implies that

$$\lim_{\rho\to\infty} h_\rho = h - \left(\Sigma'_{XW}\Sigma_{XW}\right)^{-1} F'\left(F\left(\Sigma'_{XW}\Sigma_{XW}\right)^{-1} F'\right)^{-1} Fh,$$

which directly gives

$$\lim_{\rho\to\infty} Fh_\rho = 0.$$

Therefore, as implied by our general theorem, as ρ diverges to ∞, the full fairness constraint is imposed.

Remark 7.1 Previous theorems enable us to understand the asymptotic behavior of the fair regularized IV estimator. When α goes to zero, but ρ is fixed, this estimator converges towards a function h_ρ which differs from the original function $h_\dagger$. Interestingly, we point out that the fairness constraint enables one to obtain a fair solution, but the latter does not coincide with the fair approximation of the true function, $h_\dagger$. Instead, the fair solution is obtained by considering the set of approximate solutions that satisfy the fairness constraint.

Remark 7.2 The theorem requires an additional assumption denoted by **[A6]**. This assumption aims at controlling the regularity of the *unfair* part of the function $h_\dagger$. It is analogous to a source condition imposed on the part of the solution which does not lie in the kernel of the operator F, namely $P^\perp h_\dagger$. This condition is obviously fulfilled if $h_\dagger$ is fair since $P^\perp h_\dagger = 0$.

Remark 7.3 The smoothness assumptions we impose in this paper are source conditions with regularity smaller than 2. Such restrictions come from the choice of standard Tikhonov's regularization method. Other regularization approaches, such as Landwebers's iteration or iterated Tikhonov's regularization, would enable to deal with more regular functions without changing the results presented in this work (Florens, Racine & Centorrino, 2018).

***Proof (Theorem** (7.3))* Note that the fair estimator can be decomposed into a bias and a variance term that will be studied separately

$$\begin{aligned}\hat{h}_{\alpha,\rho} &= (\alpha \mathrm{Id} + \rho F^* F + K^* K)^{-1} K^* \hat{r} \\ &= (\alpha \mathrm{Id} + \rho F^* F + K^* K)^{-1} K^* r + (\alpha \mathrm{Id} + \rho F^* F + K^* K)^{-1} K^* U_n \\ &= (B) + (V).\end{aligned}$$

Then the bias term can be decomposed as

$$\begin{aligned}(B) &= [(\alpha \mathrm{Id} + \rho F^* F + K^* K)^{-1} - (\rho F^* F + K^* K)^{-1}] K^* r + (\rho F^* F + K^* K)^{-1} K^* r \\ &= (B_1) + (B_2).\end{aligned}$$

The operator $(\alpha \mathrm{Id} + \rho F^* F + K^* K)^{-1}$ can be written as

$$\begin{aligned}(\alpha \mathrm{Id} + \rho F^* F + K^* K)^{-1} &= (R_\alpha^{-1}(K) + \rho F^* F)^{-1} \\ &= (\mathrm{Id} + \rho R_\alpha(K) F^* F)^{-1} R_\alpha(K)\end{aligned}$$

Note that condition [A4] ensures that

$$L_\alpha := R_\alpha(K) F^* F = (K^* K + \alpha \mathrm{Id})^{-1} F^* F$$

is a well-defined operator on $\mathcal{E}$. Moreover, condition [A2] ensures that $R_\alpha(K)$ is one-to-one. Hence, the kernel of the operator L_α is the kernel of F. Using the Tikhonov approximation (7.6), we thus have

$$\begin{aligned}(B1) &= [(\mathrm{Id} + \rho L_\alpha)^{-1} R_\alpha(K) - (\mathrm{Id} + \rho L)^{-1} (K^* K)^{-1}] K^* r \\ &= (\mathrm{Id} + \rho L_\alpha)^{-1} (h_\alpha - h_\dagger) + [(\mathrm{Id} + \rho L_\alpha)^{-1} - (\mathrm{Id} + \rho L)^{-1}] h_\dagger\end{aligned}$$

We will study each term separately.

• Since $\|(\mathrm{Id} + \rho L_\alpha)^{-1}\|$ is bounded, we get that the first term is of the same order as $h_\alpha - h_\dagger$. Hence, under the source condition in [A3], we have that

$$\|(\mathrm{Id} + \rho L_\alpha)^{-1} (h_\alpha - h_\dagger)\|^2 = O(\alpha^\beta).$$

- Using that for two operators

$$A^{-1} - B^{-1} = A^{-1}(B-A)B^{-1}$$

we obtain for the second term that

$$\left((\mathrm{Id}+\rho L_\alpha)^{-1} - (\mathrm{Id}+\rho L)^{-1}\right) h_\dagger = \rho(\mathrm{Id}+\rho L_\alpha)^{-1}(L-L_\alpha)(\mathrm{Id}+\rho L)^{-1} h_\dagger.$$

Note that $(L-L_\alpha)Ph_\dagger = 0$ and $(\mathrm{Id}+\rho L)^{-1}Ph_\dagger = Ph_\dagger$.

Hence we can replace $h_\dagger$ in the last expression by the projection onto the orthogonal space of $\mathcal{N}(F)$. Namely, $P^\perp h_\dagger$. Hence

$$\|\left((\mathrm{Id}+\rho L_\alpha)^{-1} - (\mathrm{Id}+\rho L)^{-1}\right) h_\dagger\|^2 = O\left(\rho^2\|L-L_\alpha\|^2\|(\mathrm{Id}+\rho L)^{-1}P^\perp h_\dagger\|^2\right).$$

We have that $(\mathrm{Id}+\rho L)^{-1}P^\perp h_\dagger\|^2 = O(1/\rho^2)$. Then

$$L - L_\alpha = \alpha(\alpha\mathrm{Id}+K^*K)^{-1}(K^*K)^{-1}F^*F.$$

Under Assumption **[E6]**, We obtain that $(K^*K)^{-1}F^*FP^\perp h_\dagger$ is of regularity γ so

$$\|(L-L_\alpha)P^\perp h_\dagger\|^2 = O\left(\alpha^\gamma\right).$$

Hence we can conclude that

$$\|\left((\mathrm{Id}+\rho T_\alpha)^{-1} - (\mathrm{Id}+\rho T)^{-1}\right) h_\dagger\|^2 = O\left(\alpha^\gamma\right).$$

The second term (B_2) is such that $(B_2) = (\rho F^*F + K^*K)^{-1}K^*r$. We can write

$$\begin{aligned}(B_2) &= \left((K^*K)^{1/2}(\mathrm{Id}+\rho(K^*K)^{-1/2}F^*F(K^*K)^{-1/2})(K^*K)^{1/2}\right)^{-1} K^*Kh_\dagger \\ &= (K^*K)^{-1/2}(\mathrm{Id}+\rho T)^{-1}(K^*K)^{1/2}h_\dagger,\end{aligned}$$

where $T := (K^*K)^{-1/2}F^*F(K^*K)^{-1/2}$ is a self-adjoint operator well-defined using Assumption **[A5]**. Let

$$h_\rho = (K^*K)^{-1/2}(\mathrm{Id}+\rho T)^{-1}(K^*K)^{1/2}h_\dagger.$$

- Note first that h_ρ converges when $\rho \to +\infty$ to the projection of $\psi := (K^*K)^{1/2}h_\dagger$ onto $\mathrm{Ker}(T)$. We can write the SVD of T as λ_j^2 and e_j for $j \geq 1$. So we get that

$$\begin{aligned}(\mathrm{Id}+\rho T)^{-1}\psi &= \sum_{j\geq 1} \frac{1}{1+\rho\lambda_j^2} < \psi, e_j > \\ &= \sum_{j\geq 1, \lambda_j\neq 0} \frac{1}{1+\rho\lambda_j^2} < \psi, e_j > e_j + \sum_{j\geq 1, \lambda_j = 0} < \psi, e_j > e_j.\end{aligned}$$

The last quantity converges when $\rho \to +\infty$ towards the projection of ψ onto the kernel of T. Applying the operator $(K^*K)^{-1/2}$ does not change the limit since K^*K is one to one.

- Note then that the kernel of the operator T can be identified as follows

$$\begin{aligned}\{\psi \in \mathrm{Ker}(T)\} &= \{\psi, \quad F(K^*K)^{-1/2}\psi = 0\} \\ &= \{\psi, \quad (K^*K)^{-1/2}\psi \in \mathrm{Ker}(F)\} \\ &= \{\psi = (K^*K)^{1/2}h, \quad h \in \mathrm{Ker}(F)\}.\end{aligned}$$

Hence h_ρ converges towards the projection of $(K^*K)^{1/2}h_\dagger$ onto the functions $(K^*K)^{1/2}h$ with $h \in \mathrm{Ker}(F)$.

- Characterization of the projection. Note that the projection can be written as

$$\begin{aligned}&\arg\min_{h\in\mathrm{Ker}(F)} \|(K^*K)^{1/2}h_\dagger - (K^*K)^{1/2}h\|^2 \\ &= \arg\min_{h\in\mathrm{Ker}(F)} \|(K^*K)^{1/2}(h_\dagger - h)\|^2 \\ &= \arg\min_{h\in\mathrm{Ker}(F)} < (K^*K)^{1/2}(h_\dagger - h), (K^*K)^{1/2}(h_\dagger - h) > \\ &= \arg\min_{h\in\mathrm{Ker}(F)} < h_\dagger - h, (K^*K)(h_\dagger - h) > \\ &= \arg\min_{h\in\mathrm{Ker}(F)} \|K(h_\dagger - h)\|^2 \\ &= \arg\min_{h\in\mathrm{Ker}(F)} \|r - Kh\|^2 \\ &= h_{K_F}\end{aligned}$$

as defined previously.

- Usual bounds enable us to prove that

$$\|h_\rho - h_{K_F}\|^2 = O\left(\frac{1}{\rho^2}\right).$$

Using all previous bounds, we can write

$$\|(B) - Ph_\dagger\|^2 = O(\frac{1}{\rho^2} + \alpha^\beta + \alpha^\gamma). \tag{7.12}$$

Finally, we prove that the variance term (V) is such that

$$\|(\alpha \mathrm{Id} + \rho F^*F + K^*K)^{-1}K^*U_n\|^2 = O\left(\frac{1}{\alpha\delta_n}\right)$$

Using previous notations, we obtain

$$\begin{aligned}\|(\alpha \mathrm{Id} + \rho F^*F + K^*K)^{-1}K^*U_n\| &= \|(\mathrm{Id} + \rho L_\alpha)^{-1}(\alpha \mathrm{Id} + K^*K)^{-1}K^*U_n\| \\ &\le \|(\mathrm{Id} + \rho L_\alpha)^{-1}\|\|(\alpha \mathrm{Id} + K^*K)^{-1}K^*\|\|U_n\|\end{aligned}$$

$$\leq \|(\mathrm{Id} + \rho L_\alpha)^{-1}\| \frac{1}{\alpha} \frac{1}{\delta_n^{1/2}}.$$

Using that $(\mathrm{Id} + \rho L_\alpha)^{-1}$ is bounded leads to the desired result.
Both bounds prove the final result for the theorem. □

Choosing the fairness constraint implies modifying the usual estimator. The following theorem quantifies, for fixed parameters ρ and α, the deviation of the fair estimator (7.10) with respect to the unfair solution of the linear inverse problem.

Theorem 7.4 (Price for fairness)

$$\|h_\alpha - h_{\alpha,\rho}\| = O\left(\frac{\rho}{\alpha^2}\right)$$

Proof

$$\begin{aligned} &\|h_\alpha - h_{\alpha,\rho}\| \\ &\leq \|(\alpha \mathrm{Id} + K^*K)^{-1}K^*r - (\alpha \mathrm{Id} + \rho F^*F + K^*K)^{-1}K^*r\| \\ &\leq \|[(\alpha \mathrm{Id} + K^*K)^{-1} - (\alpha \mathrm{Id} + \rho F^*F + K^*K)^{-1}]K^*r\|. \end{aligned}$$

Using that for two operators

$$A^{-1} - B^{-1} = A^{-1}(B - A)B^{-1}$$

we obtain

$$\|h_\alpha - h_{\alpha,\rho}\| \leq \|(\alpha \mathrm{Id} + \rho F^*F + K^*K)^{-1}\rho F^*F(\alpha \mathrm{Id} + K^*K)^{-1}\| \tag{7.13}$$

Now using that

$$\begin{aligned} \|(\alpha \mathrm{Id} + K^*K)^{-1}\| &\leq \frac{1}{\alpha} \\ (\alpha \mathrm{Id} + \rho F^*F + K^*K)^{-1} &\leq \frac{1}{\alpha} \end{aligned}$$

and since

$$\|K^*r\| \leq M$$

leads to the result. □

The previous theorem suggests that in a decision-making process, the stakeholder's choice can be cast as a function of the parameter ρ. For fixed α, as the parameter ρ diverges to ∞, the fully fair solution is imposed. However, there is also a loss of accuracy in prediction, which increases as ρ diverges. Imposing an*approximately fair* solution can help balance the benefit of fairness, which in certain situations can have a clear economic and reputational cost, with the statistical loss associated with a worse prediction. This trade-off could also be used to determine an optimal choice for the parameter ρ. We illustrate this procedure in Section 7.6.

7.5 Estimation with an Exogenous Binary Sensitive Attribute

We discuss the estimation and the finite sample implementation of our method in the simple case when S is an exogenous binary random variable (for instance, gender or race), and $Z \in \mathbb{R}^p$ only contains continuous endogenous regressors. This framework can be easily extended to the case when S is an endogenous multivariate categorical variable and to include additional exogenous components in Z (Hall & Horowitz, 2005; Centorrino, Fève & Florens, 2017; Centorrino & Racine, 2017). Our statistical model can be written as

$$Y = h_0(Z) + h_1(Z)S + U = \mathbf{S}'h(Z) + U, \tag{7.14}$$

where $h = [h_0 \quad h_1]'$, and $\mathbf{S} = [1 \quad S]'$.

This model is a varying coefficient model (see, among others, Hastie & Tibshirani, 1993; Fan & Zhang, 1999; Li, Huang, Li & Fu, 2002). Adopting the terminology that is used in this literature, we refer to $\mathbf{S}$ as the 'linear' variables (or predictors), and to the Z's as the 'smoothing' variables (or covariates) (Fan & Zhang, 2008). When Z is endogenous, Centorrino and Racine (2017) have studied the identification and estimation of this model with instrumental variables. That is, we assume there is a random vector $W \in \mathbb{R}^q$, such that $E[\mathbf{S}U|W] = 0$, and

$$E[\mathbf{S}\mathbf{S}'h(Z)|W] = 0 \quad \Rightarrow \quad h = 0, \tag{7.15}$$

where equalities are intended almost surely.[3] The *completeness* condition in equation (7.15) is necessary for identification, and it is assumed to hold. As proven in Centorrino and Racine (2017), this condition is implied by the injectivity of the conditional expectation operator (see our Assumption A2), and by the matrix $E[\mathbf{S}\mathbf{S}'|z, w]$ being full rank for almost every (z, w).

We would like to obtain a nonparametric estimator of the functions $\{h_0, h_1\}$ when a fairness constraint is imposed. We use the following operator's notations

$$\begin{aligned}
(K_s h)(w) &= E[\mathbf{S}\mathbf{S}'h(Z)|W = w] \\
(K_s^* \psi)(z) &= E[\mathbf{S}\mathbf{S}'\psi(W)|Z = z] \\
(K^* \psi)(z) &= E[\psi(W)|Z = z],
\end{aligned}$$

for every $h \in L^2(Z)$, and $\psi \in L^2(W)$.

When no fairness constraint is imposed, the regularized approximation to the pair $\{h_0, h_1\}$ is given by

$$h_\alpha = \underset{h \in L^2(Z)}{\arg\min} \|K_s h - r\|^2 + \alpha \|h\|^2, \tag{7.16}$$

where $\|h\|^2 = \|h_0\|^2 + \|h_1\|^2$. That is

[3] Notice that the moment conditions $E[\mathbf{S}U|W] = 0$ are implied by the assumption that $E[U|W, S] = 0$, although they allow to exploit the semiparametric structure of the model and reduce the curse of dimensionality (see Centorrino & Racine, 2017)

$$h_\alpha = \left(\alpha I + K_s^* K_s\right)^{-1} K_s^* r, \tag{7.17}$$

with $r(w) = E\left[\mathbf{S}Y|W = w\right]$.

As in Centorrino and Racine (2017), the quantities in equation (7.17) can be replaced by consistent estimators. Let $\{(Y_i, X_i, W_i), i = 1, \ldots, n\}$ be an iid sample from the joint distribution of (Y, X, W). We denote by

$$\mathbf{Y}_n = \begin{bmatrix} Y_1 \\ Y_2 \\ \vdots \\ Y_n \end{bmatrix}, \qquad \mathbf{S}_n = \begin{bmatrix} I_n \; diag(S_1, S_2, \ldots, S_n) \end{bmatrix},$$

the $n \times 1$ vector, which stacks the observations of the dependent variable and the $n \times 2n$ matrix of predictors, where I_n is the identity matrix of dimension n, and $diag(S_1, S_2, \ldots, S_n)$ is a $n \times n$ diagonal matrix whose diagonal elements are equal to the sample observations of the sensitive attribute S. Similarly, we let

$$\mathbf{D}_{1,n} = \begin{bmatrix} S_1 \\ S_2 \\ \vdots \\ S_n \end{bmatrix}, \text{ and } \mathbf{D}_{0,n} = \begin{bmatrix} 1 - S_1 \\ 1 - S_2 \\ \vdots \\ 1 - S_n \end{bmatrix},$$

two $n \times 1$ vectors stacking the sample observations of S and $1 - S$.

Finally, let $C(\cdot)$ a univariate kernel function, such that $C(\cdot) \geq 0$, and $\int C(u)du = 1$, and $\mathbf{C}(\cdot)$ be a multivariate product kernel. That is, for a vector $\mathbf{u} = \begin{bmatrix} u_1 \; u2 \; \ldots \; u_p \end{bmatrix}'$, with $p \geq 1$, $\mathbf{C}(\mathbf{u}) = C(u_1) \times C(u_2) \times \cdots \times C(u_p)$.

As detailed in Centorrino et al. (2017), the operators K and K^* can be approximated by finite-dimensional matrices of kernel weights. In particular, we have that

$$\underbrace{\hat{K}}_{n \times n} = \left[\mathbf{C}\left(\frac{W_i - W_j}{a_W}\right)\right]_{i,j=1}^{n} \text{ and } \underbrace{\widehat{K^*}}_{n \times n} = \left[\mathbf{C}\left(\frac{Z_i - Z_j}{a_Z}\right)\right]_{i,j=1}^{n},$$

where a_W and a_Z are bandwidth parameters, chosen in such a way that $a_W, a_Z \to 0$, as $n \to \infty$. Therefore,

$$\begin{aligned} \hat{r} &= vec\left((I_2 \otimes \hat{K})\mathbf{S}_n'\mathbf{Y}_n\right) \\ \hat{K}_s &= (I_2 \otimes \hat{K})\mathbf{S}_n'\mathbf{S}_n \\ \widehat{K^*}_s &= (I_2 \otimes \widehat{K^*})\mathbf{S}_n'\mathbf{S}_n, \end{aligned}$$

in a way that

$$\hat{h}_\alpha = \left[\hat{h}_{0,\alpha}\ \hat{h}_{1,\alpha}\right] = (vec(I_n)'\otimes I_n)\left(I_n \otimes \left(\alpha I + \widehat{K^*}_s\hat{K}_s\right)^{-1}\widehat{K^*}_s\hat{r}\right). \tag{7.18}$$

As explained above, the fairness constrain can be characterized by a linear operator F_j, such that $F_j h = 0$, where $j = \{1,2\}$. In the case of Definition 7.1, and exploiting the binary nature of S, the operator F_1 can be approximated by

$$\underbrace{F_{1,n}}_{2n\times 2n} = \begin{bmatrix} \mathbf{0}_n & \mathbf{0}_n \\ \iota_n\left[\left(\mathbf{D}_{1,n}'\mathbf{D}_{1,n}\right)^{-1}\mathbf{D}_{1,n}' - \left(\mathbf{D}_{0,n}'\mathbf{D}_{0,n}\right)^{-1}\mathbf{D}_{0,n}'\right] & \iota_n\left(\mathbf{D}_{1,n}'\mathbf{D}_{1,n}\right)^{-1}\mathbf{D}_{1,n}' \end{bmatrix},$$

where ι_n is a $n\times 1$ vector of ones, and $\mathbf{0}_n$ is a $n\times n$ matrix of zeroes.

In the case of definition 7.2, the fairness operator can be approximated by

$$\underbrace{F_{2,n}}_{2n\times 2n} = \begin{bmatrix} \mathbf{0}_n & \mathbf{0}_n \\ \mathbf{0}_n & I_n \end{bmatrix},$$

In both cases, when the function $h \in \mathcal{F}_j$, we obviously have that $F_j vec(h) = 0$, with $j = \{1,2\}$.

As detailed in Section 7.4, and for $j = \{1,2\}$, the estimator consistent with the fairness constraint can be obtained in several ways

1) By projecting the unconstrained estimator in (7.18) onto the null space of F_j. Let $P_{j,n}$ be the estimator of such projection, then we have that

$$\hat{h}_{\alpha,F,j} = (vec(I_n)'\otimes I_n)\left(I_n \otimes P_{j,n} vec(\hat{h}_\alpha)\right), \tag{7.19}$$

2) By restricting the conditional expectation operator to project onto the null space of F_j. Let

$$\hat{K}_{F,j,s} = \hat{K}_s P_{j.n}, \text{ and } \widehat{K^*}_{F,j,s} = P_{j,n}\widehat{K^*}_s,$$

then

$$\hat{h}_{\alpha,K_F,j} = (vec(I_n)'\otimes I_n)\left(I_n \otimes \left(\alpha I + \widehat{K^*}_{F,j,s}\hat{K}_{F,j,s}\right)^{-1}\widehat{K^*}_{F,j,s}\hat{r}\right), \tag{7.20}$$

3) By modifying the objecting function to include an additional term which penalizes deviations from fairness. That is, we let

$$\hat{h}_{\alpha,\rho,j} = \operatorname*{arg\,min}_{h\in\mathcal{F}_j} ||\hat{K}_s h - \hat{r}||^2 + \alpha||h||^2 + \rho||F_{j,n}h||^2,$$

in a way that

$$\hat{h}_{\alpha,\rho,j} = \left(\alpha I_n + \rho F_{j,n}F_{j,n} + \widehat{K^*}_s\hat{K}_s\right)^{-1}\widehat{K^*}_s\hat{r}. \tag{7.21}$$

For $\rho = 0$, this estimator is equivalent to the unconstrained estimator, $\hat{h}_\alpha$, and, for ρ sufficiently large, it imposes the full fairness constraint.

To implement the estimators above, we need to select several smoothing, $\{a_W, a_Z\}$, and regularization, $\{\alpha, \rho\}$, parameters. For the choice of the tuning parameters $\{a_W, a_Z, \alpha\}$, we follow Centorrino (2016) and use a sequential leave-one-out cross-validation approach. We instead select the regularization parameter ρ, for $j = \{1,2\}$ as

$$\rho_j^* = \arg\min_{\rho} ||\hat{h}_{\alpha,\rho,j} - \hat{h}_\alpha||^2 + \varsigma ||F_{j,n}\hat{h}_{\alpha,\rho,j}||^2, \tag{7.22}$$

with $\varsigma > 0$ a constant. The first term of this criterion function is a statistical loss that we incur when we impose the fairness constraint. The second term instead represents the distance of our estimator to full fairness. The smaller the norm of the second term, the closer we are to obtaining a *fair* estimator. For instance, if our unconstrained estimator, $\hat{h}_\alpha$ is *fair*, then the second term will be identically zero for any value of ρ, while the first term will be zero for $\rho = 0$, and then would increase as $\rho \to \infty$. The constant ς serves as a subjective weight for fairness. In principle, one could set $\varsigma = 1$. Values of ς higher than 1 imply that the decision-maker considers deviations from fairness to be costly and thus prefers them to be penalized more heavily. The opposite is true for values of $\varsigma < 1$.

7.6 An Illustration

We consider the following illustration of the model described in the previous Section. We generate a random vector $\tau = (\tau_1, \tau_2)'$ from a bivariate normal distribution with mean $(0, 0.5)'$ and covariance matrix equal to

$$\Sigma_\tau = \begin{bmatrix} 1 & 2\sin(\pi/12) \\ 2\sin(\pi/12) & 1 \end{bmatrix}.$$

Then, we fix

$$\begin{aligned} W &= -1 + 2\Phi(\tau_1) \\ S &= B(\Phi(\tau_2)), \end{aligned}$$

where $B(\cdot)$ is a Bernoulli distribution with probability parameter equal to $\Phi(\tau_2)$, and Φ is the cdf of a standard normal distribution.

We then let η and U to be independent normal random variables with mean 0 and variances equal to 0.16 and 0.25, respectively, and we generate

$$Z = -1 + 2\Phi\left(W - 0.5S - 0.5WS + 0.5U + \eta\right),$$

and

$$Y = h_0(Z) + h_1(Z)S + U,$$

where $h_0(Z) = 3Z^2$, and

$$h_1(Z) = 1 - 5Z^3.$$

In this illustration, the random variable Z can be thought to be an observable characteristic of the individual, while S could be a sensitive attribute related, for instance, to gender or ethnicity. Notice that the true regression function is *not fair* in the sense of either Definition 7.1 or Definition 7.2. This reflects the fact that real data may contain a bias with respect to the sensitive attribute, which is often the case in practice. We fix the sample size at $n = 1000$, and we use Epanechnikov kernels for estimation.

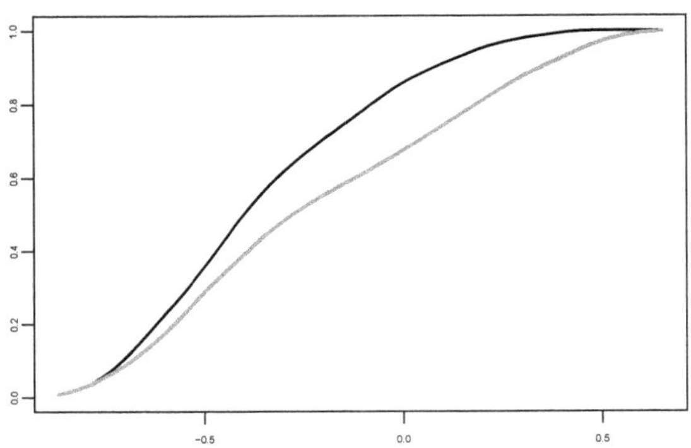

Fig. 7.2: Empirical CDF of the endogenous regressor Z, conditional of the sensitive attribute S. CDF of $Z|S = 0$, solid gray line; CDF of $Z|S = 1$, solid black line.

In Figure 7.2, we plot the empirical cumulative distribution function (CDF) of Z given $S = 0$ (solid gray line), and of Z given $S = 1$ (solid black line). We can see that the latter stochastically dominates the former. This can be interpreted as the fact that systematic differences in group characteristics can generate systematic differences in the outcome, Y, even when the sensitive attribute S is not directly taken into account.

We compare the unconstrained estimator, $\hat{h}_\alpha$, with the *fairness-constrained* estimators in the sense of Definitions 7.1 and 7.2.

In Figures 7.3 and 7.4, we plot the estimators of the functions $\{h_0, h_1\}$, under the fairness constraints in Definitions 7.1 and 7.2, respectively. Notice that, as expected, the estimator which imposes approximate fairness through the penalization parameter ρ lays somewhere in between the unconstrained estimator and the estimators which impose full fairness.

In Figure 7.5, we depict the objective function in equation (7.22) for the optimal choice of ρ, using both Definition 7.1 (left panel) and Definition 7.2 (right panel). The optimal value of ρ is obtained in our case by fixing $\varsigma = 1$ (solid black line).

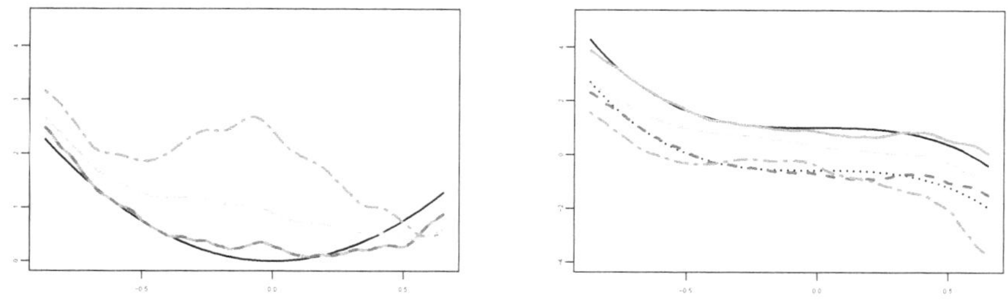

(a) $h_0(x) = 3x^2$ (b) $h_1(x) = 1 - 5x^3$

Fig. 7.3: Estimation using the definition of fairness in 7.1. Solid black line, true function; dotted black line, true function with fairness constraint; solid gray line, $\hat{h}_\alpha$; dashed gray line, $\hat{h}_{\alpha,F}$; dashed-dotted gray line, $\hat{h}_{\alpha,K_F}$; dashed-dotted light-gray line, $\hat{h}_{\alpha,\rho}$.

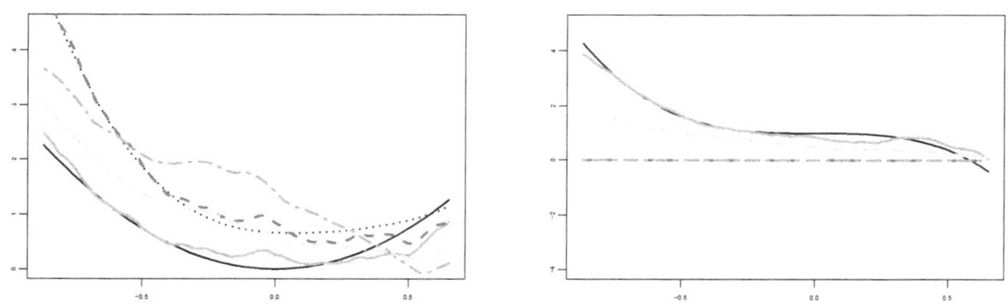

(a) $h_0(x) = 3x^2$ (b) $h_1(x) = 1 - 5x^3$

Fig. 7.4: Estimation using the definition of fairness in 7.2. Solid black line, true function; dotted black line, true function with fairness constraint; solid gray line, $\hat{h}_\alpha$; dashed gray line, $\hat{h}_{\alpha,F}$; dashed-dotted gray line, $\hat{h}_{\alpha,K_F}$; dashed-dotted light-gray line, $\hat{h}_{\alpha,\rho}$.

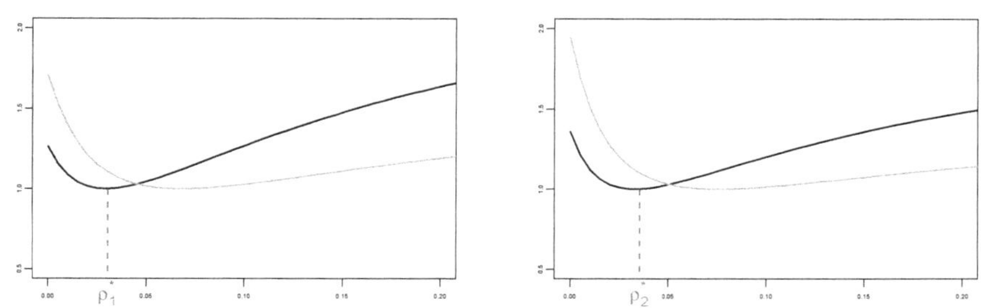

(a) Definition 7.1 (b) Definition 7.2

Fig. 7.5: Choice of the optimal value of ρ.

However, if a decision-maker wished to impose more fairness, this could be achieved by setting $\varsigma > 1$. For illustrative purposes, we also report the objective function when $\varsigma = 2$ (solid gray line). It can be seen that this leads to a larger value of ρ^*, but also that the objective function tends to flatten out.

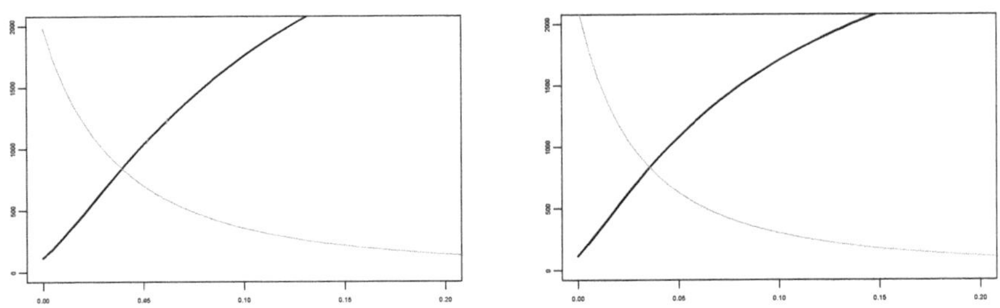

(a) Definition 7.1 (b) Definition 7.2

Fig. 7.6: Cost and benefit of fairness as a function of the penalization parameter ρ.

We also present in Figure 7.6 the trade-off between the statistical loss (solid black line), $||\hat{h}_{\alpha,\rho,j} - \hat{h}_\alpha||^2$, which can be interpreted as the cost of imposing a fair solution, and the benefit of fairness (solid gray line), which is measured by the squared norm of $F_{n,j}\hat{h}_{\alpha,\rho,j}$, when $j = \{1,2\}$ to reflect both Definitions 7.1 (left panel) and 7.2 (right panel). In both cases, we fix $\varsigma = 1$. The upward-sloping line is the squared deviation from the unconstrained estimator, which increases with ρ. The downward-sloping curve is the norm of the projection of the estimator onto the space of fair functions, which converges to zero as ρ increases.

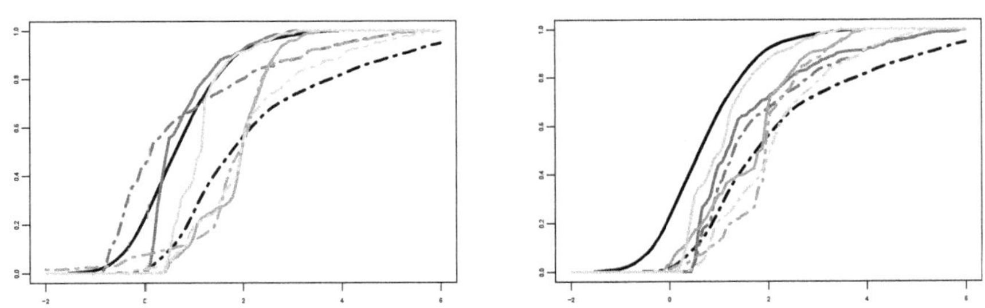

(a) Definition 7.1 (b) Definition 7.2

Fig. 7.7: Density of the predicted values from the constrained models. Solid lines represent group $S = 0$, and dashed-dotted lines group $S = 1$. Black lines are the densities of the observed data; dark-gray lines are from constrained model 1; gray lines from constrained model 2; and light-gray from constrained model 3.

Finally, it is interesting to assess how the different definitions of fairness and the different implementations affect the distribution of the predicted values. This prediction is made *in-sample* as its goal is not to assess the predictive properties of our estimator but rather to determine how the different definitions of fairness and the various ways to impose the fairness constraint in estimation affect the distribution of the model predicted values.

The black lines in Figure 7.7 represent the empirical CDF of the dependent variable Y for $S = 0$ (solid black line), and $S = 1$ (dashed-dotted black line). This is compared with the predictions using estimators 1 (dark-gray lines), 2 (gray lines), and 3 (light-gray lines). In the data, the distribution of Y given $S = 1$ stochastically dominates the distribution of Y given $S = 0$.

Notice that in the case of fairness as defined in 7.1, the estimator which modifies the conditional expectation operator to project directly onto the space of fair functions seems to behave best in terms of fairness, as the distribution of the predicted values for groups 0 and 1 are very similar. The estimator, which imposes approximate fairness, obviously lies somewhere in between the data and the previous estimator. The projection of the unconstrained estimator onto the space of fair functions does not seem to deliver an appropriate distribution of the predicted values. What happens is that this estimator penalizes people in group 1 with low values of Z, to maintain fairness on the average while preserving a substantial difference in the distribution of the two groups.

Differently, in the case of fairness, as defined in 7.2, the projection of the unconstrained estimator seems to behave best. However, this may be because the distribution of Z given $S = 0$ and $S = 1$ are substantially similar. If, however, there is more difference in the observable characteristics by group, this estimator may not behave as intended.

7.7 Conclusions

In this chapter, we consider the issue of estimating a structural econometrics model when a fairness constraint is imposed on the solution. We focus our attention on models when the function is the solution to a linear inverse problem, and the fairness constraint is imposed on the included covariates and can be expressed as a linear restriction on the function of interest. We also discuss how to construct an *approximately fair* solution to a linear functional equation and how this notion can be implemented to balance accurate predictions with the benefits of a fair machine learning algorithm. We further present regularity conditions under which the fair approximation converges towards the projection of the true function onto the null space of the fairness operator. Our leading example is a nonparametric instrumental variable model, in which the fairness constraint is imposed. We detail the example of such a model when the sensitive attribute is binary and exogenous (Centorrino & Racine, 2017).

The framework introduced in this chapter can be extended in several directions. The first significant extension would be to consider models in which the function $h_\dagger$ is the solution to a nonlinear equation. The latter can arise, for instance, when the conditional mean independence restriction is replaced with full independence between the instrumental variable and the structural error term (Centorrino, Fève & Florens, 2019; Centorrino & Florens, 2021). Moreover, one can potentially place fairness restrictions directly on the decision algorithm or on the distribution of predicted values. These restrictions usually imply that the fairness constraint is nonlinear, and a different identification and estimation approach should be employed.
In this work, we restrict to fairness as a *group* notion and do not consider fairness at an individual level as in Kusner, Loftus, Russell and Silva (2017), or De Lara, González-Sanz, Asher and Loubes (2021). This framework could enable a deeper understanding of fairness in econometrics from a causal point of view.
Finally, the fairness constraint imposed in this paper is limited to the regression function, h. However, other constraints may be imposed directly on the functional equation. For instance, on the selection of the instrumental variables, which will be the topic of future work.

Acknowledgements Jean-Pierre Florens acknowledges funding from the French National Research Agency (ANR) under the Investments for the Future program (Investissements d'Avenir, grant ANR-17-EURE-0010).

References

Angwin, J., Larson, J., Mattu, S. & Kirchner, L. (2016). Machine bias risk assessments in criminal sentencing. *ProPublica, May, 23.*

Barocas, S. & Selbst, A. D. (2016). Big data's disparate impact. *Calif. L. Rev., 104*, 671.

Besse, P., del Barrio, E., Gordaliza, P., Loubes, J.-M. & Risser, L. (2021). A survey of bias in machine learning through the prism of statistical parity. *The American Statistician*, 1–11.

Carrasco, M., Florens, J.-P. & Renault, E. (2007). Linear inverse problems in structural econometrics estimation based on spectral decomposition and regularization. In J. Heckman & E. Leamer (Eds.), *Handbook of econometrics* (p. 5633-5751). Elsevier.

Centorrino, S. (2016). Data-Driven Selection of the Regularization Parameter in Additive Nonparametric Instrumental Regressions. *Mimeo - Stony Brook University.*

Centorrino, S., Fève, F. & Florens, J.-P. (2017). Additive Nonparametric Instrumental Regressions: a Guide to Implementation. *Journal of Econometric Methods, 6*(1).

Centorrino, S., Fève, F. & Florens, J.-P. (2019). Nonparametric Instrumental Regressions with (Potentially Discrete) Instruments Independent of the Error

Term. *Mimeo - Stony Brook University.*
Centorrino, S. & Florens, J.-P. (2021). Nonparametric estimation of accelerated failure-time models with unobservable confounders and random censoring. *Electronic Journal of Statistics*, *15*(2), 5333 – 5379.
Centorrino, S. & Racine, J. S. (2017). Semiparametric Varying Coefficient Models with Endogenous Covariates. *Annals of Economics and Statistics*(128), 261–295.
Chouldechova, A. (2017). Fair prediction with disparate impact: A study of bias in recidivism prediction instruments. *Big data*, *5*(2), 153–163.
Chzhen, E., Denis, C., Hebiri, M., Oneto, L. & Pontil, M. (2020). Fair regression with wasserstein barycenters. *arXiv preprint arXiv:2006.07286*.
Darolles, S., Fan, Y., Florens, J. P. & Renault, E. (2011). Nonparametric Instrumental Regression. *Econometrica*, *79*(5), 1541–1565.
De-Arteaga, M., Romanov, A., Wallach, H., Chayes, J., Borgs, C., Chouldechova, A., . . . Kalai, A. T. (2019). Bias in bios: A case study of semantic representation bias in a high-stakes setting. In *Proceedings of the Conference on Fairness, Accountability, and Transparency* (pp. 120–128).
De Lara, L., González-Sanz, A., Asher, N. & Loubes, J.-M. (2021). Transport-based counterfactual models. *arXiv preprint arXiv:2108.13025*.
Engl, H. W., Hanke, M. & Neubauer, A. (1996). *Regularization of inverse problems* (Vol. 375). Springer Science & Business Media.
Fan, J. & Zhang, W. (1999). Statistical Estimation in Varying Coefficient Models. *Ann. Statist.*, *27*(5), 1491–1518.
Fan, J. & Zhang, W. (2008). Statistical Methods with Varying Coefficient Models. *Statistics and Its Interface*, *1*, 179–195.
Florens, J. P., Heckman, J. J., Meghir, C. & Vytlacil, E. (2008). Identification of Treatment effects using Control Functions in Models with Continuous, Endogenous Treatment and Heterogenous Effects. *Econometrica*, *76*(5), 1191–1206.
Florens, J. P., Mouchart, M. & Rolin, J. (1990). *Elements of Bayesian Statistics*. M. Dekker.
Florens, J.-P., Racine, J. & Centorrino, S. (2018). Nonparametric Instrumental Variable Derivative Estimation. *Jounal of Nonparametric Statistics*, *30*(2), 368-391.
Friedler, S. A., Scheidegger, C. & Venkatasubramanian, S. (2021, mar). The (im)possibility of fairness: Different value systems require different mechanisms for fair decision making. *Commun. ACM*, *64*(4), 136–143. Retrieved from https://doi.org/10.1145/3433949 doi: 10.1145/3433949
Gordaliza, P., Del Barrio, E., Fabrice, G. & Loubes, J.-M. (2019). Obtaining fairness using optimal transport theory. In *International conference on machine learning* (pp. 2357–2365).
Hall, P. & Horowitz, J. L. (2005). Nonparametric Methods for Inference in the Presence of Instrumental Variables. *Annals of Statistics*, *33*(6), 2904–2929.
Hastie, T. & Tibshirani, R. (1993). Varying-Coefficient Models. *Journal of the Royal Statistical Society. Series B (Methodological)*, *55*(4), 757-796.

Hoda, H., Loi, M., Gummadi, K. P. & Krause, A. (2018). A moral framework for understanding of fair ml through economic models of equality of opportunity. *Machine Learning*(a).

Hu, L. & Chen, Y. (2020). Fair classification and social welfare. In *Proceedings of the 2020 conference on fairness, accountability, and transparency* (pp. 535–545).

Jiang, R., Pacchiano, A., Stepleton, T., Jiang, H. & Chiappa, S. (2020). Wasserstein fair classification. In *Uncertainty in artificial intelligence* (pp. 862–872).

Kasy, M. & Abebe, R. (2021). Fairness, Equality, and Power in Algorithmic Decision-Making. In *Proceedings of the 2021 ACM Conference on Fairness, Accountability, and Transparency* (pp. 576–586). New York, NY, USA: Association for Computing Machinery.

Kusner, M. J., Loftus, J., Russell, C. & Silva, R. (2017). Counterfactual fairness. *Advances in neural information processing systems, 30.*

Lee, M. S. A., Floridi, L. & Singh, J. (2021). Formalising trade-offs beyond algorithmic fairness: lessons from ethical philosophy and welfare economics. *AI and Ethics*, 1–16.

Le Gouic, T., Loubes, J.-M. & Rigollet, P. (2020). Projection to fairness in statistical learning. *arXiv e-prints*, arXiv–2005.

Li, Q., Huang, C. J., Li, D. & Fu, T.-T. (2002). Semiparametric Smooth Coefficient Models. *Journal of Business & Economic Statistics*, *20*(3), 412-422.

Loubes, J.-M. & Rivoirard, V. (2009). Review of rates of convergence and regularity conditions for inverse problems. *Int. J. Tomogr. Stat*, *11*(S09), 61–82.

McIntyre, F. & Simkovic, M. (2018). Are law degrees as valuable to minorities? *International Review of Law and Economics*, *53*, 23–37.

Menon, A. K. & Williamson, R. C. (2018, 2). The cost of fairness in binary classification. In S. A. Friedler & C. Wilson (Eds.), *Proceedings of the 1st conference on fairness, accountability and transparency* (Vol. 81, pp. 107–118). New York, NY, USA: PMLR.

Natterer, F. (1984). Error bounds for tikhonov regularization in hilbert scales. *Applicable Analysis*, *18*(1-2), 29–37.

Newey, W. K. & Powell, J. L. (2003). Instrumental Variable Estimation of Nonparametric Models. *Econometrica*, *71*(5), 1565–1578.

Oneto, L. & Chiappa, S. (2020). Fairness in machine learning. In L. Oneto, N. Navarin, A. Sperduti & D. Anguita (Eds.), *Recent trends in learning from data: Tutorials from the inns big data and deep learning conference (innsbddl2019)* (pp. 155–196). Cham: Springer International Publishing. doi: 10.1007/978-3-030-43883-8_7

Rambachan, A., Kleinberg, J., Mullainathan, S. & Ludwig, J. (2020). *An economic approach to regulating algorithms* (Tech. Rep.). Cambridge, MA: National Bureau of Economic Research.

Risser, L., Sanz, A. G., Vincenot, Q. & Loubes, J.-M. (2019). Tackling algorithmic bias in neural-network classifiers using wasserstein-2 regularization. *arXiv preprint arXiv:1908.05783.*

Chapter 8
Graphical Models and their Interactions with Machine Learning in the Context of Economics and Finance

Ekaterina Seregina

Abstract Many economic and financial systems, including financial markets, financial institutions, and macroeconomic policy making can be modelled as systems of interacting agents. Graphical models, which are the main focus of this chapter, are a means of estimating the relationships implied by such systems. The main goals of this chapter are (1) acquainting the readers with graphical models; (2) reviewing the existing research on graphical models for economic and finance problems; (3) reviewing the literature that merges graphical models with other machine learning methods in economics and finance.

8.1 Introduction

Technological advances have made large data sets available for scientific discovery. Extracting information about interdependence between many variables in rich data sets plays an important role in various applications, such as portfolio management, risk assessment, forecast combinations, classification, as well as running generalized least squares regressions on large cross-sections, and choosing an optimal weighting matrix in the general method of moments (GMM). The goal of estimating variable dependencies can be formulated as a search for a covariance matrix estimator that contains pairwise relationships between variables. However, as shown in this chapter, in many applications what is required is not a covariance matrix, but its inverse, which is known as a precision matrix. Instead of direct pairwise dependencies, precision matrix contains the information about partial pairwise dependencies conditional on the remaining variables.

Originating from the literature on high-dimensional statistics and machine learning, graphical models search for the estimator of precision matrix. Building on Hastie, Tibshirani and Friedman (2001); Pourahmadi (2013) and Bishop (2006), we review

Ekaterina Seregina✉
Colby College, Waterville, ME, USA e-mail: eseregin@colby.edu

F. Chan, L. Mátyás (eds.), *Econometrics with Machine Learning*, Advanced Studies in Theoretical and Applied Econometrics 53, https://doi.org/10.1007/978-3-031-15149-1_8

the terminology used in the literature on graphical models, such as vertices, edges, directed and undirected graphs, partial correlations, sparse and fully connected graphs. We explain the connection between entries of the precision matrix, partial correlations, graph sparsity and the implications of graph structure for common economic and finance problems. We further review several prominent approaches for estimating a high-dimensional graph: Graphical LASSO (Friedman, Hastie & Tibshirani, 2007), nodewise regression (Meinshausen & Bühlmann, 2006) and CLIME (T. Cai, Liu & Luo, 2011); as well as several computational methods to recover the entries of precision matrix including stochastic gradient descent and the alternating direction method of multipliers (ADMM).

Having introduced readers to graphical models, we proceed by reviewing the literature that applies this tool to tackle economic and finance problems. Such applications include portfolio construction, see Brownlees, Nualart and Sun (2018), Barigozzi, Brownlees and Lugosi (2018), Koike (2020), Callot, Caner, Önder and Ulasan (2019) among others, and forecast combination for macroeconomic forecasting Lee and Seregina (2021a). We elaborate on what is common between such applications and why sparse precision matrix estimator is desirable in these settings. Further, we highlight the drawbacks encountered by early models that used graphical models in economic or finance context. Noticeably, such drawbacks are caused by the lack of reconciliation between models borrowed from statistical literature and stylized economic facts observed in practice. The chapter proceeds by reviewing most recent studies that propose solutions to overcome these drawbacks.

In the concluding part of the chapter, we aspire to demonstrate that graphical models are not a stand-alone tool and there are several promising directions how they can be integrated with other machine learning methods in the context of economics and finance. Along with reviewing the existing research that has attempted to tackle the aforementioned issue, we outline several directions that have been examined to a lesser extent but which we deem to be worth exploring.

8.1.1 Notation

Throughout the chapter $\mathcal{S}_p$ denote the set of all $p \times p$ symmetric matrices, and $\mathcal{S}_p^{++}$ denotes the set of all $p \times p$ positive definite matrices. For any matrix $\mathbf{C}$, its (i,j)-th element is denoted as c_{ij}. Given a vector $\mathbf{u} \in \mathbb{R}^d$ and parameter $a \in [1,\infty)$, let $\|\mathbf{u}\|_a$ denote ℓ_a-norm. Given a matrix $\mathbf{U} \in \mathcal{S}_p$, let $\Lambda_{\max}(\mathbf{U}) \equiv \Lambda_1(\mathbf{U}) \geq \Lambda_2(\mathbf{U}) \geq \ldots \geq \Lambda_{\min}(\mathbf{U}) \equiv \Lambda_p(\mathbf{U})$ be the eigenvalues of $\mathbf{U}$, and $\text{eig}_K(\mathbf{U}) \in \mathbb{R}^{K\times p}$ denote the first $K \leq p$ normalized eigenvectors corresponding to $\Lambda_1(\mathbf{U}),\ldots,\Lambda_K(\mathbf{U})$. Given parameters $a,b \in [1,\infty)$, let $|||\mathbf{U}|||_{a,b} \equiv \max_{\|\mathbf{y}\|_a=1}\|\mathbf{U}\mathbf{y}\|_b$ denote the induced matrix-operator norm. The special cases are $|||\mathbf{U}|||_1 \equiv \max_{1\leq j\leq N}\sum_{i=1}^N |u_{i,j}|$ for the ℓ_1/ℓ_1-operator norm; the operator norm (ℓ_2-matrix norm) $|||\mathbf{U}|||_2^2 \equiv \Lambda_{\max}(\mathbf{U}\mathbf{U}')$ is equal to the maximal singular value of $\mathbf{U}$; $|||\mathbf{U}|||_\infty \equiv \max_{1\leq j\leq N}\sum_{i=1}^N |u_{j,i}|$ for the ℓ_∞/ℓ_∞-operator norm. Finally, $\|\mathbf{U}\|_{\max} \equiv \max_{i,j}|u_{i,j}|$ denotes the element-wise maximum, and $|||\mathbf{U}|||_F^2 \equiv \sum_{i,j} u_{i,j}^2$ denotes the Frobenius matrix norm. Additional more specific notations are introduced throughout the chapter.

8.2 Graphical Models: Methodology and Existing Approaches

A detailed coverage of the terminology used in the network theory literature was provided in Chapter 6. We start with a brief summary to refresh and empasize the concepts that we use for studying graphical models. A *graph* consists of a set of *vertices* (nodes) and a set of *edges* (arcs) that join some pairs of the vertices. In graphical models, each vertex represents a random variable, and the graph visualizes the joint distribution of the entire set of random variables. Figure 8.1 shows a simplified example of a graph for five random variables, and Figure 8.2 depicts a larger network that has several color-coded clusters.

The graphs can be separated into *undirected graphs*-where the edges have no directional arrows, and *directed graphs*-where the edges have a direction associated with them. Directed graphs are useful for expressing causal relationship between random variables (see Pearl, 1995; Verma & Pearl, 1990 among others), whereas undirected graphs are used to study the conditional dependencies between variables. As shown later, the edges in a graph are parameterized by *potentials* (values) that encode the strength of the conditional dependence between the random variables at the corresponding vertices. *Sparse graphs* have a relatively small number of edges. Among the main challenges in working with graphical models are choosing the structure of the graph (*model selection*) and estimation of the edge parameters from the data.

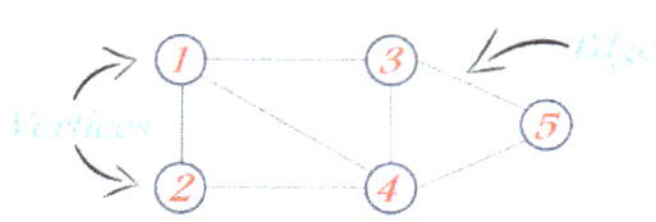

Fig. 8.1: A sample network: zoom in

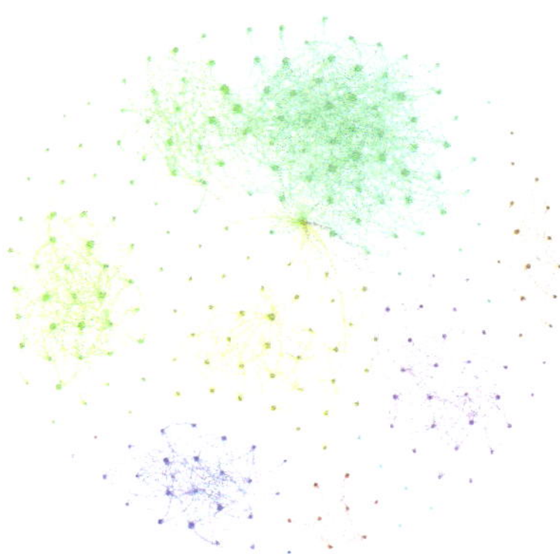

Fig. 8.2: A sample network: zoom out

Let $X = (X_1, \ldots, X_p)$ be random variables that have a multivariate Gaussian distribution with an expected value μ and covariance matrix $\mathbf{\Sigma}$, $X \sim \mathcal{N}(\mu, \mathbf{\Sigma})$. We note that even though the normality assumption is not required for estimating graphical models, it helps us illustrate the relationship between partial correlations and the entries of precision matrix.

The precision matrix $\mathbf{\Sigma}^{-1} \equiv \mathbf{\Theta}$ contains information about pairwise covariances between the variables conditional on the rest, which is known as "partial covariances". For instance, if θ_{ij}, which is the ij-th element of the precision matrix, is zero, then the variables i and j are conditionally independent, given the other variables. In order to gain more insight about Gaussian graphical models, let us partition $X = (Z, Y)$, where $Z = (X_1, \ldots, X_{p-1})$ and $Y = X_p$. Given the length of the sample size T, let $\mathbf{x}_t = (x_{1t}, \ldots, x_{pt})$, $\mathbf{X} = (\mathbf{x}'_1, \ldots, \mathbf{x}'_p)$, $\mathbf{z}_t = (x_{1t}, \ldots, x_{(p-1)t})$, $y_t = x_{pt}$, $\mathbf{Z} = (\mathbf{z}'_1, \ldots, \mathbf{z}'_{p-1})$, and $\mathbf{y} = (y_1, \ldots, y_T)'$ denote realizations of the respective random variables, where $t = 1, \ldots, T$. Then with the partitioned covariance matrix given in (8.1), we can write the conditional distribution of Y given Z:

$$\mathbf{\Sigma} = \begin{pmatrix} \mathbf{\Sigma}_{ZZ} & \boldsymbol{\sigma}_{ZY} \\ \boldsymbol{\sigma}'_{ZY} & \sigma_{YY} \end{pmatrix}, \tag{8.1}$$

$$Y|Z = z \sim \mathcal{N}\left(\mu_Y + (z - \boldsymbol{\mu}_Z)'\mathbf{\Sigma}_{ZZ}^{-1}\boldsymbol{\sigma}_{ZY}, \sigma_{YY} - \boldsymbol{\sigma}'_{ZY}\mathbf{\Sigma}_{ZZ}^{-1}\boldsymbol{\sigma}_{ZY}\right). \tag{8.2}$$

Note that the conditional mean in (8.2) is determined by the regression coefficient of population multiple linear regression of Y on Z, denoted as $\boldsymbol{\beta}_{Y|Z} = (\beta_1, \ldots, \beta_{p-1})'$, which can be expressed as the familiar OLS estimator introduced in Chapter 1 $\boldsymbol{\beta}_{Y|Z} = \mathbf{\Sigma}_{ZZ}^{-1}\boldsymbol{\sigma}_{ZY}$ with a Gram Matrix $\mathbf{\Sigma}_{ZZ} = \mathbf{Z}'\mathbf{Z}$ and $\boldsymbol{\sigma}_{ZY} = \mathbf{Z}'\mathbf{y}$.

If $\beta_j = 0$ for any $j = 1, \ldots, p-1$, then Y and Z_j are conditionally independent, given the rest. Therefore, the regression coefficients $\boldsymbol{\beta}_{Y|Z}$ determine the *conditional (in)dependence structure* of the graph. We reiterate that independence in this case is a function of normality, when the distribution is no longer normal $\beta_j = 0$ is not sufficient for independence. Let us partition $\mathbf{\Theta}$ in the same way as in (8.1):

$$\mathbf{\Theta} = \begin{pmatrix} \mathbf{\Theta}_{ZZ} & \boldsymbol{\theta}_{ZY} \\ \boldsymbol{\theta}'_{ZY} & \theta_{YY} \end{pmatrix}.$$

Remark 8.1 Let $\mathbf{M}$ be an $(m+1) \times (m+1)$ matrix partitioned into a block form:

$$\mathbf{M} = \begin{pmatrix} \underbrace{\mathbf{A}}_{m \times m} & \underbrace{\mathbf{b}}_{m \times 1} \\ \mathbf{b}' & c \end{pmatrix}.$$

We can use the following standard formula for partitioned inverses (see Cullen, 1990; Eves, 2012 among others):

$$\mathbf{M}^{-1} = \begin{pmatrix} \left(\mathbf{A} - \frac{1}{c}\mathbf{b}\mathbf{b}'\right)^{-1} & -\frac{1}{k}\mathbf{A}^{-1}\mathbf{b} \\ -\frac{1}{k}\mathbf{b}'\mathbf{A}^{-1} & \frac{1}{k} \end{pmatrix} = \begin{pmatrix} \mathbf{A}^{-1} + \frac{1}{k}\mathbf{A}^{-1}\mathbf{b}\mathbf{b}'\mathbf{A}^{-1} & -\frac{1}{k}\mathbf{A}^{-1}\mathbf{b} \\ -\frac{1}{k}\mathbf{b}'\mathbf{A}^{-1} & \frac{1}{k} \end{pmatrix}, \tag{8.3}$$

where $k = c - \mathbf{b}'\mathbf{A}^{-1}\mathbf{b}$.

Now apply (8.3) and use $\mathbf{\Sigma\Theta} = \mathbf{I}$, where $\mathbf{I}$ is an identity matrix, to get:

$$\begin{pmatrix} \mathbf{\Theta}_{ZZ} & \boldsymbol{\theta}_{ZY} \\ \boldsymbol{\theta}'_{ZY} & \theta_{YY} \end{pmatrix} = \begin{pmatrix} \mathbf{\Theta}_{ZZ} & -\theta_{YY}\mathbf{\Sigma}_{ZZ}^{-1}\boldsymbol{\sigma}_{ZY} \\ \boldsymbol{\theta}'_{ZY} & (\sigma_{YY} - \boldsymbol{\sigma}_{ZY}\mathbf{\Sigma}_{ZZ}^{-1}\boldsymbol{\sigma}_{ZY})^{-1} \end{pmatrix}, \tag{8.4}$$

where $1/\theta_{YY} = \sigma_{YY} - \boldsymbol{\sigma}'_{ZY}\mathbf{\Sigma}_{ZZ}^{-1}\boldsymbol{\sigma}_{ZY}$. From (8.4):

$$\boldsymbol{\theta}_{ZY} = -\theta_{YY}\mathbf{\Sigma}_{ZZ}^{-1}\boldsymbol{\sigma}_{ZY} = -\theta_{YY}\boldsymbol{\beta}_{Y|Z},$$

therefore,

$$\boldsymbol{\beta}_{Y|Z} = \frac{-\boldsymbol{\theta}_{ZY}}{\theta_{YY}}.$$

Hence, zero elements in $\boldsymbol{\beta}_{Y|Z}$ correspond to zeroes in $\boldsymbol{\theta}_{ZY}$ and mean that the corresponding elements of Z are conditionally independent of Y, given the rest. Therefore, $\mathbf{\Theta}$ contains all the conditional dependence information for the multivariate Gaussian model.

Let $\mathbf{W}$ be the estimate of $\mathbf{\Sigma}$. In practice, $\mathbf{W}$ can be any pilot estimator of covariance matrix. Given a sample $\{\mathbf{x}_t\}_{t=1}^{T}$, let $\mathbf{S} = (1/T)\sum_{t=1}^{T}(\mathbf{x}_t - \bar{\mathbf{x}}_t)(\mathbf{x}_t - \bar{\mathbf{x}}_t)'$ denote the sample covariance matrix, which can be used as a choice for $\mathbf{W}$. Also, let $\widehat{\mathbf{D}}^2 \equiv \text{diag}(\mathbf{W})$. We can write down the Gaussian log-likelihood (up to constants) $l(\mathbf{\Theta}) = \log\det(\mathbf{\Theta}) - \text{trace}(\mathbf{W\Theta})$. When $\mathbf{W} = \mathbf{S}$, the maximum likelihood estimator of $\mathbf{\Theta}$ is $\widehat{\mathbf{\Theta}} = \mathbf{S}^{-1}$.

In the high-dimensional settings it is necessary to regularize the precision matrix, which means that some edges will be zero. In the following subsections we discuss two most widely used techniques to estimate sparse high-dimensional precision matrices.

8.2.1 Graphical LASSO

The first approach to induce sparsity in the estimation of precision matrix is to add penalty to the maximum likelihood and use the connection between the precision matrix and regression coefficients to maximize the following *weighted penalized log-likelihood* (Janková & van de Geer, 2018):

$$\widehat{\mathbf{\Theta}}_\lambda = \arg\min_{\mathbf{\Theta}} \text{trace}(\mathbf{W\Theta}) - \log\det(\mathbf{\Theta}) + \lambda\sum_{i\neq j}\widehat{d}_{ii}\widehat{d}_{jj}|\theta_{ij}|, \tag{8.5}$$

over positive definite symmetric matrices, where $\lambda \geq 0$ is a penalty parameter and $\widehat{d}_{ii}, \widehat{d}_{jj}$ are the i-th and j-th diagonal entries of $\widehat{\mathbf{D}}$. The subscript λ in $\widehat{\boldsymbol{\Theta}}_{\lambda}$ means that the solution of the optimization problem in (8.5) will depend upon the choice of the tuning parameter. More details on the latter are provided in Janková and van de Geer (2018); Lee and Seregina (2021b) that describe how to choose the shrinkage intensity in practice. In order to simplify notation, we will omit the subscript.

The objective function in (8.5) extends the family of linear shrinkage estimators of the first moment studied in Chapter 1 to linear shrinkage estimators of the inverse of the second moments. Instead of restricting the number of regressors for estimating conditional mean, equation (8.5) restricts the number of edges in a graph by shrinking some off-diagonal entries of precision matrix to zero. We draw readers' attention to the following: first, shrinkage occurs adaptively with respect to partial covariances normalized by individual variances of both variables; second, only off-diagonal partial correlations are shrunk to zero since the goal is to identify variables with strongest pairwise conditional dependencies.

One of the most popular and fast algorithms to solve the optimization problem in (8.5) is called the Graphical LASSO (GLASSO), which was introduced by Friedman et al. (2007). Define the following partitions of $\mathbf{W}$, $\mathbf{S}$ and $\boldsymbol{\Theta}$:

$$\mathbf{W} = \begin{pmatrix} \underbrace{\mathbf{W}_{11}}_{(p-1)\times(p-1)} & \underbrace{\mathbf{w}_{12}}_{(p-1)\times 1} \\ \mathbf{w}'_{12} & w_{22} \end{pmatrix}, \mathbf{S} = \begin{pmatrix} \underbrace{\mathbf{S}_{11}}_{(p-1)\times(p-1)} & \underbrace{\mathbf{s}_{12}}_{(p-1)\times 1} \\ \mathbf{s}'_{12} & s_{22} \end{pmatrix}, \boldsymbol{\Theta} = \begin{pmatrix} \underbrace{\boldsymbol{\Theta}_{11}}_{(p-1)\times(p-1)} & \underbrace{\boldsymbol{\theta}_{12}}_{(p-1)\times 1} \\ \boldsymbol{\theta}'_{12} & \theta_{22} \end{pmatrix}.$$

Let $\boldsymbol{\beta} \equiv -\boldsymbol{\theta}_{12}/\theta_{22}$. The idea of GLASSO is to set $\mathbf{W} = \mathbf{S} + \lambda \mathbf{I}$ in (8.5) and combine the gradient of (8.5) with the formula for partitioned inverses to obtain the following ℓ_1-regularized quadratic program

$$\widehat{\boldsymbol{\beta}} = \underset{\boldsymbol{\beta} \in \mathbb{R}^{p-1}}{\arg\min} \left\{ \frac{1}{2} \boldsymbol{\beta}' \mathbf{W}_{11} \boldsymbol{\beta} - \boldsymbol{\beta}' \mathbf{s}_{12} + \lambda \|\boldsymbol{\beta}\|_1 \right\}, \tag{8.6}$$

As shown by Friedman et al. (2007), (8.6) can be viewed as a LASSO regression, where the LASSO estimates are functions of the inner products of $\mathbf{W}_{11}$ and s_{12}. Hence, (8.5) is equivalent to p coupled LASSO problems. Once we obtain $\widehat{\boldsymbol{\beta}}$, we can estimate the entries $\boldsymbol{\Theta}$ using the formula for partitioned inverses.

The LASSO penalty in (8.5) can produce a sparse estimate of the precision matrix. However, as was pointed out in Chapter 1 when discussing linear shrinkage estimators of the first moment, it produces substantial biases in the estimates of nonzero components. An approach to de-bias regularized estimators was proposed in Janková and van de Geer (2018). Let $\widehat{\boldsymbol{\Theta}} \equiv \mathbf{S}^{-1}$ be the maximum likelihood estimator of the precision matrix, $\boldsymbol{\Theta}_0 \equiv \boldsymbol{\Sigma}_0^{-1}$ is the true value which is assumed to exist. Janková and van de Geer (2018) show that asymptotic linearity of $\widehat{\boldsymbol{\Theta}}$ follows from the decomposition:

$$\widehat{\mathbf{\Theta}} - \mathbf{\Theta}_0 = \mathbf{\Theta}_0(\mathbf{S} - \mathbf{\Sigma}_0)\widehat{\mathbf{\Theta}} = -\mathbf{\Theta}_0(\mathbf{S} - \mathbf{\Sigma}_0)\mathbf{\Theta}_0 + \underbrace{\left(- \mathbf{\Theta}_0(\mathbf{S}\mathbf{\Sigma}_0)(\widehat{\mathbf{\Theta}} - \mathbf{\Theta}_0)\right)}_{\text{rem}_0}. \tag{8.7}$$

Note that the remainder term in (8.7) satisfies $||\text{rem}_0||_\infty = o(1/\sqrt{T})$[1], where we use the notation $||\mathbf{A}||_\infty = \max_{1 \le i,j \le p}|a_{ij}|$ for the supremum norm of a matrix $\mathbf{A}$.

Now consider the standard formulation of the Graphical LASSO in (8.5). Its' gradient is given by:

$$\widehat{\mathbf{\Theta}}^{-1} - \mathbf{S} - \lambda \cdot \widehat{\mathbf{\Gamma}} = \mathbf{0}, \tag{8.8}$$

where $\widehat{\mathbf{\Gamma}}$ is a matrix of component-wise signs of $\widehat{\mathbf{\Theta}}$:

$$\begin{aligned}
\gamma_{jk} &= 0 \quad \text{if } j = k, \\
\gamma_{jk} &= \text{sign}(\hat{\theta}_{jk}) \quad \text{if } \hat{\theta}_{jk} \neq 0, \\
\gamma_{jk} &= \in [-1, 1] \quad \text{if } \hat{\theta}_{jk} = 0.
\end{aligned}$$

Post-multiply (8.8) by $\widehat{\mathbf{\Theta}}$:

$$\mathbf{I} - \mathbf{S}\widehat{\mathbf{\Theta}} - \lambda\widehat{\mathbf{\Gamma}}\widehat{\mathbf{\Theta}} = \mathbf{0},$$

therefore,

$$\mathbf{S}\widehat{\mathbf{\Theta}} = \mathbf{I} - \lambda\widehat{\mathbf{\Gamma}}\widehat{\mathbf{\Theta}}.$$

Consider the following decomposition:

$$\widehat{\mathbf{\Theta}} + \widehat{\mathbf{\Theta}}'\eta(\widehat{\mathbf{\Theta}}) - \mathbf{\Theta}_0 = -\mathbf{\Theta}_0(\mathbf{S} - \mathbf{\Sigma}_0)\mathbf{\Theta}_0 + \text{rem}_0 + \text{rem}_1,$$

where $\eta(\widehat{\mathbf{\Theta}}) = \lambda\widehat{\mathbf{\Gamma}}\widehat{\mathbf{\Theta}}$ is the bias term, rem_0 is the same as in (8.7), and $\text{rem}_1 = (\widehat{\mathbf{\Theta}} - \mathbf{\Theta}_0)'\eta(\widehat{\mathbf{\Theta}})$.

Proof

$$\begin{aligned}
& -\mathbf{\Theta}_0(\mathbf{S} - \mathbf{\Sigma}_0)\widehat{\mathbf{\Theta}} + (\widehat{\mathbf{\Theta}} - \mathbf{\Theta}_0)'\eta(\widehat{\mathbf{\Theta}}) \\
&= \mathbf{\Theta}_0(\mathbf{I} - \lambda\widehat{\mathbf{\Gamma}}\widehat{\mathbf{\Theta}}) + \mathbf{\Theta}_0\mathbf{\Sigma}_0\widehat{\mathbf{\Theta}} + \widehat{\mathbf{\Theta}}'\eta(\widehat{\mathbf{\Theta}}) - \mathbf{\Theta}_0'\eta(\widehat{\mathbf{\Theta}}) \\
(\mathbf{\Theta}_0 \text{ is symmetric}) \quad &= \widehat{\mathbf{\Theta}} + \widehat{\mathbf{\Theta}}'\eta(\widehat{\mathbf{\Theta}}) - \mathbf{\Theta}_0
\end{aligned}$$

Provided that the remainder terms rem_0 and rem_1 are small enough, the *de-biased estimator* can be defined as

$$\widehat{\mathbf{T}} \equiv \widehat{\mathbf{\Theta}} + \widehat{\mathbf{\Theta}}' + \eta(\widehat{\mathbf{\Theta}}) = \widehat{\mathbf{\Theta}} + \widehat{\mathbf{\Theta}}' - \widehat{\mathbf{\Theta}}'\mathbf{S}\widehat{\mathbf{\Theta}} = 2\widehat{\mathbf{\Theta}} - \widehat{\mathbf{\Theta}}'\mathbf{S}\widehat{\mathbf{\Theta}}. \tag{8.9}$$

As pointed out by Janková and van de Geer (2018), in order to control the remainder terms, we need bounds for the ℓ_1-error of $\widehat{\mathbf{\Theta}}$ and to control the bias term, it is sufficient

[1] See (Janková & van de Geer, 2018) for the proof.

to control the upper bound $\left\|\eta(\widehat{\mathbf{\Theta}})\right\|_{\infty} = \left\|\lambda\widehat{\mathbf{\Gamma}}\widehat{\mathbf{\Theta}}\right\|_{\infty} \leq \lambda\left\|\widehat{\mathbf{\Theta}}\right\|_{1}$, where given a matrix $\mathbf{U}$ we used $|||\mathbf{U}|||_{\infty} \equiv \max_{1\leq j\leq N}\sum_{i=1}^{N}|u_{j,i}|$ and $|||\mathbf{U}|||_{1} \equiv \max_{1\leq j\leq N}\sum_{i=1}^{N}|u_{i,j}|$.

8.2.2 Nodewise Regression

An alternative approach to induce sparsity in the estimation of precision matrix in equation (8.5) is to solve for $\widehat{\mathbf{\Theta}}$ one column at a time via linear regressions, replacing population moments by their sample counterparts $\mathbf{S}$. When we repeat this procedure for each variable $j = 1,\ldots,p$, we will estimate the elements of $\widehat{\mathbf{\Theta}}$ column by column using $\{\mathbf{x}_t\}_{t=1}^{T}$ via p linear regressions. Meinshausen and Bühlmann (2006) use this approach to incorporate sparsity into the estimation of the precision matrix. Instead of running p coupled LASSO problems as in GLASSO, they fit p separate LASSO regressions using each variable (node) as the response and the others as predictors to estimate $\widehat{\mathbf{\Theta}}$. This method is known as the "nodewise“ regression and it is reviewed below based on van de Geer, Buhlmann, Ritov and Dezeure (2014) and Callot et al. (2019).

Let $\mathbf{x}_j$ be a $T\times 1$ vector of observations for the j-th regressor, the remaining covariates are collected in a $T\times p$ matrix $\mathbf{X}_{-j}$. For each $j = 1,\ldots,p$ we run the following LASSO regressions:

$$\widehat{\boldsymbol{\gamma}}_j = \underset{\boldsymbol{\gamma}\in\mathbb{R}^{p-1}}{\arg\min}\left(\|\mathbf{x}_j - \mathbf{X}_{-j}\boldsymbol{\gamma}\|_2^2/T + 2\lambda_j\|\boldsymbol{\gamma}\|_1\right), \tag{8.10}$$

where $\widehat{\boldsymbol{\gamma}}_j = \{\widehat{\gamma}_{j,k}; j = 1,\ldots,p, k\neq j\}$ is a $(p-1)\times 1$ vector of the estimated regression coefficients that will be used to construct the estimate of the precision matrix, $\widehat{\mathbf{\Theta}}$. Define

$$\widehat{\mathbf{C}} = \begin{pmatrix} 1 & -\widehat{\gamma}_{1,2} & \cdots & -\widehat{\gamma}_{1,p} \\ -\widehat{\gamma}_{2,1} & 1 & \cdots & -\widehat{\gamma}_{2,p} \\ \vdots & \vdots & \ddots & \vdots \\ -\widehat{\gamma}_{p,1} & -\widehat{\gamma}_{p,2} & \cdots & 1 \end{pmatrix}.$$

For $j = 1,\ldots,p$, define the optimal value function

$$\hat{\tau}_j^2 = \|\mathbf{x}_j - \mathbf{X}_{-j}\widehat{\boldsymbol{\gamma}}_j\|_2^2/T + 2\lambda_j\|\widehat{\boldsymbol{\gamma}}_j\|_1$$

and write

$$\widehat{\mathbf{T}}^2 = \operatorname{diag}(\hat{\tau}_1^2,\ldots,\hat{\tau}_p^2).$$

The approximate inverse is defined as

$$\widehat{\mathbf{\Theta}}_{\lambda_j} = \widehat{\mathbf{T}}^{-2}\widehat{\mathbf{C}}.$$

Similarly to GLASSO, the subscript λ_j in $\widehat{\mathbf{\Theta}}_{\lambda_j}$ means that the estimated $\mathbf{\Theta}$ will depend upon the choice of the tuning parameter: more details are provided in Callot et al. (2019) which discusses how to choose shrinkage intensity in practice. The subscript is omitted to simplify the notation.

8.2.3 CLIME

A different approach to recover the entries of precision matrix was motivated by the compressed sensing and high-dimensional linear regression literature: instead of using ℓ_1-MLE estimators, it proceeds by using a method of constrained ℓ_1 minimization for inverse covariance matrix estimation (CLIME, T. Cai et al., 2011). To illustrate the motivation of CLIME and its connection with GLASSO, recall from (8.5) that when $\mathbf{W} = \mathbf{S}$, the solution $\widehat{\mathbf{\Theta}}_{\text{GL}}$ satisfies:

$$\widehat{\mathbf{\Theta}}_{\text{GL}}^{-1} - \mathbf{S} = \lambda \widehat{\mathbf{Z}}, \tag{8.11}$$

where $\widehat{\mathbf{Z}}$ is an element of the subdifferential $\partial\Big(\sum_{i\neq j} \widehat{d}_{ii}\widehat{d}_{jj}|\theta_{ij}|\Big)$, which is a partial derivative with respect to each off-diagonal entry of precision matrix. This leads T. Cai et al. (2011) to consider the following optimization problem:

$$\min ||\mathbf{\Theta}||_1 \text{ s.t. } ||\mathbf{\Theta} - \mathbf{S}||_\infty \leq \lambda,\ \mathbf{\Theta} \in \mathbb{R}^{p\times p}, \tag{8.12}$$

Notice that for a suitable choice of the tuning parameter, first-order conditions of (8.12) coincide with the (8.11), meaning that both approaches would theoretically lead to the same optimal solution for the precision matrix. However, the feasible set in (8.12) is complicated, hence, the following relaxation is proposed:

$$\min ||\mathbf{\Theta}||_1 \text{ s.t. } ||\mathbf{S}\mathbf{\Theta} - \mathbf{I}||_\infty \leq \lambda. \tag{8.13}$$

To make the solution $\widehat{\mathbf{\Theta}}$ in (8.13) symmetric, an additional symmetrization procedure is performed which selects the values with smaller magnitude from lower and upper parts of $\widehat{\mathbf{\Theta}}$.

Following T. Cai et al. (2011), Figure 8.3 illustrates the solution for recovering a 2 by 2 precision matrix $\begin{pmatrix} x & z \\ z & y \end{pmatrix}$, and only the plane $x(=y)$ is plotted versus z for simplicity. The CLIME solution $\widehat{\mathbf{\Omega}}$ is located at the tangency of the feasible set (shaded polygon) and objective function as in (8.13) (dashed diamond).The log-likelihood function of GLASSO (as in (8.5)) is represented by the dotted line.

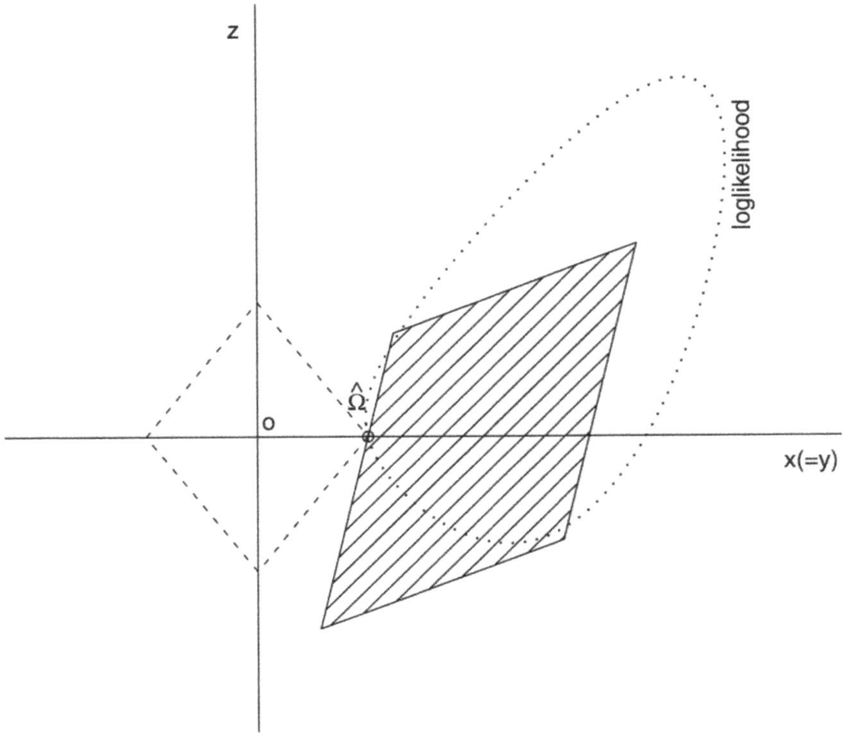

Fig. 8.3: Objective functions of CLIME (dashed diamond) and GLASSO (the dotted line) with the constrained feasible set (shaded polygon)

8.2.4 Solution Techniques

We now comment on the procedures used to obtain solutions for equations (8.6), (8.10), and (8.13). T. Cai et al. (2011) use a linear relaxation followed by the primal dual interior method approach. For GLASSO and nodewise regression, the classical solution technique proceeds by applying stochastic gradient descent to (8.6) and (8.10). Another useful technique to solve for precision matrix from (8.6) and (8.10) is through the alternating direction method of multipliers (ADMM) – a decomposition-coordination procedure in which the solutions to small local subproblems are coordinated to find a solution to a large global problem. A detailed coverage of a general ADMM method is provided in Boyd, Parikh, Chu, Peleato and Eckstein (2011), we limit the discussion below to a GLASSO-specific procedure.

We now illustrate the use of ADMM to recover the entries of precision matrix using GLASSO. Let us rewrite our objective function in (8.5):

$$\widehat{\Theta} = \arg\min_{\Theta > 0} \operatorname{trace}(\mathbf{W}\Theta) - \log\det(\Theta) + \lambda||\Theta||_1, \tag{8.14}$$

over nonnegative definite matrices (denoted as $\Theta > 0$). We now reformulate the unconstrained problem in (8.14) as a constrained problem which can be solved using ADMM:

$$\begin{aligned} &\min_{\Theta > 0} \operatorname{trace}(\mathbf{W}\Theta) - \log\det(\Theta) + \lambda||\Theta||_1 \\ &\text{s.t.} \quad \Theta = \mathbf{Z}, \end{aligned}$$

where $\mathbf{Z}$ is an auxiliary variable designed for optimization purposes to track deviations of the estimated precision matrix from the constraint. Now we can use scaled ADMM to write down the augmented Lagrangian:

$$\mathcal{L}_\rho(\Theta, \mathbf{Z}, \mathbf{U}) = \operatorname{trace}(\mathbf{S}\Theta) - \log\det(\Theta) + \lambda||\mathbf{Z}||_1 + \frac{\rho}{2}||\Theta - \mathbf{Z} + \mathbf{U}||_F^2 - \frac{\rho}{2}||\mathbf{U}||_F^2.$$

The iterative updates are:

$$\Theta^{k+1} \equiv \arg\min_{\Theta} \left(\operatorname{trace}(\mathbf{W}\Theta) - \log\det(\Theta) + \frac{\rho}{2}||\Theta - \mathbf{Z}^k + \mathbf{U}^k||_F^2\right), \tag{8.15}$$

$$\mathbf{Z}^{k+1} \equiv \arg\min_{\mathbf{Z}} \left(\lambda||\mathbf{Z}||_1 + \frac{\rho}{2}||\Theta^{k+1} - \mathbf{Z} + \mathbf{U}^k||_F^2\right), \tag{8.16}$$

$$\mathbf{U}^{k+1} \equiv \mathbf{U}^k + \Theta^{k+1} - \mathbf{Z}^{k+1},$$

where $||\cdot||_F^2$ denotes the Frobenius norm which is calculated as the square root of the sum of the squares of the entries. The updating rule in (8.16) is easily recognized to be the element-wise soft thresholding operator:

$$\mathbf{Z}^{k+1} \equiv S_{\lambda/\rho}(\Theta^{k+1} + \mathbf{U}^k),$$

where the *soft-thresholding operator* is defined as:

$$S_k(a) = \begin{cases} a - k, & \text{for } a > k \\ 0, & \text{for } |a| \leq k \\ a + k, & \text{for } a < -k \end{cases}$$

Take the gradient of the updating rule in (8.15) in order to get a closed-form solution to this updating rule:

$$\mathbf{W} - \Theta^{-1} + \rho\left(\Theta - \mathbf{Z}^k + \mathbf{U}^k\right) = 0,$$

Rearranging,

$$\rho\Theta - \Theta^{-1} = \rho\left(\mathbf{Z}^k - \mathbf{U}^k\right) - \mathbf{W}. \tag{8.17}$$

Equation (8.17) implies that $\mathbf{\Theta}$ and $\rho\left(\mathbf{Z}^k - \mathbf{U}^k\right) - \mathbf{W}$ share the same eigenvectors.[2] Let $\mathbf{Q\Lambda Q}'$ be the eigendecomposition of $\rho\left(\mathbf{Z}^k - \mathbf{U}^k\right) - \mathbf{W}$, where $\mathbf{\Lambda} = \text{diag}(\lambda_1, \ldots, \lambda_N)$, and $\mathbf{Q}'\mathbf{Q} = \mathbf{QQ}' = \mathbf{I}$. Pre-multiply (8.17) by $\mathbf{Q}'$ and post-multiply it by $\mathbf{Q}$:

$$\rho\widetilde{\mathbf{\Theta}} - \widetilde{\mathbf{\Theta}}^{-1} = \mathbf{\Lambda}. \tag{8.18}$$

Now construct a diagonal solution of (8.18):

$$\rho\tilde{\theta}_j - \frac{1}{\tilde{\theta}_j} = \lambda_j,$$

where $\tilde{\theta}_j$ denotes the j-th eigenvalue of $\widetilde{\mathbf{\Theta}}$. Solving for $\tilde{\theta}_j$ we get:

$$\tilde{\theta} = \frac{\lambda_j + \sqrt{\lambda_j^2 + 4\rho}}{2\rho}.$$

Now we can calculate $\mathbf{\Theta}$ which satisfies the optimality condition in (8.18):

$$\mathbf{\Theta} = \frac{1}{2\rho}\mathbf{Q}\left(\mathbf{\Lambda} + \sqrt{\mathbf{\Lambda}^2 + 4\rho\mathbf{I}}\right)\mathbf{Q}'.$$

Note that the computational cost of the update in (8.15) is determined by the eigenvalue decomposition of $p \times p$ matrix, which is $O(p^3)$.

As pointed out by Danaher, Wang and Witten (2014), suppose we determine that the estimated precision matrix $\widehat{\mathbf{\Theta}}$ is block diagonal with $\{b_l\}_{l=1}^B$ blocks, where each block contains p_l features. Then instead of computing the eigendecomposition of a $p \times p$ matrix, we only need to compute the eigendecomposition of matrices of dimension $p_1 \times p_1, \ldots, p_B \times p_B$. As a result, the computational complexity decreases to $\sum_{l=1}^B O(p_l^3)$.

8.3 Graphical Models in the Context of Finance

This section addresses the questions of *why* and *how* graphical models are useful for finance problems. In particular, we review the use of graphical models for portfolio allocation and asset pricing.

---

[2] $\mathbf{\Theta}\mathbf{q}_i = \frac{1}{\rho}\left(\rho\left(\mathbf{Z}_{ii}^k - \mathbf{U}_{ii}^k\right) - \mathbf{W}_{ii} + \mathbf{\Theta}^{-1}\right)\mathbf{q}_i = \theta_i\mathbf{q}_i$, where $\mathbf{q}_i$ is the eigenvector of $\mathbf{\Theta}$ corresponding to its eigenvalue θ_i. Post-multiply both parts of (8.17) by $\mathbf{q}_i$ and rearrange: $\left(\rho\left(\mathbf{Z}^k - \mathbf{U}^k\right) - \mathbf{W}\right)\mathbf{q}_i = \rho\mathbf{\Theta}\mathbf{q}_i - \mathbf{\Theta}^{-1}\mathbf{q}_i = \rho\theta_i\mathbf{q}_i - \frac{1}{\theta_i}\mathbf{q}_i = \left(\rho\theta_i - \frac{1}{\theta_i}\right)\mathbf{q}_i$. See Witten and Tibshirani (2009) for more general cases.

We start by reviewing the basic problem faced by investors that allocate their savings by investing in financial markets and forming a portfolio of financial assets. Hence, they need to choose which stocks to include in a financial portfolio and how much to invest in these stocks. Suppose we observe $i = 1,\ldots,N$ assets over $t = 1,\ldots,T$ period of time. Let $\mathbf{r}_t = (r_{1t}, r_{2t}, \ldots, r_{Nt})' \sim \mathcal{D}(\mathbf{m}, \boldsymbol{\Sigma})$ be an $N \times 1$ return vector drawn from a distribution $\mathcal{D}$ which can belong to the either sub-Gaussian or elliptical families.

The investment strategy is reflected by the choice of portfolio weights $\mathbf{w} = (w_1,\ldots,w_N)$ which indicates how much is invested in each asset. Given $\mathbf{w}$ the expected return of an investment portfolio is $\mathbf{m}'\mathbf{w}$, and the risk associated with the investment strategy $\mathbf{w}$ is $\mathbf{w}'\boldsymbol{\Sigma}\mathbf{w}$. When a weight is positive, it is said that an investor has a long position in an asset (i.e. they bought an asset), whereas negative weights correspond to short positions (they are expected to deliver an asset). When the sum of portfolio weights is equal to one, it means that an investor allocates all available budget (normalized to one) for portfolio positions; when the sum is greater than one, it means that an investors borrows additional amount on top of the initial budget; when the sum is less than one, it means that an investor keeps some money as cash.

The choice of the investment strategy depends on several parameters and constraints, including the minimum level of target return an investor desires to achieve (which we denote as μ), the maximum level of risk an investor is willing to tolerate (denoted as σ), whether an investor is determined to allocate all available budget for portfolio positions (in this case the portfolio weights are required to sum up to one), whether short-selling is allowed (i.e. whether weights are allowed to be negative). The aforementioned inputs give rise to three portfolio formulations that are discussed below.

The first two formulations originate from Markowitz mean-variance portfolio theory (Markowitz, 1952) that formulates the search for optimal portfolio weights as a trade-off of achieving the maximum desired portfolio return while minimizing the risk. Naturally, riskier strategies are associated with higher expected returns. The statistics aimed at capturing such trade-off is called Sharpe Ratio (SR), which is defined as $\mathbf{m}'\mathbf{w}/\sqrt{\mathbf{w}'\boldsymbol{\Sigma}\mathbf{w}}$. The aforementioned goal can be formulated as the following quadratic optimization problem:

$$\begin{cases} \min\limits_{\mathbf{w}} \dfrac{1}{2}\mathbf{w}'\boldsymbol{\Sigma}\mathbf{w} \\ \text{s.t. } \mathbf{w}'\boldsymbol{\iota} = 1 \\ \mathbf{m}'\mathbf{w} \geq \mu, \end{cases} \tag{8.19}$$

where $\mathbf{w}$ is an $N \times 1$ vector of assets weights in the portfolio, $\boldsymbol{\iota}$ is an $N \times 1$ vector of ones, and μ is a desired expected rate of portfolio return. The first constraint in (8.19) requires investors to have all available budget, normalized to one, invested in a portfolio. This assumption can be easily relaxed and we demonstrate the implications of this constraint on portfolio weights.

Equation (8.19) gives rise to two portfolio formulations. First, when the second constraint is not binding, then the solution to (8.19) yields the *global minimum-*

variance portfolio (GMV) weights $\mathbf{w}_G$:

$$\mathbf{w}_G = (\boldsymbol{\iota}'\boldsymbol{\Theta}\boldsymbol{\iota})^{-1}\boldsymbol{\Theta}\boldsymbol{\iota}. \tag{8.20}$$

Note that portfolio strategy in (8.19) does not depend on the target return or risk tolerance, in this sense the solution is a global minimizer of portfolio risk for all levels of portfolio return. Despite its simplicity, the importance of the minimum-variance portfolio formation strategies as a risk-management tool has been studied by many researchers. In Exhibit 1, Clarke, de Silva and Thorley (2011) provide empirical evidence of superior performance of the GMV portfolios compared to market portfolios for 1,000 largest U.S. stocks over 1968-2009. The most notable difference occurs at recessions: during the financial crisis of 2007-09 minimum-variance portfolio outperformed the market by 15-20% on average.

Below we provide a detailed proof showing how to derive an expression for a portfolio strategy $\mathbf{w}_G$ in (8.20). Let λ_1 and λ_2 denote Lagrange multipliers for the first and the second constraints in (8.19) respectively.

Proof If $\mathbf{m}'\mathbf{w} > \mu$, then $\mu = (\boldsymbol{\iota}'\Theta\boldsymbol{\iota})^{-1}\boldsymbol{\iota}'\Theta\mathbf{m}$, $\lambda_1 = (\boldsymbol{\iota}'\Theta\boldsymbol{\iota})^{-1}$, $\lambda_2 = 0$. If $\lambda_2 = 0$, the first-order condition of (8.19) yields:

$$\mathbf{w} = -\frac{1}{2}\lambda_1\boldsymbol{\Theta}\boldsymbol{\iota}, \tag{8.21}$$

Pre-multiply both sides of (8.21) by $\boldsymbol{\iota}'$ and express λ_1:

$$\lambda_1 = -2\frac{1}{\boldsymbol{\iota}'\boldsymbol{\Theta}\boldsymbol{\iota}}, \tag{8.22}$$

Plug-in (8.22) into (8.21):

$$\mathbf{w}_G = (\boldsymbol{\iota}'\boldsymbol{\Theta}\boldsymbol{\iota})^{-1}\boldsymbol{\Theta}\boldsymbol{\iota}. \tag{8.23}$$

The second portfolio formulation arises when the second constraint in (8.19) is binding (i.e. $\mathbf{m}'\mathbf{w} = \mu$), meaning that the resulting investment strategy $\mathbf{w}_{MWC}$ achieves minimum risk for a certain level of target return μ. We refer to this portfolio strategy as Markowitz Weight-Constrained (MWC) portfolio:

$$\mathbf{w}_{MWC} = (1-a_1)\mathbf{w}_{GMV} + a_1\mathbf{w}_M^*, \tag{8.24}$$

$$\mathbf{w}_M^* = (\boldsymbol{\iota}'\boldsymbol{\Theta}\mathbf{m})^{-1}\boldsymbol{\Theta}\mathbf{m}, \tag{8.25}$$

$$a_1 = \frac{\mu(\mathbf{m}'\boldsymbol{\Theta}\boldsymbol{\iota})(\boldsymbol{\iota}'\boldsymbol{\Theta}\boldsymbol{\iota}) - (\mathbf{m}'\boldsymbol{\Theta}\boldsymbol{\iota})^2}{(\mathbf{m}'\boldsymbol{\Theta}\mathbf{m})(\boldsymbol{\iota}'\boldsymbol{\Theta}\boldsymbol{\iota}) - (\mathbf{m}'\boldsymbol{\Theta}\boldsymbol{\iota})^2}.$$

This result is a well-known two-fund separation theorem introduced by Tobin (1958): MWC strategy can be viewed as holding a GMV portfolio and a proxy for the market fund $\mathbf{w}_M^*$ since the latter capture all mean-related market information. In terms of the parameters, this strategy requires an additional input asking investors to specify their target return level. Below we provide a detailed proof showing how to derive an expression for a portfolio strategy $\mathbf{w}_{MWC}$ in (8.24).

Proof Suppose the second constraint in (8.19) is binding, i.e. $\mathbf{m}'\mathbf{w} = \mu$. From the first-order condition of (8.19) with respect to $\mathbf{w}$ we get:

$$\begin{aligned}\mathbf{w} &= \boldsymbol{\Theta}(\lambda_1 \boldsymbol{\iota} + \lambda_2 \mathbf{m}) = \lambda_1 \boldsymbol{\Theta}\boldsymbol{\iota} + \lambda_2 \boldsymbol{\Theta}\mathbf{m} \\ &= \lambda_1 (\boldsymbol{\iota}'\boldsymbol{\Theta}\boldsymbol{\iota})\mathbf{w}_G + \lambda_2 (\boldsymbol{\iota}'\boldsymbol{\Theta}\mathbf{m})\mathbf{w}_M.\end{aligned} \tag{8.26}$$

Rewrite the first constraint in (8.19):

$$1 = \mathbf{w}'\boldsymbol{\Theta}\boldsymbol{\Sigma}\boldsymbol{\iota} = \lambda_1 \boldsymbol{\iota}'\boldsymbol{\Theta}\boldsymbol{\iota} + \lambda_2 \boldsymbol{\iota}'\boldsymbol{\Theta}\mathbf{m},$$

therefore, set

$$\lambda_2 \boldsymbol{\iota}'\boldsymbol{\Theta}\mathbf{m} = a_1, \text{ and } \mathbf{w}'\boldsymbol{\Theta}\boldsymbol{\Sigma}\boldsymbol{\iota} = \lambda_1 \boldsymbol{\iota}'\boldsymbol{\Theta}\boldsymbol{\iota} = 1 - a_1.$$

To solve for λ_2 combine both constraints and (8.26):

$$\begin{aligned}1 &= \mathbf{w}'\boldsymbol{\Theta}\boldsymbol{\Sigma}\boldsymbol{\iota} = \lambda_1 \boldsymbol{\iota}'\boldsymbol{\Theta}\boldsymbol{\iota} + \lambda_2 \boldsymbol{\iota}'\boldsymbol{\Theta}\mathbf{m}, \\ \mu &= \mathbf{m}'\boldsymbol{\Theta}\boldsymbol{\Sigma}\mathbf{w} = \lambda_1 \mathbf{m}'\boldsymbol{\Theta}\boldsymbol{\iota} + \lambda_2 \mathbf{m}'\boldsymbol{\Theta}\mathbf{m},\end{aligned}$$

therefore,

$$\begin{aligned}\lambda_1 &= \frac{(\mathbf{m}'\boldsymbol{\Theta}\mathbf{m}) - (\mathbf{m}'\boldsymbol{\Theta}\boldsymbol{\iota})\mu}{(\mathbf{m}'\boldsymbol{\Theta}\mathbf{m})(\boldsymbol{\iota}'\boldsymbol{\Theta}\boldsymbol{\iota}) - (\mathbf{m}'\boldsymbol{\Theta}\boldsymbol{\iota})^2}, \\ \lambda_2 &= \frac{(\boldsymbol{\iota}'\boldsymbol{\Theta}\boldsymbol{\iota})\mu - \mathbf{m}'\boldsymbol{\Theta}\boldsymbol{\iota}}{(\mathbf{m}'\boldsymbol{\Theta}\mathbf{m})(\boldsymbol{\iota}'\boldsymbol{\Theta}\boldsymbol{\iota}) - (\mathbf{m}'\boldsymbol{\Theta}\boldsymbol{\iota})^2}, \\ a_1 &= \frac{\mu(\mathbf{m}'\boldsymbol{\Theta}\boldsymbol{\iota})(\boldsymbol{\iota}'\boldsymbol{\Theta}\boldsymbol{\iota}) - (\mathbf{m}'\boldsymbol{\Theta}\boldsymbol{\iota})^2}{(\mathbf{m}'\boldsymbol{\Theta}\mathbf{m})(\boldsymbol{\iota}'\boldsymbol{\Theta}\boldsymbol{\iota}) - (\mathbf{m}'\boldsymbol{\Theta}\boldsymbol{\iota})^2}.\end{aligned}$$

It is possible to relax the constraint in (8.19) that requires portfolio weights to sum up to one: this gives rise to the Markowitz Risk-Constrained (MRC) problem which maximizes SR subject to either target risk or target return constraints, but portfolio weights are not required to sum up to one:

$$\max_{\mathbf{w}} \frac{\mathbf{m}'\mathbf{w}}{\sqrt{\mathbf{w}'\boldsymbol{\Sigma}\mathbf{w}}} \text{ s.t. (i) } \mathbf{m}'\mathbf{w} \geq \mu \text{ or(ii) } \mathbf{w}'\boldsymbol{\Sigma}\mathbf{w} \leq \sigma^2,$$

when $\mu = \sigma\sqrt{\mathbf{m}'\boldsymbol{\Theta}\mathbf{m}}$, the solution to either of the constraints is given by

$$\mathbf{w}_{MRC} = \frac{\sigma}{\sqrt{\mathbf{m}'\boldsymbol{\Theta}\mathbf{m}}}\boldsymbol{\Theta}\mathbf{m}. \tag{8.27}$$

Equation (8.27) tells us that once an investor specifies the desired return, μ, and maximum risk-tolerance level, σ, this pins down the Sharpe Ratio of the portfolio.

The objective function in (8.19) only models risk preferences of investors. If we want to incorporate investors' desire to maximize expected return while minimizing risk, we need to introduce the Markowitz mean-variance function, denoted as $M(\mathbf{m}, \boldsymbol{\Sigma})$. According to Fan, Zhang and Yu (2012), if $\mathbf{r}_t \sim \mathcal{N}(\mathbf{m}, \boldsymbol{\Sigma})$, and the utility function is given by $U(x) = 1 - \exp(-Ax)$, where A is the absolute risk-aversion parameter, maximizing the expected utility would be equivalent to maximizing $M(\mathbf{m}, \boldsymbol{\Sigma}) = \mathbf{w}'\mathbf{m} - \gamma\mathbf{w}'\boldsymbol{\Sigma}\mathbf{w}$, where $\gamma = A/2$. Hence, we can formulate the following optimization problem:

$$\begin{cases} \max_{\mathbf{w}} \mathbf{w}'\mathbf{m} - \gamma\mathbf{w}'\boldsymbol{\Sigma}\mathbf{w} \\ s.t.\ \mathbf{w}'\boldsymbol{\iota} = 1. \end{cases} \tag{8.28}$$

The solution to the above problem is analogous to the equation (8.24):

$$\begin{aligned} \mathbf{w} &= (1 - a_2)\mathbf{w}_G + a_2\mathbf{w}_M, \\ a_2 &= \gamma(\mathbf{m}'\boldsymbol{\Theta}\boldsymbol{\iota}). \end{aligned}$$

We can see from equations (8.23), (8.24), and (8.27) that in order to obtain optimal portfolio weights, one needs to get an estimate of the precision matrix, $\boldsymbol{\Theta}$. As pointed out by Zhan, Sun, Jakhar and Liu (2020), "from a graph viewpoint, estimating the covariance using historic returns models a fully connected graph between all assets. The fully connected graph appears to be a poor model in reality, and substantially adds to the computational burden and instability of the problem". In the following sections we will examine existing approaches to solving Markowitz mean-variance portfolio problem. In addition, we will also propose desirable characteristics in the estimated precision matrix that are attractive for a portfolio manager.

In practice, solution of (8.28) depends sensitively on the input vectors $\mathbf{m}$ and $\boldsymbol{\Sigma}$, and their accumulated estimation errors. Summarizing the work by Jagannathan and Ma (2003), Fan et al. (2012) show that the sensitivity of utility function to estimation errors is bounded by:

$$|M(\widehat{\mathbf{m}}, \widehat{\boldsymbol{\Sigma}}) - M(\mathbf{m}, \boldsymbol{\Sigma})| \leq ||\widehat{\mathbf{m}} - \mathbf{m}||_\infty ||\mathbf{w}||_1 + \gamma||\widehat{\boldsymbol{\Sigma}} - \boldsymbol{\Sigma}||_\infty ||\mathbf{w}||_1^2, \tag{8.29}$$

where $||\widehat{\mathbf{m}} - \mathbf{m}||_\infty$ and $||\widehat{\boldsymbol{\Sigma}} - \boldsymbol{\Sigma}||_\infty$ are the maximum componentwise estimation errors[3]. The sensitivity problem can be alleviated by considering a modified version of (8.19), known as the *optimal no-short-sale portfolio*:

$$\begin{cases} \min_{\mathbf{w}} \frac{1}{2}\mathbf{w}'\boldsymbol{\Sigma}\mathbf{w} \\ s.t.\ \mathbf{w}'\boldsymbol{\iota} = 1 \\ ||\mathbf{w}||_1 \leq c, \end{cases} \tag{8.30}$$

where $||\mathbf{w}||_1 \leq c$ is the gross-exposure constraint for a moderate c. When $c = 1$ - no short sales are allowed[4]. We do not provide a closed-form solution for the optimization problem in (8.30) since it needs to be solved numerically. Furthermore, equation (8.24) no longer holds since the portfolio frontier can not be constructed from a linear combination of any two optimal portfolios due to the restrictions on weights imposed by the gross-exposure constraint. The constraint specifying the

---

[3] The proof of (8.29) is a straightforward application of the Hölder's inequality.

[4] If $\mathbf{w}'\boldsymbol{\iota} = 1$ and $w_i \geq 0,\ \forall i$, then $||\mathbf{w}||_1 = |w_1| + \ldots |w_N| \leq 1$.

target portfolio return is also omitted since there may not exist a no-short-sale portfolio that reaches this target.

Define a measure of risk $R(\mathbf{w}, \boldsymbol{\Sigma}) = \mathbf{w}'\boldsymbol{\Sigma}\mathbf{w}$, where risk is measured by the variance of the portfolio. For risk minimization with the gross-exposure constraint we obtain:

$$|R(\mathbf{w}, \widehat{\boldsymbol{\Sigma}}) - R(\mathbf{w}, \boldsymbol{\Sigma})| \leq \|\widehat{\boldsymbol{\Sigma}} - \boldsymbol{\Sigma}\|_\infty \|\mathbf{w}\|_1^2. \tag{8.31}$$

The minimum of the right-hand side of (8.31) is achieved under no short-sale constraint $\|\mathbf{w}\|_1 = 1$. Fan et al. (2012) concluded that the optimal no-short-sale portfolio in (8.30) has smaller actual risk than that for the global minimum-variance portfolio described by weights in (8.20). However, their empirical studies showed that the optimal no-short-sale portfolio is not diversified enough and the performance can be improved by allowing some short positions.

8.3.1 The No-Short-Sale Constraint and Shrinkage

Investment strategies discussed so far did not put any sign-restrictions on portfolio weights. Jagannathan and Ma (2003) explored the connection between regularizing the sample covariance matrix and restricting portfolio weights: they showed that the solution to the short sale constrained problem using the sample covariance matrix $\mathbf{S} = \frac{1}{T}\sum_{t=1}^{T}(\mathbf{r}_t - \bar{\mathbf{r}}_t)(\mathbf{r}_t - \bar{\mathbf{r}}_t)'$ (depicted in the left column of (8.32)) coincides with the solution to the unconstrained problem (in the right column of (8.32)) if $\mathbf{S}$ is replaced by $\widehat{\boldsymbol{\Sigma}}_{JM}$:

$$\begin{cases} \min\limits_{\mathbf{w}} \mathbf{w}'\mathbf{S}\mathbf{w} \\ s.t.\ \mathbf{w}'\boldsymbol{\iota} = 1 \\ w_i \geq 0\ \forall i \end{cases} \Longrightarrow \begin{cases} \min\limits_{\mathbf{w}} \mathbf{w}'\widehat{\boldsymbol{\Sigma}}_{JM}\mathbf{w} \\ s.t.\ \mathbf{w}'\boldsymbol{\iota} = 1 \\ \widehat{\boldsymbol{\Sigma}}_{JM} = \mathbf{S} - \boldsymbol{\lambda}\boldsymbol{\iota}' - \boldsymbol{\iota}\boldsymbol{\lambda}', \end{cases} \tag{8.32}$$

where $\boldsymbol{\lambda} \in \mathbb{R}^N$ is the vector of Lagrange multipliers for the short sale constraint. (8.32) means that each of the no-short-sale constraints is equivalent to reducing the estimated covariance of the corresponding asset with other assets by a certain amount. Jagannathan and Ma (2003) interpret the estimator $\widehat{\boldsymbol{\Sigma}}_{JM}$ as a shrinkage version of the sample covariance matrix $\mathbf{S}$ and argue that it can reduce sampling error even when no-short-sale constraints do not hold in the population. In order to understand the relationship between covariance matrix and portfolio weights, consider the following unconstrained GMV problem:

$$\begin{cases} \min\limits_{\mathbf{w}} \frac{1}{2}\mathbf{w}'\mathbf{S}\mathbf{w} \\ s.t.\ \mathbf{w}'\boldsymbol{\iota} = 1. \end{cases} \tag{8.33}$$

The first-order condition of (8.33) is:

$$\sum_{i=1}^{N} w_i s_{j,i} = \lambda \geq 0, \qquad j = 1, \ldots, N \tag{8.34}$$

where $\lambda \in \mathbb{R}$ is the Lagrange multiplier. Let us denote the investment strategy obtained using a short-sale constraint as $\mathbf{w}^S$. Equation (8.34) means that at the optimum the marginal contribution of stock j to the portfolio variance is the same as the marginal contribution of stock i for any j, i. This is consistent with well-known asset pricing models such as Capital Asset Pricing Model (CAPM) and Arbitrage Pricing Theory (APT): the underlying principle behind them states that under the assumption that the markets are efficient, total risk of a financial asset can be decomposed into common and idiosyncratic parts. Common risk stems from similar drivers of volatility for all assets, such as market movements and changes in macroeconomic indicators. Since all assets are influenced by common drivers, the risk associated with this component cannot be reduced by increasing the number of stocks in the portfolio. In contrast, idiosyncratic (or asset-specific) risk differs among various assets depending on the specificity of a firm, industry, country associated with an asset. It is possible to decrease one's exposure to idiosyncratic risk by increasing the number of stocks in the portfolio (by doing so, investors reduce their exposure to a particular country or industry. This is known as diversification – naturally, higher diversification benefits can be achieved when portfolio includes assets with low or negative correlation.

Suppose stock j has higher covariance with other stocks, meaning that the j-th row of $\mathbf{S}$ has larger elements compared to other rows. This means that stock j will contribute more to the portfolio variance. Hence, to satisfy optimal condition in (8.34) we need to reduce the weight of stock j in the portfolio. If stock j has high variance and is highly correlated with other stocks, then its' weight can be negative. According to Green and Hollifield (1992), the presence of dominant factors leads to extreme negative weights even in the absence of the estimation errors.

Remark 8.2 Consider the short-sale-constrained optimization problem in (8.32). Suppose the non-negativity constraint for asset j is binding. Then its' covariances with other assets will be reduced by $\lambda_j + \lambda_i$ for all $i \neq j$, and its' variance is reduced by $2\lambda_j$. Jagannathan and Ma (2003) argue that since the largest covariance estimates are more likely caused by upward-biased estimation error, the shrinking of covariance matrix may reduce the estimation error. On the other hand, Green and Hollifield (1992) suggest that the short-sale constraint will not, in general, hold in the population when asset returns have dominant factors.

According to Jagannathan and Ma (2003), given a short-sale-constrained optimal portfolio weight $\mathbf{w}^S$ in (8.32), there exist many covariance matrix estimates that have $\mathbf{w}^S$ as their unconstrained GMV portfolio. Under the joint normality of returns, Jagannathan and Ma (2003) show that $\widehat{\boldsymbol{\Sigma}}_{JM}$ is the constrained MLE of the population covariance matrix. Let $\mathbf{r}_t = (r_{1t}, r_{2t}, \ldots, r_{Nt}) \overset{i.i.d.}{\sim} \mathcal{N}(\mathbf{m}, \boldsymbol{\Sigma})$ be an $N \times 1$ return vector, $\mathbf{S} = \frac{1}{T}\sum_{t=1}^{T}(\mathbf{r}_t - \bar{\mathbf{r}}_t)(\mathbf{r}_t - \bar{\mathbf{r}}_t)'$ is the unconstrained MLE of $\boldsymbol{\Sigma}$ and define $\boldsymbol{\Sigma}^{-1} \equiv \boldsymbol{\Theta}$. Then the log-likelihood as a function of covariance matrix becomes (up to the constants):

$$l(\mathbf{\Theta}) = -\log\det\mathbf{\Sigma} - \text{trace}(\mathbf{S\Theta})$$
$$= \log\det\mathbf{\Theta} - \text{trace}(\mathbf{S\Theta}).$$

Jagannathan and Ma (2003) show that $\widehat{\mathbf{\Sigma}}_{JM}$ constructed from the solution to the constrained GMV problem in (8.32) is the solution of the constrained ML problem (8.35), where no short sales constraint in (8.32) is translated into regularization of precision matrix in (8.35):

$$\begin{cases} \max_{\mathbf{\Theta}} \Big(\log\det\mathbf{\Theta} - \text{trace}(\mathbf{S\Theta}) \Big) \\ s.t. \sum_j \theta_{i,j} \geq 0, \end{cases} \tag{8.35}$$

where $\theta_{i,j}$ is the (i, j)-th element of the precision matrix. Even though the regularization of weights in (8.35) is translated into the regularization of $\mathbf{\Theta}$, Jagannathan and Ma (2003) solve the above problem for $\widehat{\mathbf{\Sigma}}$.

The main findings of Jagannathan and Ma (2003) can be summarized as follows:

1. No-short-sale constraint shrinks the large elements of the sample covariance matrix towards zero which has two effects. On the one hand, if an estimated large covariance is due to the sampling error, the shrinkage reduces this error. However, if the population covariance is large, the shrinkage introduces specification error. The net effect is determined by the trade-off between sampling and specification errors.
2. No-short-sale constraint deteriorates performance for factor models and shrinkage covariance estimators (such as Ledoit and Wolf (2003), which is formed from a combination of the sample covariance matrix and the 1-factor (market return) covariance matrix);
3. Under no-short-sale constraint, minimum-variance portfolios constructed using the sample covariance matrix perform comparable to factor models and shrinkage estimators;
4. GMV outperform MWC portfolio which implies that the estimates of the mean returns are very noisy

Let us now elaborate on the idea of incorporating no-short-sale constraint and, consequently, shrinking the sample covariance estimator. We will examine the case studied in Jagannathan and Ma (2003) when the asset returns are normally distributed. Consider the first-order condition in (8.34). Suppose that stock j has high covariance with other stocks. As a result, the weight of this stock can be negative and large in the absolute value. In order to reduce the impact of the j-th stock on portfolio variance, no-short-sale approach will set the weight of such asset to zero by shrinking the corresponding entries of the sample covariance matrix. The main motivation of such shrinkage comes from the assumption that high covariance is caused by the estimation error. However, as pointed out by Green and Hollifield

(1992), extreme negative weights can be a result of the dominant factors rather than the estimation errors. Hence, imposing no-short-sale assumption will fail to account for important structural patterns in the data.

Furthermore, the approach discussed by Jagannathan and Ma (2003) only studies the impact of the covariance structure. Assume we have 10 assets. Let asset $j = 1$ be highly correlated with all other assets. We can calculate partial correlation of asset $j = 1$ and $j_{-1} = 2, \ldots, 10$. Suppose we find out that once we condition on $j = 2$, the partial correlation of stock j with all assets except $j = 2$ becomes zero. That would mean that high covariance of the j-th stock with all other assets was caused by its strong relationship with asset $j = 2$. The standard approaches to the mean-variance analysis will assume that this high covariance was a result of an estimation error, and will reduce the weight of asset j in the portfolio. We claim that instead of doing that, we should exploit this relationship between assets $j = 1$ and $j = 2$. However, the covariance (therefore, correlation) matrix will not be able to detect this structure. Hence, we need another statistics, such as a matrix of partial correlations (precision matrix) to help us draw such conclusions.

8.3.2 The A-Norm Constraint and Shrinkage

DeMiguel, Garlappi, Nogales and Uppal (2009) establish the relationship between portfolio weights regularization and shrinkage estimator of covariance matrix proposed by Ledoit and Wolf (2003, 2004a, 2004b). The latter developed a linear shrinkage estimator denoted as $\widehat{\mathbf{\Sigma}}_{LW}$ which is a combination of the sample covariance matrix $\mathbf{S}$ and a low-variance target estimator $\widehat{\mathbf{\Sigma}}_{\text{target}}$:

$$\widehat{\mathbf{\Sigma}}_{LW} = \frac{1}{1+\nu}\mathbf{S} + \frac{\nu}{1+\nu}\widehat{\mathbf{\Sigma}}_{\text{target}},$$

where $\nu \in \mathbb{R}$ is a positive constant. Define $||\mathbf{w}||_A = (\mathbf{w}'\mathbf{A}\mathbf{w})^{1/2}$ to be an A-norm, where $\mathbf{A} \in \mathbb{R}^{N\times N}$ is a positive-definite matrix. DeMiguel et al. (2009) show that for each $\nu \geq 0$ there exists a δ such that the solution to the A-norm constrained GMV portfolio problem coincides with the solution to the unconstrained problem if the sample covariance matrix is replaced by $\widehat{\mathbf{\Sigma}}_{LW}$:

$$\begin{cases} \min_{\mathbf{w}} \mathbf{w}'\mathbf{S}\mathbf{w} \\ s.t.\ \mathbf{w}'\boldsymbol{\iota} = 1 \\ \mathbf{w}'\mathbf{A}\mathbf{w} \leq \delta \end{cases} \Longrightarrow \begin{cases} \min_{\mathbf{w}} \mathbf{w}'\widehat{\mathbf{\Sigma}}_{LW}\mathbf{w} \\ s.t.\ \mathbf{w}'\boldsymbol{\iota} = 1 \\ \widehat{\mathbf{\Sigma}}_{LW} = \frac{1}{1+\nu}\mathbf{S} + \frac{\nu}{1+\nu}\mathbf{A}. \end{cases} \tag{8.36}$$

If $\mathbf{A}$ is chosen to be the identity matrix $\mathbf{I}$, then there is a one-to-one correspondence between the A-norm-constrained portfolio on the left of (8.36) and the shrinkage estimator proposed in Ledoit and Wolf (2004b). If $\mathbf{A}$ is chosen to be the 1-factor (market return) covariance matrix $\widehat{\boldsymbol{\Sigma}}_F$, then there is a one-to-one correspondence with the shrinkage portfolio in Ledoit and Wolf (2003). Therefore, direct regularization of weights using A-norm constraint achieves shrinkage of the sample covariance matrix. In order to understand this result, let us consider the A-norm-constrained optimization problem in (8.36). Note that in contrast to (8.32), the A-norm shrinks the total norm of the minimum-variance portfolio weights rather than shrinking every weight. Now suppose the A-norm constraint in (8.36) binds. In this case $v > 0$ and in order to ensure the A-norm constraint is not violated, the sample covariance matrix will be forced to shrink towards $\mathbf{A}$.

Remark 8.3 When $\mathbf{A} = \mathbf{I}$, the A-norm becomes an ℓ_2-norm and the constraint becomes:

$$\sum_{i=1}^{N}\left(w_i - \frac{1}{N}\right)^2 \leq \left(\delta - \frac{1}{N}\right). \tag{8.37}$$

Equation (8.37) follows from the footnote 10 of DeMiguel et al. (2009):

$$\sum_{i=1}^{N}\left(w_i - \frac{1}{N}\right)^2 = \sum_{i=1}^{N} w_i^2 + \sum_{i=1}^{N}\frac{1}{N^2} - 2\sum_{i=1}^{N}\frac{2w_i}{N} = \sum_{i=1}^{N} w_i^2 - \frac{1}{N}.$$

Therefore, using an ℓ_2-norm constraint imposes an upper-bound on the deviations of the minimum-variance portfolio from the equally weighted portfolio. If $\delta = 1/N$ we obtain $w_i = 1/N$.

DeMiguel et al. (2009) show that empirically A-norm constrained portfolios outperform the portfolio strategies in Jagannathan and Ma (2003); Ledoit and Wolf (2003, 2004b), factor portfolios, and equally-weighted portfolio in terms of out-of-sample Sharpe ratio. They study monthly returns for 5 datasets with the number of assets less than $N = 50$ for four datasets, and $N = 500$ for the last dataset. Once the number of assets is increased, direct regularization of weights is computationally challenging. Moreover, one needs to justify the choice of the free parameter δ in the A-norm constraint.

Let us now elaborate on the idea of the A-norm constraint and shrinkage estimators, such as in Ledoit and Wolf (2003, 2004b). Note that the A-norm constraint is equivalent to constraining the total exposure of the portfolio. Since the constraint does not restrict individual weights, the resulting portfolio is not sparse. Therefore, even when some assets have negligible weights, they are still included in the final portfolio. However, as pointed out by Li (2015), when the number of assets is large, sparse portfolio rule is desired. This is motivated from two main perspectives: zero portfolio weights reduce transaction costs as well as portfolio management costs; and, since the number of historical asset returns, T, might be relatively small compared to the number of assets, N, this increases the estimation error. Hence, we need some regularization scheme which would induce sparsity in the portfolio weights.

Furthermore, Rothman, Bickel, Levina and Zhu (2008) emphasized that shrinkage estimators of the form proposed by Ledoit and Wolf (2003, 2004b) do not affect the eigenvectors of the covariance, only the eigenvalues. However, Johnstone and Lu (2009) showed that the sample eigenvectors are also not consistent in high-dimensions. Moreover, recall that in the formulas for portfolio weights we need to use an estimator of precision matrix, $\mathbf{\Theta}$. Once we shrink the eigenvalues of the sample covariance matrix, it becomes invertible and we could use it for calculating portfolio weights. However, shrinking the eigenvalues of the sample covariance matrix will not improve the spectral behavior of the estimated precision. That is, it might have exploding eigenvalues which might lead to extreme portfolio positions. In this sense, consistency of precision in ℓ_2-operator norm implies that the eigenvalues of $\widehat{\mathbf{\Theta}}$ consistently estimate the corresponding eigenvalues of $\mathbf{\Theta}$. Therefore, we need a sparse estimator of precision matrix that would be able to handle high-dimensional financial returns and have a bounded spectrum.

8.3.3 Classical Graphical Models for Finance

Graphical models were shown to provide consistent estimates of the precision matrix Friedman et al. (2007); Meinshausen and Bühlmann (2006); T. Cai et al. (2011). Goto and Xu (2015) estimated a sparse precision matrix for portfolio hedging using graphical models. They found out that their portfolio achieves significant out-of-sample risk reduction and higher return, as compared to the portfolios based on equal weights, shrunk covariance matrix, industry factor models, and no-short-sale constraints. Awoye (2016) used Graphical LASSO to estimate a sparse covariance matrix for the Markowitz mean-variance portfolio problem to improve covariance estimation in terms of lower realized portfolio risk. Millington and Niranjan (2017) conducted an empirical study that applies Graphical LASSO for the estimation of covariance for the portfolio allocation. Their empirical findings suggest that portfolios that use Graphical LASSO for covariance estimation enjoy lower risk and higher returns compared to the empirical covariance matrix. They show that the results are robust to missing observations: they remove a number of samples randomly from the training data and compare how the corruption of the data affects the risks and returns of the portfolios produced on both seen and unseen data. Millington and Niranjan (2017) also construct a financial network using the estimated precision matrix to explore the relationship between the companies and show how the constructed network helps to make investment decisions. Callot et al. (2019) use the nodewise-regression method of Meinshausen and Bühlmann (2006) to establish consistency of the estimated variance, weights and risk of high-dimensional financial portfolio. Their empirical application demonstrates that the precision matrix estimator based on the nodewise-regression outperforms the principal orthogonal complement thresholding estimator (POET) (Fan, Liao & Mincheva, 2013) and linear shrinkage (Ledoit &

Wolf, 2004b). T. T. Cai, Hu, Li and Zheng (2020) use constrained ℓ_1-minimization for inverse matrix estimation (CLIME) of the precision matrix (T. Cai et al., 2011) to develop a consistent estimator of the minimum variance for high-dimensional global minimum-variance portfolio. It is important to note that all the aforementioned methods impose some sparsity assumption on the precision matrix of excess returns.

Having originated from the literature on statistical modelling, graphical models inherit the properties and assumptions common in that literature, such as sparse environment and lack of dynamics. Natural questions are (1) whether these statistical assumptions are justified in economics and finance settings, and (2) how to augment graphical models and make them suitable for the use in economics and finance?

8.3.4 Augmented Graphical Models for Finance Applications

We start with analysing the *sparsity* assumption imposed in all graphical models: many entries of precision matrix are zero, which is a necessary condition to consistently estimate inverse covariance.

The arbitrage pricing theory (APT), developed by (Ross, 1976), postulates that the expected returns on securities should be related to their covariance with the common components or factors only. The goal of the APT is to model the tendency of asset returns to move together via factor decomposition. Assume that the return generating process ($\mathbf{r}_t$) follows a K-factor model:

$$\underbrace{\mathbf{r}_t}_{p\times 1} = \mathbf{B} \underbrace{\mathbf{f}_t}_{K\times 1} + \varepsilon_t, \quad t = 1, \ldots, T \tag{8.38}$$

where $\mathbf{f}_t = (f_{1t}, \ldots, f_{Kt})'$ are the factors, $\mathbf{B}$ is a $p \times K$ matrix of factor loadings, and ε_t is the idiosyncratic component that cannot be explained by the common factors. Without loss of generality, we assume throughout the paper that unconditional means of factors and idiosyncratic component are zero. Factors in (8.38) can be either observable, such as in (Fama & French, 1993, 2015), or can be estimated using statistical factor models. Unobservable factors and loadings are usually estimated by the principal component analysis (PCA), as studied in Connor and Korajczyk (1988); Bai (2003); Bai and Ng (2002); Stock and Watson (2002). Strict factor structure assumes that the idiosyncratic disturbances, ε_t, are uncorrelated with each other, whereas approximate factor structure allows correlation of the idiosyncratic disturbances (see Chamberlain and Rothschild (1983); Bai (2003) among others).

When common factors are present across financial returns, the precision matrix cannot be sparse because all pairs of the forecast errors are partially correlated given other forecast errors through the common factors. To illustrate this point, we generated variables that follow (8.38) with $K = 2$ and $\varepsilon_t \sim \mathcal{N}(\mathbf{0}, \boldsymbol{\Sigma}_\varepsilon)$, where $\sigma_{\varepsilon,ij} = 0.4^{|i-j|}$ is the i, j-th element of $\boldsymbol{\Sigma}_\varepsilon$. The vector of factors $\mathbf{f}_t$ is drawn from $\mathcal{N}(\mathbf{0}, \mathbf{I}_K/10)$, and the entries of the matrix of factor loadings for forecast error $j = 1, \ldots, p$, $\mathbf{b}_j$, are drawn from $\mathcal{N}(\mathbf{0}, \mathbf{I}_K/100)$. The full loading matrix is given by $\mathbf{B} = (\mathbf{b}_1, \ldots, \mathbf{b}_p)'$. Let $\widehat{K}$

denote the number of factors estimated by the PCA. We set $(T, p) = (200, 50)$ and plot the heatmap and histogram of population partial correlations of financial returns $\mathbf{r}_t$, which are the entries of a precision matrix, in Figure 8.4. We now examine the performance of graphical models for estimating partial correlations under the factor structure. Figure 8.5 shows the partial correlations estimated by GLASSO that does not take into account factors: due to strict sparsity imposed by graphical models almost all partial correlations are shrunk to zero which degenerates the histogram in Figure 8.5. This means that strong sparsity assumption on $\mathbf{\Theta}$ imposed by classical graphical models (such as GLASSO, nodewise regression, or CLIME discussed in Section 8.2) is not realistic under the factor structure.

One attempt to integrate factor modeling and high-dimensional precision estimation was made by (Fan, Liu & Wang, 2018) (Section 5.2): the authors referred to such class of models as "conditional graphical models". However, this was not the main focus of their paper which concentrated on covariance estimation through elliptical factor models. As (Fan et al., 2018) pointed out, "*though substantial amount of efforts have been made to understand the graphical model, little has been done for estimating conditional graphical model, which is more general and realistic*". One of the studies that examines theoretical and empirical performance of graphical models integrated with the factor structure in the context of portfolio allocation is (Lee & Seregina, 2021b). They develop a Factor Graphical LASSO Algorithm that decomposes precision matrix of stock returns into low-rank and sparse components, with the latter estimated using GLASSO. To have a better understanding of the framework, let us introduce some notations. First, rewrite (8.38) in matrix form:

$$\underbrace{\mathbf{R}}_{p\times T} = \underbrace{\mathbf{B}}_{p\times K}\mathbf{F} + \mathbf{E}, \tag{8.39}$$

The factors and loadings in (8.39) are estimated by solving the following minimization problem: $(\widehat{\mathbf{B}}, \widehat{\mathbf{F}}) = \arg\min_{\mathbf{B},\mathbf{F}} \|\mathbf{R} - \mathbf{BF}\|_F^2$ s.t. $\frac{1}{T}\mathbf{FF}' = \mathbf{I}_K$, $\mathbf{B}'\mathbf{B}$ is diagonal. The constraints are needed to identify the factors (Fan et al., 2018).

Given a symmetric positive semi-definite matrix $\mathbf{U}$, let $\Lambda_{\max}(\mathbf{U}) \equiv \Lambda_1(\mathbf{U}) \geq \Lambda_2(\mathbf{U}) \geq \ldots \geq \Lambda_{\min}(\mathbf{U}) \equiv \Lambda_p(\mathbf{U})$ be the eigenvalues of $\mathbf{U}$, and $\text{eig}_K(\mathbf{U}) \in \mathbb{R}^{K\times p}$ denote the first $K \leq p$ normalized eigenvectors corresponding to $\Lambda_1(\mathbf{U}), \ldots, \Lambda_K(\mathbf{U})$. It was shown (Stock & Watson, 2002) that $\widehat{\mathbf{F}} = \sqrt{T}\text{eig}_K(\mathbf{R}'\mathbf{R})$ and $\widehat{\mathbf{B}} = T^{-1}\mathbf{R}\widehat{\mathbf{F}}'$.

Given $\widehat{\mathbf{F}}, \widehat{\mathbf{B}}$, define $\widehat{\mathbf{E}} = \mathbf{R} - \widehat{\mathbf{B}}\widehat{\mathbf{F}}$. Let $\mathbf{\Sigma}_\varepsilon = T^{-1}\mathbf{EE}'$ and $\mathbf{\Sigma}_f = T^{-1}\mathbf{FF}'$ be covariance matrices of the idiosyncratic components and factors, and let $\mathbf{\Theta}_\varepsilon = \mathbf{\Sigma}_\varepsilon^{-1}$ and $\mathbf{\Theta}_f = \mathbf{\Sigma}_f^{-1}$ be their inverses. Given a sample of the estimated residuals $\{\widehat{\boldsymbol{\varepsilon}}_t = \mathbf{r}_t - \widehat{\mathbf{B}}\widehat{\mathbf{f}}_t\}_{t=1}^T$ and the estimated factors $\{\widehat{\mathbf{f}}_t\}_{t=1}^T$, let $\widehat{\mathbf{\Sigma}}_\varepsilon = (1/T)\sum_{t=1}^T \widehat{\boldsymbol{\varepsilon}}_t\widehat{\boldsymbol{\varepsilon}}_t'$ and $\widehat{\mathbf{\Sigma}}_f = (1/T)\sum_{t=1}^T \widehat{\mathbf{f}}_t\widehat{\mathbf{f}}_t'$ be the sample counterparts of the covariance matrices.

To decompose precision of financial returns into low-rank and sparse components, the authors apply Sherman-Morrison-Woodbury formula to estimate the final precision matrix of excess returns:

$$\widehat{\mathbf{\Theta}} = \widehat{\mathbf{\Theta}}_\varepsilon - \widehat{\mathbf{\Theta}}_\varepsilon\widehat{\mathbf{B}}[\widehat{\mathbf{\Theta}}_f + \widehat{\mathbf{B}}'\widehat{\mathbf{\Theta}}_\varepsilon\widehat{\mathbf{B}}]^{-1}\widehat{\mathbf{B}}'\widehat{\mathbf{\Theta}}_\varepsilon. \tag{8.40}$$

The estimated precision matrix from(8.40) is used to compute portfolio weights, risk and the Sharpe Ratio and establish consistency of these performance metrics.

Let us now revisit the motivating example at the beginning of this section: Figures 8.6-8.8 plot the heatmaps and the estimated partial correlations when precision matrix is computed using Factor GLASSO with $\widehat{K} \in \{1, 2, 3\}$ statistical factors. The heatmaps and histograms closely resemble population counterparts in Figure 8.4, and the result is not very sensitive to over- or under-estimating the number of factors $\widehat{K}$. This demonstrates that using a combination of classical graphical models and factor structure via Factor Graphical Models improves upon the performance of classical graphical models.

We continue with the lack of dynamics in the graphical models literature. First, recall that a precision matrix represents a network of interacting entities, such as corporations or genes. When the data is Gaussian, the sparsity in the precision matrix encodes the conditional independence graph - two variables are conditionally independent given the rest if and only if the entry corresponding to these variables in the precision matrix is equal to zero. Inferring the network is important for portfolio allocation problem. At the same time, the financial network changes over time, that is, the relationships between companies can change either smoothly, or abruptly (e.g. as a response to an unexpected policy shock, or in the times of economic downturns). Therefore, it is important to account for time-varying nature of stock returns. The time-varying network also implies time-varying second moments of a distribution – this naturally means that both covariance and precision matrices vary with time.

There are two streams of literature that study time-varying networks. The first one models dynamics in the precision matrix locally. Zhou, Lafferty and Wasserman (2010) develop a nonparametric method for estimating time-varying graphical structure for multivariate Gaussian distributions using an ℓ_1-penalized log-likelihood. They find out that if the covariances change smoothly over time, the covariance matrix can be estimated well in terms of predictive risk even in high-dimensional problems. Lu, Kolar and Liu (2015) introduce nonparanormal graphical models that allow to model high-dimensional heavy-tailed systems and the evolution of their network structure. They show that the estimator consistently estimates the latent inverse Pearson correlation matrix. The second stream of literature allows the network to vary with time by introducing two different frequencies. Hallac, Park, Boyd and Leskovec (2017) study time-varying Graphical LASSO with smoothing evolutionary penalty.

One of the works that combines latent factor modeling, time-variation and graphical modeling is a paper by Zhan et al. (2020). To capture the latent space distribution, the authors use PCA and autoencoders. To model temporal dependencies they employ variational autoencoders with Gaussian and Cauchy priors. A graphical model relying on GLASSO is associated to each time interval and the graph is updated when moving to the next time point.

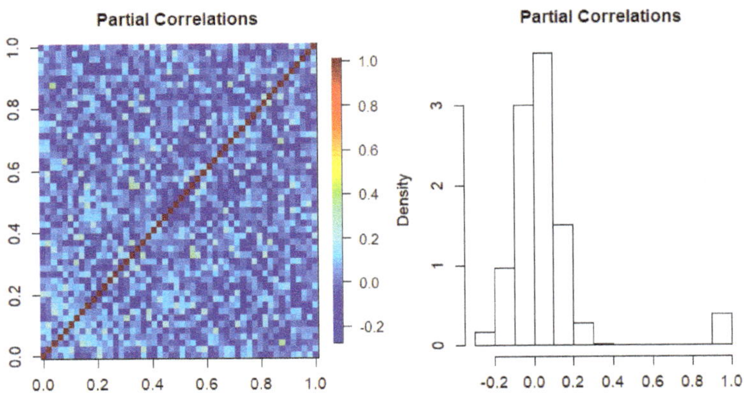

Fig. 8.4: Heatmap and histogram of population partial correlations; $T = 200$, $p = 50$, $K = 2$

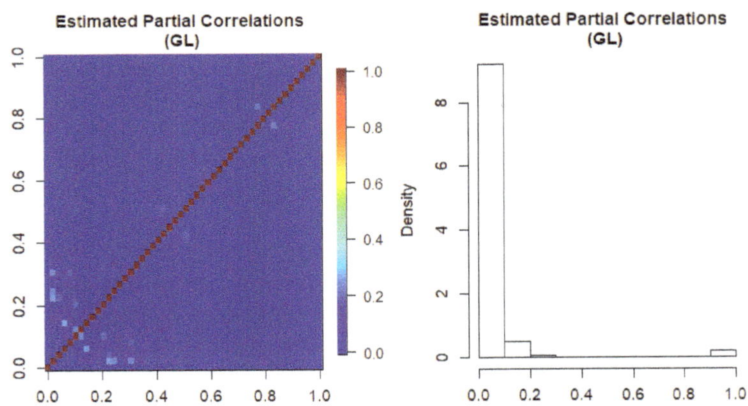

Fig. 8.5: Heatmap and histogram of sample partial correlations estimated using GLASSO with no factors; $T = 200$, $p = 50$, $q = 2$, $\hat{q} = 0$

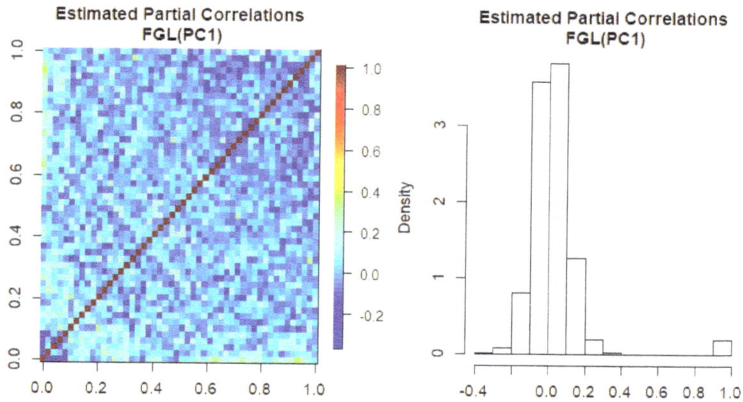

Fig. 8.6: Heatmap and histogram of sample partial correlations estimated using Factor GLASSO with 1 statistical factor; $T = 200$, $p = 50$, $K = 2$, $\hat{K} = 1$

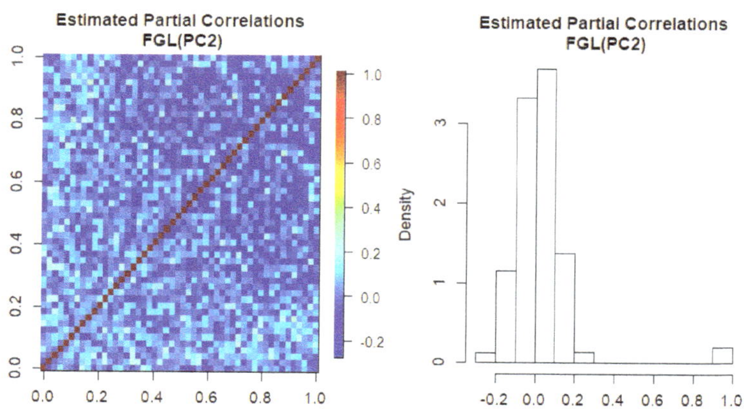

Fig. 8.7: Heatmap and histogram of sample partial correlations estimated using Factor GLASSO with 2 statistical factors; $T = 200$, $p = 50$, $K = 2$, $\hat{K} = 2$

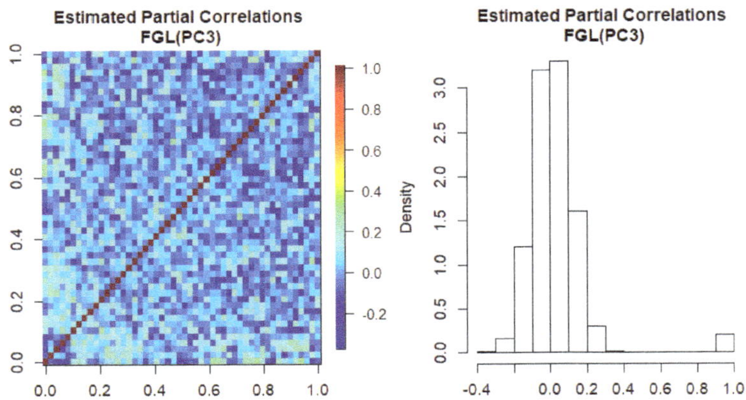

Fig. 8.8: Heatmap and histogram of sample partial correlations estimated using Factor GLASSO with 3 statistical factors; $T = 200$, $p = 50$, $K = 2$, $\hat{K} = 3$

8.4 Graphical Models in the Context of Economics

In this section we review other applications of graphical models to economic problems and revisit further extensions. We begin with a forecast combination exercise that bears a close resemblance with portfolio allocation strategies reviewed in the previous section. Then we deviate from the optimization problem of searching for the optimal weights and review other important applications which rely on the use of the estimator of precision matrix. We devote special attention to Vector Autoregressive (VAR) models that emphasize the relationship between network estimation and recovery of a sparse inverse covariance.

8.4.1 Forecast Combinations

Not surprisingly, the area of economic forecasting bears a close resemblance to portfolio allocation exercise reviewed in the previous section. Suppose we have p competing forecasts, $\widehat{\mathbf{y}}_t = (\hat{y}_{1,t}, \ldots, \hat{y}_{p,t})'$, of the variable y_t, $t = 1, \ldots, T$. Let $\mathbf{y}_t = (y_t, \ldots, y_t)'$. Define $\mathbf{e}_t = \mathbf{y}_t - \widehat{\mathbf{y}}_t = (e_{1t}, \ldots, e_{pt})'$ to be a $p \times 1$ vector of forecast errors. The forecast combination is defined as follows:

$$\widehat{y}_t^c = \mathbf{w}'\widehat{\mathbf{y}}_t$$

where $\mathbf{w}$ is a $p \times 1$ vector of weights. Define the mean-squared forecast error (MSFE) to be a measure of risk $\text{MSFE}(\mathbf{w}, \boldsymbol{\Sigma}) = \mathbf{w}'\boldsymbol{\Sigma}\mathbf{w}$. As shown in Bates and Granger (1969), the *optimal* forecast combination minimizes the variance of the combined forecast error:

$$\min_{\mathbf{w}} \text{MSFE} = \min_{\mathbf{w}} \mathbb{E}[\mathbf{w}'\mathbf{e}_t\mathbf{e}_t'\mathbf{w}] = \min_{\mathbf{w}} \mathbf{w}'\boldsymbol{\Sigma}\mathbf{w}, \text{ s.t. } \mathbf{w}'\boldsymbol{\iota}_p = 1, \tag{8.41}$$

where $\boldsymbol{\iota}_p$ is a $p \times 1$ vector of ones. The solution to (8.41) yields a $p \times 1$ vector of the optimal forecast combination weights:

$$\mathbf{w} = \frac{\boldsymbol{\Theta}\boldsymbol{\iota}_p}{\boldsymbol{\iota}_p'\boldsymbol{\Theta}\boldsymbol{\iota}_p}. \tag{8.42}$$

If the true precision matrix is known, the equation (8.42) guarantees to yield the optimal forecast combination. In reality, one has to estimate $\boldsymbol{\Theta}$. Hence, the out-of-sample performance of the combined forecast is affected by the estimation error. As pointed out by Smith and Wallis (2009), when the estimation uncertainty of the weights is taken into account, there is no guarantee that the "optimal" forecast combination will be better than the equal weights or even improve the individual forecasts. Define $a = \boldsymbol{\iota}_p'\boldsymbol{\Theta}\boldsymbol{\iota}_p/p$, and $\widehat{a} = \boldsymbol{\iota}_p'\widehat{\boldsymbol{\Theta}}\boldsymbol{\iota}_p/p$. We can write

$$\left|\frac{\text{MSFE}(\widehat{\mathbf{w}}, \widehat{\boldsymbol{\Sigma}})}{\text{MSFE}(\mathbf{w}, \boldsymbol{\Sigma})} - 1\right| = \left|\frac{\hat{a}^{-1}}{a^{-1}} - 1\right| = \frac{|a - \hat{a}|}{|\hat{a}|},$$

and

$$\left\|\widehat{\mathbf{w}} - \mathbf{w}\right\|_1 \leq \frac{a\frac{\left\|(\widehat{\boldsymbol{\Theta}}-\boldsymbol{\Theta})\boldsymbol{\iota}_p\right\|_1}{p} + |a - \widehat{a}|\frac{\left\|\boldsymbol{\Theta}\boldsymbol{\iota}_p\right\|_1}{p}}{|\widehat{a}|a}.$$

Therefore, in order to control the estimation uncertainty in the MSFE and combination weights, one needs to obtain a consistent estimator of the precision matrix $\boldsymbol{\Theta}$.

Lee and Seregina (2021a) apply the idea of the Factor Graphical LASSO described in Subsection 8.3.4 to forecast combination for macroeconomics time-series. They argue that the success of equal-weighted forecast combinations is partly due to the fact that the forecasters use the same set of public information to make forecasts, hence, they tend to make common mistakes. For example, they illustrate that in the European Central Bank's Survey of Professional forecasters of Euro-area real GDP growth, the forecasters tend to *jointly* understate or overstate GDP growth. Therefore, the authors stipulate that the forecast errors include common and idiosyncratic components, which allows the forecast errors to move together due to the common error component. Their paper provides a framework to learn from analyzing forecast errors: the authors separate unique errors from the common errors to improve the accuracy of the combined forecast.

8.4.2 Vector Autoregressive Models

The need for graphical modeling and learning the network structure among a large set of time series arises in economic problems that adhere to VAR framework. Examples of such applications include macroeconomic policy making and forecasting, and assessing connectivity among financial firms. Lin and Michailidis (2017) provide a good overview that draws the links between VAR models and networks, which we summarise below. The authors start with an observation that in many applications the components of a system can be partitioned into interacting blocks. As an example, Cushman and Zha (1997) examined the impact of monetary policy in a small open economy. The economy is modeled as one block, whereas variables in foreign economies as the other. Both blocks have their own autoregressive structure, and there is unidirectional interdependence between the blocks: the foreign block influences the small open economy, but not the other way around. Hence, there exists a linear ordering amongst blocks. Another example provided by Lin and Michailidis (2017) stems from the connection between the stock market and employment macroeconomic variables (Farmer, 2015 that focuses on the impact through a wealth effect mechanism of the former on the latter. In this case the underlying hypothesis of interest is that the stock market influences employment, but not the other way around. An extension of the standard VAR modeling introduces an additional exogenous block of variables "X" that exhibits autoregressive dynamics – such extension is referred to as VAR-X model. For instance, Pesaran, Schuermann and Weiner (2004) build a model to study regional inter-dependencies where country specific macroeconomic indicators evolve according to a VAR model, and they are influenced by key macroeconomic variables from neighbouring countries/regions (an exogenous block). Abeysinghe (2001) studies the direct and indirect impact of oil prices on the GDP growth of 12 Southeast and East Asian economies, while controlling for such exogenous variables as the country's consumption and investment expenditures along with its trade balance.

Let us now follow Lin and Michailidis (2017) to formulate the aforementioned VAR setup as a recursive linear dynamical system comprising two blocks of variables:

$$\mathbf{x}_t = \mathbf{A}\mathbf{x}_{t-1} + \mathbf{u}_t, \tag{8.43}$$

$$\mathbf{z}_t = \mathbf{B}\mathbf{x}_{t-1} + \mathbf{C}\mathbf{z}_{t-1} + \mathbf{v}_t, \tag{8.44}$$

where $\mathbf{x}_t \in \mathbb{R}^{p_1}$ and $\mathbf{z}_t \in \mathbb{R}^{p_2}$ are the variables in groups 1 and 2, respectively. Matrices $\mathbf{A}$ and $\mathbf{C}$ capture the temporal intra-block dependence, while matrix $\mathbf{B}$ captures the inter-block dependence. Note that the block of $\mathbf{x}_t$ variables acts as an exogenous effect to the evolution of the $\mathbf{z}_t$ block. Furthermore, $\mathbf{z}_t$ is Granger-caused by $\mathbf{X}_t$. Noise processes $\{\mathbf{u}_t\}$ and $\{\mathbf{v}_t\}$ capture additional contemporaneous intra-block dependence of $\mathbf{x}_t$ and $\mathbf{z}_t$. In addition, the noise processes are assumed to follow zero mean Gaussian distributions:

$$\mathbf{u}_t \sim \mathcal{N}(\mathbf{0}, \mathbf{\Sigma}_u) \quad \mathbf{v}_t \sim \mathcal{N}(\mathbf{0}, \mathbf{\Sigma}_v),$$

where $\mathbf{\Sigma}_u$ and $\mathbf{\Sigma}_v$ are covariance matrices. The parameters of interest are transition matrices $\mathbf{A} \in \mathbb{R}^{p_1 \times p_1}$, $\mathbf{B} \in \mathbb{R}^{p_2 \times p_1}$, $\mathbf{C} \in \mathbb{R}^{p_2 \times p_2}$, and the covariance matrices $\mathbf{\Sigma}_u$ and $\mathbf{\Sigma}_v$. Lin and Michailidis (2017) assume that the matrices $\mathbf{A}$, $\mathbf{C}$, $\mathbf{\Theta}_u \equiv \mathbf{\Sigma}_u^{-1}$, and $\mathbf{\Theta}_v \equiv \mathbf{\Sigma}_v^{-1}$ are sparse, whereas $\mathbf{B}$ can be either sparse or low-rank.

We now provide an overview of the estimation procedure used to obtain the ML estimates of the aforementioned transition matrices and precision matrices, based on Lin and Michailidis (2017). First, let us introduce some notations: the "response" matrices from time 1 to T are defined as:

$$\mathcal{X}^T = [x_1\ x_2\ \dots\ x_T]' \quad \mathcal{Z}^T = [z_1\ z_2\ \dots\ z_T]',$$

where $\{x_0, \dots, x_T\}$ and $\{z_0, \dots, z_T\}$ is centered time series data. Further, define the "design" matrices from time 0 to T-1 as:

$$\mathcal{X} = [x_0\ x_1\ \dots\ x_{T-1}]' \quad \mathcal{Z}^T = [z_0\ z_1\ \dots\ z_{T-1}]'.$$

The error matrices are denoted as $\mathcal{U}$ and $\mathcal{V}$. The authors proceed by formulating optimization problems using penalized log-likelihood functions to recover $\mathbf{A}$ and $\mathbf{\Theta}_u$:

$$(\widehat{\mathbf{A}}, \widehat{\mathbf{\Theta}}_u) = \underset{\mathbf{A}, \mathbf{\Theta}_u}{\arg\min} \Big[\operatorname{tr}\Big(\mathbf{\Theta}_u (\mathcal{X}^T - \mathcal{X}\mathbf{A}')'(\mathcal{X}^T - \mathcal{X}\mathbf{A}')/T\Big) - \log|\mathbf{\Theta}_u| + \lambda_A \|\mathbf{A}\|_1 + \rho_u \|\mathbf{\Theta}_u\|_{1,\text{off}} \Big], \tag{8.45}$$

as well as $\mathbf{B}$, $\mathbf{C}$, and $\mathbf{\Theta}_v$:

$$(\widehat{\mathbf{B}}, \widehat{\mathbf{C}}, \widehat{\mathbf{\Theta}}_v) = \underset{\mathbf{B}, \mathbf{C}, \mathbf{\Theta}_v}{\arg\min} \Big[\operatorname{tr}\Big(\mathbf{\Theta}_v (\mathcal{Z}^T - \mathcal{X}\mathbf{B}' - \mathcal{Z}\mathbf{C}')'(\mathcal{Z}^T - \mathcal{X}\mathbf{B}' - \mathcal{Z}\mathbf{C}')/T\Big) - \log|\mathbf{\Theta}_v| + \lambda_B \mathcal{R}(\mathbf{B}) + \lambda_C \|\mathbf{C}\|_1 + \rho_v \|\mathbf{\Theta}_v\|_{1,\text{off}}, \Big], \tag{8.46}$$

where $\|\mathbf{\Theta}_u\|_{1,\text{off}} = \sum_{i \neq j} |\theta_{u,ij}|$, and $\lambda_A, \lambda_B, \lambda_C, \rho_u, \rho_v$ are tuning parameters controlling the regularization strength. The regularizer $\mathcal{R}(\mathbf{B}) = \|\mathbf{B}\|_1$ if $\mathbf{B}$ is assumed to be sparse, and $\mathcal{R}(\mathbf{B}) = |||\mathbf{B}|||_* = \sum_{i=1}^{min(p_2, p_1)}$(singular values$_i$ of $\mathbf{B}$) if $\mathbf{B}$ is assumed to be low-rank, where $|||\cdot|||_*$ is a nuclear norm. To solve (8.45) and (8.46), Lin and Michailidis (2017) develop two algorithms that iterate between estimating transition matrices by minimizing the regularized sum of squared residuals keeping the precision matrix fixed, and estimating precision matrices using GLASSO keeping the transition matrices fixed. The finite sample error bounds for the obtained estimates are also established.

In their empirical application, the authors extend the model of Farmer (2015): they analyze the temporal dynamics of the log-returns of stocks with large market capitalization and key macroeconomic variables. In the context of the aforementioned notations, the $\mathbf{x}_t$ block consists of the stock log-returns, whereas the $\mathbf{Z}_t$ block consists of the macroeconomic variables. The authors' findings are consistent with the previous empirical results documenting increased connectivity during the crisis periods and strong impact of the stock market on total employment arguing that the stock market provides a plausible explanation for the great recession.

Basu, Li and Michailidis (2019) suggest that the sparsity assumption on (8.43) may not be sufficient: fir instance, returns on assets tend to move together in a more concerted manner during financial crisis periods. The authors proceed by studying high-dimensional VAR models where the transition matrix **A** exhibits a more complex structure: it is low rank and/or (group) sparse, which can be formulated as the following optimization problem:

$$\mathbf{x}_t = \mathbf{A}\mathbf{x}_{t-1} + \mathbf{u}_t, \quad \mathbf{u}_t \sim \mathcal{N}(\mathbf{0}, \boldsymbol{\Sigma}_u)$$
$$\mathbf{A} = \mathbf{L}^* + \mathbf{R}^*, \quad \text{rank}(\mathbf{L}^*) = r,$$

where $\mathbf{L}^*$ is the low-rank component and $\mathbf{R}^*$ is either sparse ($\mathbf{S}^*$), or group-sparse ($\mathbf{G}^*$). The authors assume that the number of non-zero elements in a sparse case (as defined by the cardinality of $\mathbf{S}^*$ shown as an ℓ_0-norm: $||\cdot||_0$) is $||\mathbf{S}^*||_0 = s$, and $||\mathbf{G}^*||_{2,0} = g$ which denotes the number of nonzero groups in $\mathbf{G}^*$ in the group sparse case. The goal is to estimate $\mathbf{L}^*$ and $\mathbf{R}^*$ accurately based on $T \ll p^2$. To overcome an inherent identifiability issue in the estimation of sparse and low-rank components, Basu, Li and Michailidis (2019) impose a well-known incoherence condition which is sufficient for exact recovery of $\mathbf{L}^*$ and $\mathbf{R}^*$ by solving the following convex program:

$$(\widehat{\mathbf{L}}, \widehat{\mathbf{R}}) = \underset{\mathbf{L}\in\boldsymbol{\Omega},\mathbf{R}}{\arg\min}\, l(\mathbf{L}, \mathbf{R}) \tag{8.47}$$
$$l(L, R) \equiv \frac{1}{2}\left\|\mathcal{X}^T - \mathcal{X}(\mathbf{L}+\mathbf{R})\right\|_F^2 + \lambda_N ||\mathbf{L}||_* + \mu_N ||\mathbf{R}||_\diamond,$$

where $\boldsymbol{\Omega} = \{\mathbf{L} \in \mathbb{R}^{p\times p} : ||\mathbf{L}||_{\max} \leq \alpha/p\}$ (for sparse) or $\boldsymbol{\Omega} = \{\mathbf{L} \in \mathbb{R}^{p\times p} : ||\mathbf{L}||_{2,\max} \equiv \max_{k=1,\dots,K}\left\|(\mathbf{L})_{G_k}\right\|_F \leq \beta/\sqrt{K}\}$ (for group sparse); $||\cdot||_\diamond$ represents $||\cdot||_1$ or $||\cdot||_{2,1}$ depending on sparsity or group sparsity of **R**. The parameters α and β control the degree of non-identifiability of the matrices allowed in the model class. To solve the optimization problem in (8.47), Basu, Li and Michailidis (2019) develop an iterative algorithm based on the gradient descent, which they call "Fast Network Structure Learning".

In their empirical application, the authors employ the proposed framework to learn Granger causal networks of asset pricing data obtained from CRSP and WRDS. They examine the network structure of realized volatilities of financial institutions representing banks, primary broker/dealers and insurance companies. Two main findings can be summarized as follows: (1) they document increased connectivity pattern during the crisis periods; (2) significant sparsity of the estimated sparse component provides a further scope for better examining specific firms that are key drivers in the volatility network.

Another example of the use of graphical models for economic problems is to measure systemic risk. For instance, network connectivity of large financial institutions can be used to identify systematically important institutions based on the centrality of their role in a network. Basu, Das, Michailidis and Purnanandam (2019) propose a system-wide network measure that uses GLASSO to estimate connectivity among many firms using a small sample size. The authors criticise a pairwise approach of learning network structures drawing the analogy of the latter with the omitted variable bias in standard regression models which leads to inconsistently estimated model parameters. To overcome the aforementioned limitation of the pairwise approach and correctly identify the interconnectedness structure of the system, Basu, Das et al. (2019) fit a sparse high-dimensional VAR model and develop a testing framework for Granger causal effects obtained from regularized estimation of large VAR models.

They consider the following model of stock returns of p firms:

$$\mathbf{x}_t = \mathbf{A}\mathbf{x}_{t-1} + \boldsymbol{\varepsilon}_t,\ \boldsymbol{\varepsilon}_t \sim \mathcal{N}(\mathbf{0}, \boldsymbol{\Sigma}_{\varepsilon}),\ \boldsymbol{\Sigma}_{\varepsilon} = \text{diag}(\sigma_1^2, \ldots, \sigma_p^2),\ \sigma_j^2 > 0\ \forall j,$$

which is a simplified version of (8.43). The paper assumes that the financial network is sparse: in other words, they require the true number of interconnections between the firms to be very small. To recover partial correlations between stock returns, Basu, Das et al. (2019) use a debiased Graphical LASSO as in (8.9).

In their empirical exercise, the authors use monthly returns for three financial sectors: banks, primary broker/dealers and insurance companies. They focus on three systemically important events: the Russian default and LCTM bankruptcy in late 1998, the dot-com bubble accompanied with the growth of mortgage-backed securities in 2002, and the global financial crisis of 2007-09. The paper finds connectivity, measured by either the count of neighbours or distance between nodes, increases before and during systemically important events. Furthermore, the approach developed in the paper allows tracing the effect of a negative shock on a firm on the entire network by tracing its effects through the direct linkages. Finally, using an extensive simulation exercise, the authors show that debiased GLASSO outperforms competing methods in terms of the estimation and detection accuracy.

Finally, non-stationarity is another distinctive feature of economic time series. Basu and Rao (2022) take a further step in extending graphical models: they develop a nonparametric framework for high-dimensional VARs based on GLASSO for non-stationary multivariate time series. In addition to the concepts of conditional dependence/independence commonly used in the graphical modelling literature, the authors introduce the concepts of conditional stationarity/non-stationarity. The paper demonstrates that the non-stationary graph structure can be learned from finite-length time series in the Fourier domain. They conclude with numerical experiments showing the feasibility of the proposed method.

8.5 Further Integration of Graphical Models with Machine Learning

Given that many economic phenomena can be visualised as networks, graphical models can serve as a useful tool to infer the structure of such networks. In this chapter we aimed at reviewing recent advances in graphical modelling that have made the latter more suitable for finance and economics applications. Tracing back historical development of commonly used econometric approaches, including non-parametric and Bayesian ones, they started with a simplified environment, such as a low-dimensional setup, and developed into more complex frameworks that deal with high dimensions, missing observations, non-stationarity etc. As these developments were growing, more parallels evolved suggesting that machine learning methods can be viewed as complements rather than substitutes of the existing approaches. Our stand is that the literature on graphical modelling is still getting integrated into economic and finance applications, and a further development would be to establish common grounds for this class of models with other machine learning methods. We wrap up the chapter with a discussion on the latter.

To start off, instead of investing in all available stocks (as was implicitly assumed in Section 8.3), several works focus on selecting and managing a subset of financial instruments such as stocks, bonds, and other securities. Such technique is referred to as a sparse portfolio. This stream of literature integrates graphical models with shrinkage techniques (such as LASSO) and reinforcement learning. To illustrate, Seregina (2021) proposes a framework for constructing a sparse portfolio in high dimensions using nodewise regression. The author reformulates Markowitz portfolio allocation exercise as a constrained regression problem, where the portfolio weights are shrunk to zero. The paper finds that in contrast to non-sparse counterparts, sparse portfolios are robust to recessions and can be used as hedging vehicles during such times. Furthermore, they obtain the oracle bounds of sparse weight estimators and provide guidance regarding their distribution. Soleymani and Paquet (2021) take a different approach for constructing sparse portfolios: they develop a graph convolutional reinforcement learning framework, DeepPocket, whose objective is to exploit the time-varying interrelations between financial instruments. Their framework has three ingredients: (1) a restricted stacked autoencoder (RSAE) for feature extraction and dimensionality reduction. Concretely, the authors map twelve features including opening, closing, low and high prices, and financial indicators to obtain a lower dimensional representation of the data. They use the latter and apply (2) a graph convolutional network (GCN) to acquire interrelations among financial instruments: a GCN is a generalization of convolutional neural networks (CNN) to a data that has a graph structure. The output of the GCN is passed to (3) a convolutional network for each of the actor and the critic to enforce investment policy and estimate the return on investment. The model is trained using historical data. The authors evaluate model performance over three distinct investment periods including during the COVID-19 recession: they find superior performance of DeepPocket compared to market indices and equally-weighted portfolio in terms of the return on investment.

GCNs provide an important step in merging the information contained in graphical models with neural networks, which opens further ground to use graphical methods in classification, link prediction, community detection, and graph embedding. Convolution operator (which is defined as the integral of the product of the two functions after one is reversed and shifted) is essential since it has proven efficient at extracting complex features, and it represents the backbone of many deep learning models. Graph kernels that use a "kernel trick" (Kutateladze, 2022) serve as an example of a convolution filter used for GCNs.

Fellinghauer, Buhlmann, Ryffel, von Rhein and Reinhardt (2013) use random forests (Breiman, Friedman, Olshen & Stone, 2017) in combination with nodewise regression to extend graphical approaches to models with mixed-type data. They call the proposed framework Graphical Random Forest (GRaFo). In order to determine which edges should be included in the graphical model, the edges suggested by the individual regressions need to be ranked such that a smaller rank indicates a better candidate for inclusion. However, if variables are mixed-type, a global ranking criterion is difficult to find. For instance, continuous and categorical response variables are not directly comparable. To overcome this issue, the authors use random forests for performing the individual nonlinear regressions and obtain the ranking scheme from random forests' variable importance measure. GRaFo demonstrates promising performance on the two-health related data set for studying the interconnection of functional health components, personal and environmental factors; and for identifying which risk factors may be associated with adverse neurodevelopment after open-heart surgery.

Finally, even though not yet commonly used in economic applications, a few studies have combined graphical models with neural networks for classification tasks. Ji and Yao (2021) develop a CNN-based model with Graphical LASSO (CNNGLasso) to extract sparse topological features for brain disease classification. Their approach can be summarized as a three-step procedure: (1) they develop a novel Graphical LASSO model to reveal the sparse connectivity patterns of the brain network by estimating the sparse inverse covariance matrices (SICs). In this model, the Cholesky composition is performed on each SIC for ensuring its positive definite property, and SICs are divided into several groups according to the classification task, which can interpret the difference between patients and normal controls while maintaining the inter-subject variability. (2) the filters of the convolutional layer are multiplied with the estimated SICs in an element-wise manner, which aims at avoiding the redundant features in the high level topological feature extraction. (3) the obtained sparse topological features are used to classify patients with brain diseases from normal controls. Ji and Yao (2021) use a CNN as a choice of a neural network, whereas for economic application alternative models such as RNNs (Dixon & London, 2021) and LSTMs (Zhang et al., 2019) were shown to demonstrate good performance for time-series modelling and financial time-series prediction.

Acknowledgements The author would like to express her sincere gratitude to Chris Zhu (ychris.zhu@gmail.com) who helped create a GitHub repository with toy examples of several machine learning methods from the papers reviewed in this chapter. Please visit Seregina and Zhu (2022) for further details.

References

Abeysinghe, T. (2001). Estimation of direct and indirect impact of oil price on growth. *Economics letters*, *73*(2), 147–153.

Awoye, O. A. (2016). *Markowitz minimum variance portfolio optimization using new machine learning methods* (Unpublished doctoral dissertation). University College London.

Bai, J. (2003). Inferential theory for factor models of large dimensions. *Econometrica, 71*(1), 135–171. Retrieved from https://doi.org/10.1111/1468-0262.00392

Bai, J. & Ng, S. (2002). Determining the number of factors in approximate factor models. *Econometrica, 70*(1), 191–221. Retrieved from https://doi.org/10.1111/1468-0262.00273 doi: 10.1111/1468-0262.00273

Barigozzi, M., Brownlees, C. & Lugosi, G. (2018). Power-law partial correlation network models. *Electronic Journal of Statistics, 12*(2), 2905–2929. Retrieved from https://doi.org/10.1214/18-EJS1478

Basu, S., Das, S., Michailidis, G. & Purnanandam, A. (2019). A system-wide approach to measure connectivity in the financial sector. *Available at SSRN 2816137.*

Basu, S., Li, X. & Michailidis, G. (2019). Low rank and structured modeling of high-dimensional vector autoregressions. *IEEE Transactions on Signal Processing, 67*(5), 1207-1222. doi: 10.1109/TSP.2018.2887401

Basu, S. & Rao, S. S. (2022). Graphical models for nonstationary time series. *arXiv preprint arXiv:2109.08709.*

Bates, J. M. & Granger, C. W. J. (1969). The combination of forecasts. *Operations Research, 20*(4), 451–468. Retrieved from http://www.jstor.org/stable/3008764

Bishop, C. M. (2006). *Pattern recognition and machine learning (information science and statistics).* Berlin, Heidelberg: Springer-Verlag.

Boyd, S., Parikh, N., Chu, E., Peleato, B. & Eckstein, J. (2011, January). Distributed optimization and statistical learning via the alternating direction method of multipliers. *Found. Trends Mach. Learn., 3*(1), 1–122. Retrieved from http://dx.doi.org/10.1561/2200000016

Breiman, L., Friedman, J. H., Olshen, R. A. & Stone, C. J. (2017). *Classification and regression trees.* Routledge.

Brownlees, C., Nualart, E. & Sun, Y. (2018). Realized networks. *Journal of Applied Econometrics, 33*(7), 986-1006. Retrieved from https://onlinelibrary.wiley.com/doi/abs/10.1002/jae.2642

Cai, T., Liu, W. & Luo, X. (2011). A constrained l1-minimization approach to sparse precision matrix estimation. *Journal of the American Statistical Association, 106*(494), 594–607.

Cai, T. T., Hu, J., Li, Y. & Zheng, X. (2020). High-dimensional minimum variance portfolio estimation based on high-frequency data. *Journal of Econometrics, 214*(2), 482-494.

Callot, L., Caner, M., Önder, A. O. & Ulasan, E. (2019). A nodewise regression approach to estimating large portfolios. *Journal of Business & Economic Statistics, 0*(0), 1-12. Retrieved from https://doi.org/10.1080/07350015.2019.1683018

Chamberlain, G. & Rothschild, M. (1983). Arbitrage, factor structure, and mean-variance analysis on large asset markets. *Econometrica, 51*(5), 1281–1304. Retrieved from http://www.jstor.org/stable/1912275

Clarke, R., de Silva, H. & Thorley, S. (2011). Minimum-variance portfolio composition. *The Journal of Portfolio Management*, *37*(2), 31–45. Retrieved from https://jpm.pm-research.com/content/37/2/31

Connor, G. & Korajczyk, R. A. (1988). Risk and return in an equilibrium APT: Application of a new test methodology. *Journal of Financial Economics*, *21*(2), 255–289. Retrieved from http://www.sciencedirect.com/science/article/pii/0304405X88900621

Cullen, C. (1990). *Matrices and linear transformations*. Courier Corporation, 1990. Retrieved from https://books.google.com/books?id=fqUTMxPsjt0C

Cushman, D. O. & Zha, T. (1997). Identifying monetary policy in a small open economy under flexible exchange rates. *Journal of Monetary economics*, *39*(3), 433–448.

Danaher, P., Wang, P. & Witten, D. M. (2014). The joint graphical LASSO for inverse covariance estimation across multiple classes. *Journal of the Royal Statistical Society: Series B (Statistical Methodology)*, *76*(2), 373-397. Retrieved from https://rss.onlinelibrary.wiley.com/doi/abs/10.1111/rssb.12033

DeMiguel, V., Garlappi, L., Nogales, F. J. & Uppal, R. (2009). A generalized approach to portfolio optimization: Improving performance by constraining portfolio norms. *Management Science*, *55*(5), 798–812. Retrieved from http://www.jstor.org/stable/40539189

Dixon, M. & London, J. (2021). Financial forecasting with alpha-rnns: A time series modeling approach. *Frontiers in Applied Mathematics and Statistics*, *6*, 59. Retrieved from https://www.frontiersin.org/article/10.3389/fams.2020.551138

Eves, H. (2012). *Elementary matrix theory*. Courier Corporation, 2012. Retrieved from https://books.google.com/books?id=cMLCAgAAQBAJ

Fama, E. F. & French, K. R. (1993). Common risk factors in the returns on stocks and bonds. *Journal of Financial Economics*, *33*(1), 3–56. Retrieved from http://www.sciencedirect.com/science/article/pii/0304405X93900235

Fama, E. F. & French, K. R. (2015). A five-factor asset pricing model. *Journal of Financial Economics*, *116*(1), 1–22. Retrieved from http://www.sciencedirect.com/science/article/pii/S0304405X14002323

Fan, J., Liao, Y. & Mincheva, M. (2013). Large covariance estimation by thresholding principal orthogonal complements. *Journal of the Royal Statistical Society: Series B*, *75*(4), 603–680.

Fan, J., Liu, H. & Wang, W. (2018, 08). Large covariance estimation through elliptical factor models. *The Annals of Statistics*, *46*(4), 1383–1414. Retrieved from https://doi.org/10.1214/17-AOS1588

Fan, J., Zhang, J. & Yu, K. (2012). Vast portfolio selection with gross-exposure constraints. *Journal of the American Statistical Association*, *107*(498), 592-606. Retrieved from https://doi.org/10.1080/01621459.2012.682825 (PMID: 23293404)

Farmer, R. E. (2015). The stock market crash really did cause the great recession. *Oxford Bulletin of Economics and Statistics*, *77*(5), 617–633.

Fellinghauer, B., Buhlmann, P., Ryffel, M., von Rhein, M. & Reinhardt, J. D. (2013). Stable graphical model estimation with random forests for discrete, continuous, and mixed variables. *Computational Statistics and Data Analysis*, *64*, 132-152. Retrieved from https://www.sciencedirect.com/science/article/pii/S0167947313000789

Friedman, J., Hastie, T. & Tibshirani, R. (2007, 12). Sparse inverse covariance estimation with the Graphical LASSO. *Biostatistics*, *9*(3), 432-441. Retrieved from https://doi.org/10.1093/biostatistics/kxm045

Goto, S. & Xu, Y. (2015). Improving mean variance optimization through sparse hedging restrictions. *Journal of Financial and Quantitative Analysis*, *50*(6), 1415–1441. doi: 10.1017/S0022109015000526

Green, R. C. & Hollifield, B. (1992). When will mean-variance efficient portfolios be well diversified? *The Journal of Finance*, *47*(5), 1785–1809. Retrieved from http://www.jstor.org/stable/2328996

Hallac, D., Park, Y., Boyd, S. & Leskovec, J. (2017). Network inference via the time-varying graphical LASSO. In *Proceedings of the 23rd acm sigkdd international conference on knowledge discovery and data mining* (pp. 205–213). New York, NY, USA: ACM. Retrieved from http://doi.acm.org/10.1145/3097983.3098037

Hastie, T., Tibshirani, R. & Friedman, J. (2001). *The elements of statistical learning*. New York, NY, USA: Springer New York Inc.

Jagannathan, R. & Ma, T. (2003). Risk reduction in large portfolios: Why imposing the wrong constraints helps. *The Journal of Finance*, *58*(4), 1651-1683. Retrieved from https://onlinelibrary.wiley.com/doi/abs/10.1111/1540-6261.00580

Janková, J. & van de Geer, S. (2018). Chapter 14: Inference in high-dimensional graphical models. In (p. 325 - 351). CRC Press.

Ji, J. & Yao, Y. (2021). Convolutional neural network with graphical LASSO to extract sparse topological features for brain disease classification. *IEEE/ACM Transactions on Computational Biology and Bioinformatics*, *18*(6), 2327-2338. doi: 10.1109/TCBB.2020.2989315

Johnstone, I. M. & Lu, A. Y. (2009). Sparse principal components analysis. *arXiv preprint arXiv:0901.4392*.

Koike, Y. (2020). De-biased graphical LASSO for high-frequency data. *Entropy*, *22*(4), 456.

Kutateladze, V. (2022). The kernel trick for nonlinear factor modeling. *International Journal of Forecasting*, *38*(1), 165-177. Retrieved from https://www.sciencedirect.com/science/article/pii/S0169207021000741

Ledoit, O. & Wolf, M. (2003). Improved estimation of the covariance matrix of stock returns with an application to portfolio selection. *Journal of Empirical Finance*, *10*(5), 603 - 621. Retrieved from http://www.sciencedirect.com/science/article/pii/S0927539803000070

Ledoit, O. & Wolf, M. (2004a). Honey, I shrunk the sample covariance matrix. *The Journal of Portfolio Management*, *30*(4), 110–119. Retrieved from https://jpm.iijournals.com/content/30/4/110

Ledoit, O. & Wolf, M. (2004b). A well-conditioned estimator for large-dimensional covariance matrices. *Journal of Multivariate Analysis*, *88*(2), 365 - 411. Retrieved from http://www.sciencedirect.com/science/article/pii/S0047259X03000964

doi: https://doi.org/10.1016/S0047-259X(03)00096-4
Lee, T.-H. & Seregina, E. (2021a). Learning from forecast errors: A new approach to forecast combinations. *arXiv:2011.02077*.
Lee, T.-H. & Seregina, E. (2021b). Optimal portfolio using factor graphical LASSO. *arXiv:2011.00435*.
Li, J. (2015). Sparse and stable portfolio selection with parameter uncertainty. *Journal of Business & Economic Statistics*, *33*(3), 381-392. Retrieved from https://doi.org/10.1080/07350015.2014.954708
Lin, J. & Michailidis, G. (2017). Regularized estimation and testing for high-dimensional multi-block vector-autoregressive models. *Journal of Machine Learning Research*, *18*(117), 1-49. Retrieved from http://jmlr.org/papers/v18/17-055.html
Lu, J., Kolar, M. & Liu, H. (2015). Post-regularization inference for time-varying nonparanormal graphical models. *J. Mach. Learn. Res.*, *18*, 203:1-203:78.
Markowitz, H. (1952). Portfolio selection*. *The Journal of Finance*, *7*(1), 77-91. Retrieved from https://onlinelibrary.wiley.com/doi/abs/10.1111/j.1540-6261.1952.tb01525.x
Meinshausen, N. & Bühlmann, P. (2006, 06). High-dimensional graphs and variable selection with the LASSO. *Ann. Statist.*, *34*(3), 1436–1462. Retrieved from https://doi.org/10.1214/009053606000000281
Millington, T. & Niranjan, M. (2017, 10). Robust portfolio risk minimization using the graphical LASSO. In (p. 863-872). doi: 10.1007/978-3-319-70096-0_88
Pearl, J. (1995). Causal diagrams for empirical research. *Biometrika*, *82*(4), 669–688. Retrieved from http://www.jstor.org/stable/2337329
Pesaran, M. H., Schuermann, T. & Weiner, S. M. (2004). Modeling regional interdependencies using a global error-correcting macroeconometric model. *Journal of Business & Economic Statistics*, *22*(2), 129–162.
Pourahmadi, M. (2013). *High-dimensional covariance estimation: With high-dimensional data*. John Wiley and Sons, 2013. Retrieved from https://books.google.com/books?id=V3e5SxlumuMC
Ross, S. A. (1976). The arbitrage theory of capital asset pricing. *Journal of Economic Theory*, *13*(3), 341–360. Retrieved from http://www.sciencedirect.com/science/article/pii/0022053176900466
Rothman, A. J., Bickel, P. J., Levina, E. & Zhu, J. (2008). Sparse permutation invariant covariance estimation. *Electron. J. Statist.*, *2*, 494–515. Retrieved from https://doi.org/10.1214/08-EJS176
Seregina, E. (2021). A basket half full: Sparse portfolios. *arXiv:2011.04278*.
Seregina, E. & Zhu, C. (2022). *Chapter 8: GitHub Repository*. https://github.com/ekat92/Book-Project-Econometrics-and-MLhttps://github.com/ekat92/Book-Project-Econometrics-and-ML.
Smith, J. & Wallis, K. F. (2009). A simple explanation of the forecast combination puzzle. *Oxford Bulletin of Economics and Statistics*, *71*(3), 331–355.
Soleymani, F. & Paquet, E. (2021, Nov). Deep graph convolutional reinforcement learning for financial portfolio management – deeppocket. *Expert Systems*

with Applications, *182*, 115127. Retrieved from http://dx.doi.org/10.1016/j.eswa.2021.115127

Stock, J. H. & Watson, M. W. (2002). Forecasting using principal components from a large number of predictors. *Journal of the American Statistical Association*, *97*(460), 1167–1179. Retrieved from https://doi.org/10.1198/016214502388618960

Tobin, J. (1958, 02). Liquidity Preference as Behaviour Towards Risk. *The Review of Economic Studies*, *25*(2), 65-86. Retrieved from https://doi.org/10.2307/2296205

van de Geer, S., Buhlmann, P., Ritov, Y. & Dezeure, R. (2014, 06). On asymptotically optimal confidence regions and tests for high-dimensional models. *The Annals of Statistics*, *42*(3), 1166–1202. Retrieved from https://doi.org/10.1214/14-AOS1221

Verma, T. & Pearl, J. (1990). Causal networks: Semantics and expressiveness. In *Uncertainty in artificial intelligence* (Vol. 9, p. 69 - 76). North-Holland. Retrieved from http://www.sciencedirect.com/science/article/pii/B9780444886507500111

Witten, D. M. & Tibshirani, R. (2009). Covariance-regularized regression and classification for high dimensional problems. *Journal of the Royal Statistical Society. Series B (Statistical Methodology)*, *71*(3), 615–636. Retrieved from http://www.jstor.org/stable/40247591

Zhan, N., Sun, Y., Jakhar, A. & Liu, H. (2020). Graphical models for financial time series and portfolio selection. In *Proceedings of the first acm international conference on ai in finance* (pp. 1–6).

Zhang, X., Liang, X., Zhiyuli, A., Zhang, S., Xu, R. & Wu, B. (2019, jul). AT-LSTM: An attention-based LSTM model for financial time series prediction. *IOP Conference Series: Materials Science and Engineering*, *569*(5), 052037. Retrieved from https://doi.org/10.1088/1757-899x/569/5/052037

Zhou, S., Lafferty, J. & Wasserman, L. (2010, 1st Sep). Time varying undirected graphs. *Machine Learning*, *80*(2), 295–319. Retrieved from https://doi.org/10.1007/s10994-010-5180-0

Chapter 9
Poverty, Inequality and Development Studies with Machine Learning

Walter Sosa-Escudero, Maria Victoria Anauati and Wendy Brau

Abstract This chapter provides a hopefully complete 'ecosystem' of the literature on the use of machine learning (ML) methods for poverty, inequality, and development (PID) studies. It proposes a novel taxonomy to classify the contributions of ML methods and new data sources used in this field. Contributions lie in two main categories. The first is making available better measurements and forecasts of PID indicators in terms of frequency, granularity, and coverage. The availability of more granular measurements has been the most extensive contribution of ML to PID studies. The second type of contribution involves the use of ML methods as well as new data sources for causal inference. Promising ML methods for improving existent causal inference techniques have been the main contribution in the theoretical arena, whereas taking advantage of the increased availability of new data sources to build or improve the outcome variable has been the main contribution in the empirical front. These inputs would not have been possible without the improvement in computational power.

9.1 Introduction

No aspect of empirical academic inquiry remains unaffected by the so-called 'data science revolution', and development economics is not an exception. Nevertheless, the

Walter Sosa-Escudero ✉
Universidad de San Andres, CONICET and Centro de Estudios para el Desarrollo Humano (CEDH-UdeSA) Buenos Aires, Argentina, e-mail: wsosa@udesa.edu.ar

Maria Victoria Anauati
Universidad de San Andres CONICET and CEDH-UdeSA, Buenos Aires, Argentina, e-mail: vanauati@udesa.edu

Wendy Brau
Universidad de San Andres and CEDH-UdeSA, Buenos Aires, Argentina, e-mail: wbrau@udesa .edu.ar

F. Chan, L. Mátyás (eds.), *Econometrics with Machine Learning*, Advanced Studies in Theoretical and Applied Econometrics 53, https://doi.org/10.1007/978-3-031-15149-1_9

combination of big data (tentatively defined as mostly observational data that arise from interacting with interconnected devices) and machine learning (ML henceforth) methods made its entrance at a moment when Economics was still embracing the *credibility revolution* in empirical analysis (Angrist and Pischke (2010)): the adoption of experimental or quasi-experimental datasets and statistical tools that allowed researchers to identify causal effects cleanly. The Nobel Prizes awarded in a lapse of only two years to Banerjee, Duflo and Kremer (2019), and to Angrist, Imbens and Card (2021) are a clear accolade to this approach. In such a context, the promises of big data/ML bring back memories of the correlation fallacies that the experimental approach tried to avoid. Consequently, Economics is a relatively new comer to the data science revolution that has already permeated almost every academic and professional field.

Aside from these considerations, development economics –and its immediate connection with policy– is a field that benefits enormously from detailed descriptions, measurements and predictions in the complex, heterogeneous and multidimensional contexts under which it operates. Additionally, the field has been particularly successful in exploiting observational data to find causal channels, either through meticulous institutional or historic analysis that isolate exogenous variation in such datasets (as in quasi experimental studies) and/or by using tools specifically aimed at dealing with endogeneities, such as difference-in-difference, instrumental variables or discontinuity design strategies. The tension between these concerns and opportunities may explain why, though relatively late, the use of ML techniques in development studies virtually exploded in the last years.

This chapter provides a complete picture of the use of ML for poverty, inequality development (PID) studies. The rate at which such studies have accumulated very recently makes it impossible to offer an exhaustive panorama that is not prematurely obsolete. Hence, the main goal of this chapter is to provide a useful taxonomy of the contribution of ML methods in this field that helps understand the main advantages and limitations of this approach.

Most of the chapter is devoted to two sections. The first one focuses on the use of ML to provide better measurements and forecasts of poverty, inequality and other development indicators. Monitoring and measurement is a crucial aspect of development studies, and the availability of big data and ML provides an invaluable opportunity to either improve existing indicators or illuminate aspects of social and economic behavior that are difficult or impossible to reach with standard sources like household surveys, census or administrative records. The chapter reviews such improvements, in terms of combining standard data sources with *non-traditional* ones, like satellite images, cell phone usage, social media and other *digital fingerprints*. The use of such data sets and modern ML techniques led to dramatic improvements in terms of more granular measurements (either temporal or geographically) or being able to reach otherwise difficult regions like rural areas or urban slums. The construction of indexes –as an application of non-supervised methods like modern versions of principal component analysis (PCA)– and the use of regularization tools to solve difficult missing data problems are also an important part of the use of ML in development studies and are reviewed in Section 9.2. The number of relevant articles on the subject is copious. This chapter focuses on a subset of them that help provide a complete picture of the use of ML in development studies. The Electronic Online Supplement contains a complete list of all articles reviewed in this chapter. Similarly, Table 9.3 details the ML methods used in the articles reviewed.

The second part of this chapter focuses on the still emerging field of causal inference with ML methods (see Chapter 3). Contributions come from the possibility of exploiting ML to unveil heterogeneities in treatment effects, improve treatment design, build better counterfactuals and to combine observational and experimental datasets that further facilitate clean identification of causal effects. Table 9.1 describes the main contributions of these two branches and highlights the key articles of each.

A final section explores advantages arising directly from the availability of more computing power, in terms of faster algorithms or the use of computer based strategies to either generate or interact with data, like bots or modern data visualization tools.

9.2 Measurement and Forecasting

ML help improve measurements and forecasts of poverty, inequality, and development (PID) in three ways (see Table 9.1). First, ML tools can be used to combine different data sources to improve data availability in terms time frequency and spatial disaggregation (*granularity*) or extension (*coverage*). They are mainly supervised ML algorithms –both for regression and classification– that are used for this purpose: the response variable to be predicted is a PID indicator. Therefore, improving the granularity or frequency of PID indicators implies predicting unobserved data points using ML. Second, ML methods can be used to reduce data dimensionality, which is useful to build indexes and characterize groups (using mainly unsupervised ML algorithms), or to select a subset of relevant variables (through both supervised and non-supervised methods) in order to design shorter and cheaper surveys. Finally, ML can solve data problems like missing observations in surveys or the lack of panel data.

The lack of detailed and high quality data has been a timely deterrent of income and wealth distribution studies. Household surveys, the most common source for these topics, provide reliable measures of income and consumption but at a low frequency or with a long delay, and at low levels of disaggregation. Census data, which attempt to solve the concern of disaggregation, usually have poor information on income or consumption. In addition, surveys and censuses are costly, and for many relevant issues they are available at a low frequency, a problem particularly relevant in many developing countries.

New non-traditional data sources, such as satellite images, digital fingerprints from cell phone calls records, social media, or the 'Internet of Things', can be fruitfully combined with traditional surveys, census and administrative data, with the aim of improving poverty predictions in terms of their availability, frequency and granularity. In this section, we first focus on the contributions in terms of improving availability and time frequency of estimates (Section 9.2.1). Second, we focus on spatial granularity (Section 9.2.2). In the Electronic Online Supplement, Table 9.1

Table 9.1: Taxonomy on ML contributions to PID studies

Category	Contributions	Key papers
Better measurements and forecasts	Combining data sources to improve data availability, frequency, and granularity; dimensionality reduction; data imputation.	Elbers et al. (2003), Blumenstock et al. (2015), Chi et al. (2022), Jean et al. (2016), Caruso et al. (2015).
Causal inference	Heterogeneous Treatment Effects; optimal treatment assignment; handling high- dimensional data and debiased ML; machine-building counterfactuals; leveraging new data sources; combining observational and experimental data.	Chowdhury et al. (2021), Chernozhukov et al. (2018a, 2018b), Athey and Wager (2021), Banerjee et al. (2021a, 2021b), Belloni et al. (2017), Athey et al. (2020), Ratledge et al. (2021), Huang et al. (2015).

provides an exhaustive list of studies that aim at improving the availability, frequency and granularity of poverty, inequality, and development indicators, along with their main characteristics (the scope of the paper, the data sources, the contribution, the ML methods used). Next, we review PID studies that aim at reducing data dimensionality (Section 9.2.3) and, finally, we center on PID studies that use ML methods to solve missing data problems (Section 9.2.4).

9.2.1 Combining Sources to Improve Data Availability

New data sources share several relevant characteristics. They are: (1) passively collected data, (2) relatively easy to obtain at a significantly lower cost than data from traditional sources, (3) updated frequently, and thus useful for generating nearly real-time estimates of regional vulnerability and for providing early warning and real-time monitoring of vulnerable populations. Therefore, one of the most immediate applications of data science to PID studies is to provide a source of inexpensive and more frequent estimations that complement or supplement national statistics. Non-traditional data sources in general do not replace official statistics. On the contrary, traditional ones often act as the *ground truth data*, i.e., the measure that is known to be real or true, used as response variable to train and test the performance of the predictive models.

Early work explored the potential offered by data from *nighttime lights*: satellite photographs taken at night that capture light emitted from the Earth's surface. Sutton et al. (2007) is one of first studies to apply a spatial analytic approach to the patterns in the nighttime imagery to proxy economic activity. They predict GDP at the state level for China, India, Turkey, and the United States for the year 2000. Elvidge et al. (2009) calculate a poverty index using population counts and the brightness of nighttime lights for all countries in the world. Both studies use cross-sectional data to predict poverty in a given year. Henderson et al. (2012) is the first study that measures real income growth using panel data. They calculate a brightness score for each pixel of a satellite image, and then aggregate these scores over a region to obtain an indicator that can be used as a proxy for economic activity under the assumption that lighting is a normal good. By examining cross-country GDP growth rates between 1992 and 2008, they develop a statistical framework that combines the growth in this light measure for each country with estimates of GDP growth from the World Development Indicators. Results show that the light-GDP elasticity lies between 0.28 and 0.32; this result is used to predict income growth for a set of countries with very low capacity national statistical agencies.

Since then, the use of satellite images as a proxy for local economic activity has increased significantly (see Donaldson & Storeygard, 2016, and Bennett & Smith, 2017, for a review). They lead to reliable poverty measures on a more frequent basis than that in traditional data sources, and they complement official statistics when there are measurement errors or when data are not available at disaggregated levels. They are used to track the effectiveness of poverty-reduction efforts in specific localities and to monitor poverty across time and space. For other examples, see Michalopoulos and Papaioannou (2014), Hodler and Raschky (2014), and Kavanagh et al. (2016), described in the Electronic Online Supplement, Table 9.1.

Despite the promise of satellite imagery as a proxy for economic output, Chen and Nordhaus (2015) show that for time-series analysis estimations of economic growth from multi-temporal nighttime imagery are not sufficiently accurate. In turn, Jean et al. (2016) warned that nightlight data is less effective at distinguishing differences in economic activity in areas at the bottom end of the income distribution, where satellite images appear uniformly dark. They suggest extracting information from daytime satellite imagery, which has a much higher resolution.

Satellite imagery are a type of *remote sensing data*, that is, data on the physical characteristics of an area which are detected, monitored and collected by measuring its radiation at a distance. Apart from satellite images, numerous studies resort to the use of digital fingerprints left by cell phone transaction records, which are increasingly ubiquitous even in very poor regions. Past history of mobile phone use can be employed to infer socioeconomic status in the absence of official statistics. Eagle et al. (2010) analyze the mobile and landline network of a large proportion of the population in the UK coupled with the Multiple Deprivation Index, a composite measure of relative prosperity of the community. Each residential landline number is associated with the rank of the Multiple Deprivation Index of the exchange area in which it is located. They map each census region to the telephone exchange area with greatest spatial overlap. Results show that the diversity of individuals' relationships is strongly correlated with the economic development of communities. Soto et al. (2011) and Frias-Martinez and Virseda (2012) extend the analysis to several countries in Latin America. A key article on improving data availability by combining traditional data with *call detail records* seems to be the widely cited work by Blumenstock et al.

(2015). They propose a methodology to predict poverty at the individual level based on the intensity of cell phone usage.

Naturally, there are other data sources from the private sector which help predict development indicators. An important article is Chetty et al. (2020a), which builds a public database that tracks spending, employment, and other outcomes at a high frequency (daily) and granular level (disaggregated by ZIP code, industry, income group, and business size) that uses anonymized data from private companies. The authors use this database to explore the heterogeneity of COVID-19's impact on the U.S. economy. Another example is Farrell et al. (2020), who use administrative banking data in combination with zip code-level characteristics to provide an estimate of gross family income using gradient boosting machines, understood as ensemble of many classification or regression trees (see Chapter 2), in order to improve their predictive performance. Boosting is a popular method to improve the accuracy or predictions by locally retraining the model to handle errors from previous training stages (see chapter 4 and Hastie et al., 2009 for more details on boosting).

Network data is also an important source of rich information for PID studies with ML. UN Global (2016) explores the network structure of the international postal system to produce indicators for countries' socioeconomic profiles, such as GDP, Human Development Index or poverty, analyzing 14 million records of dispatches sent between 187 countries over 2010-2014. Hristova et al. (2016) measure the position of each country in six different global networks (trade, postal, migration, international flights, IP and digital communications) and build proxies for a number of socioeconomic indicators, including GDP per capita and Human Development Index.

Several studies explore how data from *internet interactions and social media* can provide proxies for economic indicators at a fine temporal resolution. This type of data stands out for its low cost of acquisition, wide geographical coverage and real-time update. However, one disadvantage is that access to the web is usually limited in low-income areas, which conditions its use for the prediction of socioeconomic indicators in the areas where it is most needed. As many of the social media data come in the form of text, natural language processing techniques (NLP) are necessary for processing them. NLP is a subfield of ML that analyses human language in speech and text (a classic reference is the book by Jurafsky & Martin, 2014). For example, Quercia et al. (2012) explore Twitter users in London communities, and study the relationship between sentiment expressed in tweets and community socioeconomic well-being. To this end, a word count sentiment score is calculated by counting the number of positive and negative words. There are different dictionaries annotating the sentiment of words, i.e., whether they are positive or negative. The authors use a commonly used dictionary called Linguistic Inquiry Word Count that annotates 2,300 English words. An alternative to using a dictionary would be manually annotating the sentiment of a subset of tweets, and then use that subset to train an algorithm that learns to predict the sentiment label from text features. Then, by averaging the sentiment score of users in each community, they calculate the gross community happiness. Lansley and Longley (2016) use an unsupervised learning algorithm to classify geo-tagged tweets from Inner London into 20 distinctive topic groupings and find that users' socioeconomic characteristics can be inferred from their behaviors on Twitter. Liu et al. (2016) analyze nearly 200 million users' activities over 2009-2012 in the largest social network in China (Sina Microblog) and explore the relationship between online activities and socioeconomic indices.

Another important branch of studies focuses on tracking *labor market* indicators. Ettredge et al. (2005) find a significant association between the job-search variables and the official unemployment data for the U.S. Askitas and Zimmermann (2009) use Google keyword searches to predict unemployment rates in Germany. González-Fernández and González-Velasco (2018), use a similar methodology in Spain. Antenucci et al. (2014) use data from Twitter to create indexes of job loss, job search, and job posting using PCA. Llorente et al. (2015) also use data from Twitter to infer city-level behavioral measures, and then uncover their relationship with unemployment.

Finally, there several studies that combine multiple non-traditional data sources. For instance, satellite imagery data provide information about physical properties of the land, which are cost-effective but relatively coarse in urban areas. By contrast, call records from mobile phones have high spatial resolution in cities though insufficient in rural areas due to the sparsity of cellphone towers. Thus, recent studies show that their combination can produce better predictions. Steele et al. (2017) is a relevant reference. They use remote sensing data, call detail records and traditional survey-based data from Bangladesh to provide the first systematic evaluation of the extent to which different sources of input data can accurately estimate different measures of poverty, namely the Wealth Index, Progress out of Poverty Index and reported household income. They use hierarchical bayesian geostatistical models to construct highly granular maps of poverty for these indicators. Chi et al. (2022) use data from satellites, mobile phone networks, topographic maps, as well as aggregated and de-identified connectivity data from Facebook to build micro-estimates of wealth and poverty at 2.4km resolution for low and middle-income countries. They first use deep learning methods to convert the raw data to a set of quantitative features of each village. Deep learning algorithms such as neural networks are ML algorithms organized in layers: an input layer, an output layer, and hidden layers connecting the input and output layers. The input layer takes in the initial raw data, the hidden layers processes the data using nonlinear functions with certain parameters and outputs it to the next layer, and the output layer is connected to the last hidden layer and presents the prediction outcome (for a reference see Chapters 4 and 6 as well as the book by Goodfellow, Bengio & Courville, 2016). Then, they use these features to train a supervised ML model that predicts the relative and absolute wealth of each all populated 2.4km grid cells (see Section 9.2.2 for more details on the methodology and results of this study. See also Njuguna & McSharry, 2017 and Pokhriyal & Jacques, 2017 for further examples).

Many data sources, such as phone call records, are often proprietary. Therefore, the usefulness of the ML methods and their reproducibility depends on the accessibility to these data. The work by Chetty et al. (2014), Chetty et al. (2018) or Blumenstock et al. (2015) are examples of successful partnerships between governments, researchers and the private sector. However, according to Lazer et al. (2020), access to data from private companies is rarely available and when it is, the access is generally established on an ad-hoc basis. On their articles on the contributions of big data to development and social sciences, both Blumenstock (2018a) and Lazer et al. (2020) call for fostering collaboration among scientists, development experts, governments, civil society and the private sector for establishing clear guidelines for data-sharing. Meanwhile, works that use publicly available data become particularly valuable. For example, Jean et al. (2016) use satellite imagery that are publicly available and Rosati et al. (2020) use only open-source data, many of which are collected via scraping (see Section 9.2.2).

Moreover, these data often require dealing with privacy issues, as they are often highly disaggregated and have sensitive personal information. Blumenstock (2018a) states that a pitfall on the uses of big data for development is that there are few data regulation laws and checks and balances to control access to sensitive data in developing countries. However, Lazer et al. (2020) note that there are emerging examples of methodologies and models that facilitate data analysis while preserving privacy and keeping sensitive data secure. To mention a few, Blumenstock et al. (2015) emphasize that they use an anonymized phone call records database and request for informed consent to merge it with a follow-up phone survey database, which solicited no personally identifying information. Chetty et al. (2020a) use anonymized data from different private companies and took some to protect their confidentiality (such as excluding outliers, reporting percentage changes relative to a baseline rather than reporting levels of each series, or combining data from multiple firms).

9.2.2 More Granular Measurements

This section explores contributions of big data and ML in terms of data geolocation, visualization techniques, and methods for data interpolation that facilitate access, improve the granularity, and extended the coverage of development indicators to locations where standard data are scarce or non-existent.

9.2.2.1 Data Visualization and High-Resolution Maps

One of the main contributions of ML to PID studies is the construction of high resolution maps that help design more focused policies. The use of poverty maps has become more widespread in the last two decades. A compilation of many of these efforts can be found in (Chi et al., 2022). Other important development indicators in areas such as health, education, or child-labor have also benefited from this approach, as in Bosco et al. (2017), Graetz et al. (2018), or ILO-ECLAC (2018).

Bedi et al. (2007b) describes applications of poverty maps. They are a powerful communication tool, as they summarize a large volume of data in a visual format that is easy to understand while preserving spatial relationships. For instance, Chetty et al. (2020) made available an interactive atlas of children outcomes such as earnings distribution and incarceration rates by parental income, race and gender at the census tract level in the United States, by linking census and federal income tax returns data. Figure 9.1b shows a typical screenshot of their atlas. Soman et al. (2020) map an index of access to street networks (as a proxy for slums) worldwide, and Chi et al. (2022) map estimates of wealth and poverty around the world (see Table 9.1 in the Electronic Online Supplement for more details).

Perhaps the main application of poverty maps is in program targeting, i.e., determining eligibility more precisely. Elbers et al. (2007) quantify the impact on poverty of adopting a geographically-targeted budget using poverty maps for three countries (Ecuador, Madagascar and Cambodia). Their simulations show that the gains from increased granularity in targeting are important. For example, in Cambodia, the poverty reduction that can be achieved using 54.5% of the budget with a uniform transfer to each of the six provinces is the same as that achievable using only 30.8% of the budget but targeting each of the 1594 communes. Finally, maps can also be used for planning government policies at the sub-regional level, for analyzing the relationship of poverty or other development indicators between neighboring areas, for studying their geographic determinants or for evaluating the impact of policies (Aiken, Bedoya, Coville & Blumenstock, 2020; Bedi, Coudouel & Simler, 2007a).

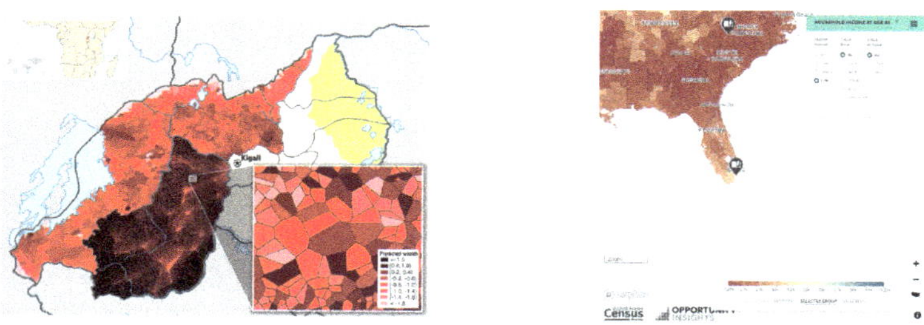

(a) Blumenstock et al. (2015)

(b) Source: Chetty et al. (2020)

Fig. 9.1: Examples of visualizations of poverty and inequality maps

9.2.2.2 Interpolation

Spatial interpolation exploits the correlation between indices of population well-being or development and geographical, socioeconomic or infrastructure features to predict the value of indicators where survey data are not available (Bosco et al., 2017). Researchers have developed several methods to use and combine both traditional and non-traditional sources of data to estimate poverty or other measures of development at a more granular level (such as the ELL method and others that use ML; see Table 9.1 in the Electronic Online Supplement).

The theoretical work of Elbers et al. (2003), coming from small area estimation statistics and before 'Data Science' was even a term, is seminal in this area. Their *ELL method* combines traditional survey and census data to improve granularity. Many variables are usually available in surveys with higher temporal frequency but at an aggregated or sparse spatial scale, and therefore are integrated with the finer scale census data to infer the high resolution of a variable of interest. The ELL method estimates the joint distribution of the variable of interest (avilable in the more frequent surveys) and a vector of covariates, restricting the set of explanatory variables to those that can be linked to observations in the census.

ML may improve the ELL method considerably, in two ways. Firstly, as supervised ML techniques focus on predicting accurately out-of-sample and allow more flexibility, they could improve the performance of linear imputation methods and of program targeting when using poverty proxies treating them as a missing value that can be predicted (Sohnesen & Stender, 2017; McBride & Nichols, 2018). Secondly, the ELL method relies on the availability of census data, while ML methods often leverage non-traditional data for poverty predictions. They are useful for preprocessing the data, given that the new data available are often unstructured (i.e., they are not organized in a predefined data format such as tables or graphs), unlabelled (i.e., each observation is not linked with a particular value of the response variable), or high-dimensional (e.g., image data come in the form of pixels arranged in at least two dimensions). In particular, deep learning methods have been widely adopted to extract a subset of numerical variables from this kind of data, as in Jean et al. (2016) or Chi et al. (2022). In general, this subset of numerical variables are not interpretable, and are only useful for prediction.

Most studies train supervised models to predict well-being indicators from traditional data sources based on features from non-traditional ones. The work by Blumenstock et al. (2015) is seminal in this approach. They take advantage of the fact that cell phone data are available at an individual level to study the distribution of wealth in Rwanda at a much finer level of disaggregation than official statistics allow. The authors merge an anonymized cell phone call data set from 1.5 million cell phone users and a follow-up phone socioeconomic survey of a geographically stratified random sample of 856 individual subscribers. Then, they predict poverty and wealth of individual subscribers and create regional estimates for 12,148 distinct cells, instead of just 30 districts that the census allows (see Figure 9.1a for a visualization of their wealth index). They start by estimating the first principal component of several survey responses related to wealth to construct a wealth index. This wealth index serves as the dependent variable they want to predict from the subscriber's historical patterns of phone use. Secondly, they use feature engineering to generate the features of phone use. They employ a combinatorial method that automatically generates thousands of metrics from the phone logs. Then, they train an elastic net regularization model to predict wealth from the features on phone usage (Chapter 1 explains the elastic net as well as others shrinkage estimators). Finally, they validate their estimations by comparing them with data collected by the Rwandan government.

Jean et al. (2016) is another seminal article. They use diurnal satellite imagery and nighttime lights data to estimate cluster-level expenditures or assets in Nigeria, Tanzania, Uganda, Malawi, and Rwanda. Clusters are roughly equivalent to villages in rural areas or wards in urban areas. Using diurnal satellite imagery overcomes the limitation of nightlights for tracking the livelihoods of the very poor due to the fact that luminosity is low and shows little variation in areas with populations living near and below the poverty line. In contrast, daytime satellite images can provide more information, such as the roof material of houses or the distance to roads. Like Blumenstock et al. (2015), their response variable is an asset index computed as the first principal component of the Demographic and Health Surveys' responses to questions about asset ownership. They also use expenditure data from the World Bank's Living Standards Measurement Study (LSMS) surveys. The authors predict these measures of poverty obtained from a traditional data source using features extracted from a non-traditional data source: diurnal satellite imagery. However, these images require careful preprocessing to extract relevant features for poverty prediction, as they are unstructured and unlabelled. To overcome these challenges, Jean et al. (2016) use Convolutional Neural Networks (CNNs) and a three-step transfer learning approach. The three steps are the following:

1. Pretraining a CNN on ImageNet, a large dataset with labelled images. CNNs are a particular type of neural network, widely used for image classification tasks, that has at least one *convolution* layer. Briefly, each image can be thought of as a matrix or array, where each pixel is represented with a numerical value. Then, the convolution layer consists of performing matrix products between the original matrix and other matrices called *convolution filters*, which are useful for determining whether a local or *low-level* feature (e.g., certain shape or edges) is present in an image. Therefore, in this first step the model learns to identify low-level image features.
2. *Fine-tuning* the model for a more specific task: training it to estimate nighttime light intensities, which proxy economic activity, from the input daytime satellite imagery. Therefore, this second step is useful for extracting image features which are relevant for poverty prediction, if some of the features that explain variation in nightlights are also predictive of economic outcomes.
3. Training Ridge regression models to predict cluster-level expenditures and assets using the features obtained in the previous steps.

One of the main contributions of the work by Jean et al. (2016) is that the satellite images used are publicly available. In the same venue, Rosati et al. (2020) map health vulnerability at the census block level in Argentina using only open-source data. They create an index of health vulnerability using dimensionality reduction techniques). Bosco et al. (2017) also take advantage of the availability of geolocated household survey data. They apply bayesian learning methods and neural networks to predict and map literacy, stunting and the use of modern contraceptive methods from a combination of other demographic and health variables, and find that the accuracy of these methods in producing high resolution maps disaggregated by gender is overall high but varies substantially by country and by dependent variable.

At the end of the road, and probably reflecting the state of the art in the field, is Chi et al. (2022). They obtain more than 19 million micro-estimates of wealth (both absolute and relative measures) that cover the populated surface of all 135 low and middle-income countries at a 2.4km resolution. Wealth information from

Table 9.2: ML methods for data interpolation in three key papers

Paper	Response variable	Features	Prediction
Blumenstock et al. (2015)	1st principal component from survey questions	Data engineering from phone calls data	Elastic net Regularization
Jean et al. (2016)	1st principal component from survey questions	CNN and transfer learning from satellite images	Ridge regularization
Chi et al. (2022)	1st principal component from survey questions	CNN from satellite images + features from phone calls and social media	Gradient boosting

traditional face-to-face Demographic and Health Surveys covering more than 1.3 million households in 56 different countries is the ground truth data. Once again, as in the previous works, a relative wealth index is calculated by taking the first principal component of 15 questions related to assets and housing characteristics. In turn, the authors combine a variety of non-traditional data sources –satellites, mobile phone networks, topographic maps and aggregated and de-identified connectivity data from Facebook– to build features for each micro-region. In the case of satellite imagery, the authors follow Jean et al. (2016) and use CNN to get 2048 features from each image, and then keep their first 100 principal components. From these features they train a gradient boosted regression tree to predict measurements of household wealth collected through surveys, and find that including different data sources improves the performance of their model, and that features related to mobile connectivity are among the most predictive ones.

The algorithms of Chi et al. (2022) have also been implemented for aid-targeting during the Covid-19 crisis by the governments of Nigeria and Togo. For the case of Togo, Aiken et al. (2021) quantify that ML methods reduce errors of exclusion by 4-21% relative to other geographic options considered by the government at the time (such as making all individuals within the poorest prefectures or poorest cantons eligible, or targeting informal workers). Likewise, Aiken et al. (2020) conclude that supervised learning methods leveraging mobile phone data are useful for targeting beneficiaries of an antipoverty program in Afghanistan. The errors of inclusion and exclusion when identifying the phone-owning ultra-poor are similar to when using survey-based measures of welfare, and methods combining mobile phone data and survey data are more accurate than methods using any one of the data sources. However, they highlight that its utility is limited by incomplete mobile phone penetration among the program beneficiaries.

Table 9.2 outlines the methods used in the three key papers leveraging ML techniques for interpolation in PID studies.

9.2.2.3 Extended Regional Coverage

Rural areas and informal settlements (slums) are typically underrepresented in official statistics. Many studies using ML or non-traditional data sources have contributed to extend data coverage to these areas.

Several of the interpolation studies mentioned before estimate development indicators in rural areas (see the Electronic Online Supplement, Table 9.1 for a detailed description of these studies). For example, Chi et al. (2022) show that their model can differentiate variation in wealth within rural areas, and Aiken et al. (2021) specifically designed their model to determine eligibility in a rural assistance program. The model in Engstrom et al. (2017), using day-time and night-time satellite imagery is even more accurate in rural areas than in urban ones. Watmough et al. (2019) use remote sensor satellite data to predict rural poverty at the household level.

The detection of high poverty areas such as slums (i.e., urban poverty) has been widely explored. It has its origins in spatial statistics and geographical information systems literature, and the body of work has increased since the availability of high-resolution remote sensing data, satellite and street view images and georeferenced data from crowd-sourced maps, as well as new ML methods to process them. Kuffer et al. (2016) and Mahabir et al. (2018) review studies using remote sensing data for slum mapping.

A variety of methods have been employed to detect slums. For instance, image texture analysis extracts features based on the shape, size and tonal variation within an image. The most frequently used method is object-based image analysis (commonly known as OBIA), a set of techniques to segment the image into meaningful objects by grouping adjacent pixels. Texture analysis might serve as input for OBIA as in Kohli et al. (2012) and Kohli et al. (2016), and OBIA can also be combined with georeferenced data (GEOBIA techniques). More recently, ML techniques have been added to the toolkit for slum mapping, and they have been highly accurate (Kuffer et al., 2016; Mahabir et al., 2018). Supervised ML algorithms use a subset of data labelled as slums to learn which combination of data features are relevant to identify slums, in order to minimize the error when predicting if there are slums in unlabeled areas. Roughly, two different supervised ML approaches have been used. The first one is to use a two-step strategy (Soman et al., 2020): preprocessing the data to extract an ex-ante and ad-hoc set of features, and then training a model to classify areas into slums or not. Different techniques can be combined for feature extraction, such as texture analysis or GEOBIA for images, and different algorithms can be used for the classification tasks. Some of the works using this approach are Baylé (2016), Graesser et al. (2012), Wurm et al. (2017), Owen and Wong (2013), Huang et al. (2015), Schmitt et al. (2018), Dahmani et al. (2014) and Khelifa and Mimoun (2012). The second and more recent approach is using deep learning techniques, where feature extraction takes place automatically and jointly with classification, instead of being defined ad-hoc and ex-ante. For example, Maiya and Babu (2018) and Wurm et al. (2017) use CNNs.

9.2.2.4 Extrapolation

The evidence on the possibility of extrapolating the results of a model trained in a certain location to a different one is mixed. For example, Bosco et al. (2017) find a large variability in the accuracy of the models across countries in their study. This means that not all the geolocated socioeconomic variables are equally useful for predicting education and health indicators in every country, so a model that works well in one country cannot be expected to work well in another. In this respect, Blumenstock (2018b) find that models trained using mobile phone data from Afghanistan are quite inaccurate to predict the wealth in Rwanda and vice-versa. The multi-data supervised models of Chi et al. (2022) are also generalizable between countries and they find that ML models trained in one country are more accurate to predict wealth in other countries when applied to neighboring countries and to countries with similar observable characteristics.

Regarding slum detection, given that the morphological structure of slums varies in different parts of the world and at different growth stages, models often need to be re-trained for different contexts (Taubenböck, Kraff & Wurm, 2018; Soman et al., 2020). As a result, slum mapping tends to be oriented to specific areas (Mahabir et al., 2018), although the possibility of generalizing results varies across methods. According to Kuffer et al. (2016), texture-based methods are among the most robust across cities and imagery. Soman et al. (2020) develop a methodology based on topological features from open source map data with the explicit purpose of generalizing results across the world.

In sum, there are two kinds of factors that hinder extrapolation. First, the heterogeneity of the relationship between observable characteristics and development indicators across different locations. Second, there may be noise inherent to the data collection process, such as satellite images taken at different times of the day or in different seasons, although the increased collection frequency of satellite images may help to tackle it. However, domain generalization techniques, which have been increasingly applied in other fields (Dullerud et al., 2021; Zhuo & Tan, 2021; Kim, Kim, Kim, Kim & Kim, 2019; Gulrajani & Lopez-Paz, 2020), are still under-explored in poverty and development studies. One example is the work by Wald et al. (2021), that introduces methods to improve multi-domain calibration by training or modifying an existing model that achieve better performance on unseen domains, and apply their methods to the dataset collected by Yeh et al. (2020).

9.2.3 Dimensionality Reduction

Much of the literature points towards the multidimensional nature of welfare (Sen, 1985), which translates almost directly into that of poverty or deprivation. Even when there is an agreement on the multidimensionality of well-being, there remains the problem of deciding how many dimensions are relevant. This problem of *dimensionality reduction* is important not only for a more accurate prediction of poverty but also for a more precise identification of which variables are relevant in order to design shorter surveys with lower non-response rates and lower costs.

Non-supervised ML techniques, such as clustering and factor analysis, can contribute towards that direction. Traditionally, research has addressed the multidimensionality of welfare by first reducing the dimensionality of the original welfare space using factor methods, and finally proceeding to identify the poor based on this reduced set of variables. For instance, Gasparini et al. (2013) apply factor analysis to 12 variables in the *Gallup World Poll* for Latin American countries, concluding that three factors are necessary for representing welfare (related to income, subjective well-being and basic needs). In turn, Luzzi et al. (2008) apply factor analysis to 32 variables in the Swiss Household Panel, concluding that four factors suffice (summarizing financial, health, neighborhood and social exclusion conditions). Then, they use these factors for cluster analysis in order to identify the poor.

Instead, Caruso et al. (2015) propose a novel methodology that first identifies the poor and then explores the dimensionality of welfare. They first identify the poor by applying clustering methods on a rather large set of attributes and then reduce the dimension of the original welfare space by finding the smallest set of attributes that can reproduce as accurately as possible the poor/non-poor classification obtained in the first stage, based on a 'blinding' variable selection method as in Fraiman et al. (2008). The reduced set of variables identified in the second stage is a strict subset of the variables originally in the welfare space, and hence readily interpretable. Therefore, their methodology overcomes one of the limitations of PCA and factor analysis: they result in an index that is by construction a linear combination of all the original features. To solve this problem Gasparini et al. (2013) and Luzzi et al. (2008) use rotation techniques. However, rotations cannot guarantee that enough variables have a zero loading in each component to render them interpretable. In turn, Merola and Baulch (2019) suggest using sparse PCA to obtain more interpretable asset indexes from household survey data in Vietnam and Laos. Sparse PCA techniques embed regularization methods into PCA so as to obtain principal components with sparse loadings, that is, each principal component is a combination of only a subset of the original variables.

Edo et al. (2021) aim at identifying the middle class. They propose a method for building multidimensional well-being quantiles from an unidimensional well-being index obtained with PCA. Then, they reduce the dimensionality of welfare using the 'blinding' variable selection method of Fraiman et al. (2008). Others use supervised ML methods to select a subset of relevant variables. Thoplan (2014) apply random forests to identify the key variables that predict poverty in Mauritius, and Okiabera (2020) use this same algorithm to identify key determinants of poverty in Kenya. Mohamud and Gerek (2019) use the wrapper feature selector in order to find a set of features that allows classifying a household into four possible poverty levels.

Another application is poverty targeting, which generally implies ranking or classifying households from poorest to wealthiest and selecting the program beneficiaries. Proxy Means Tests are common tools for this task. They consist of selecting from a large set of potential observables a subset of household characteristics that can account for a substantial amount of the variation in the dependent variable. For that, stepwise regressions are in general used, and the best performing tool is selected following the criteria of the best in-sample performance. Once the Proxy Means Tests has been estimated from a sample, the tool can be applied to the subpopulation selected for intervention to rank or classify households according to Proxy Means Tests score. This process involves the implementation of a household survey in the targeted subpopulation so as to assign values for each of the household characteristics identified during the tool development. ML methods –in particular, ensemble methods– have been shown to outperform Proxy Mean Tests. McBride and Nichols (2018) show that regression forests and quantile regression forests algorithms can substantially improve the out-of-sample performance of Proxy Mean Tests. These methods have the advantage of selecting the variables that offer the greatest predictive accuracy without the need to resort to stepwise regression and/or running multiple model specifications.

9.2.4 Data Imputation

In addition to improving the temporal and spatial frequency of estimates, ML techniques have been exploited to solve other missing data problems. Rosati (2017) compares the performance of an ensemble of LASSO regression models against the traditional 'hot-deck' imputation method for missing data in Argentina's Permanent Household Survey. Chapter 1 in this book discusses the LASSO in detail.

ML methods can also be used to compensate for the lack of panel data, a very promising area of research that brings together econometrics, ML and PID studies. The exploitation of panel data is at the heart of the contributions of econometrics, but they are often not available, or suffer from non-random attrition problems. Adding to the literature on synthetic panels for welfare dynamics (Dang, Lanjouw, Luoto & McKenzie, 2014, which also builds on Elbers et al., 2003), Lucchetti (2018) uses LASSO to estimate economic mobility from cross-sectional data. Later on, Lucchetti et al. (2018) propose to combine LASSO with predictive mean matching (LASSO-PMM). Although this methodology does not substitute panel data, Lucchetti et al. (2018)'s findings are sufficiently encouraging to suggest that estimating economic mobility using LASSO-PMM may approximate actual welfare indicators in settings where cross-sections are routinely collected, but where panel data are unavailable.

Another example is Feigenbaum (2016) who uses ML models and text comparison to link individuals across datasets that lack clean identifiers and which are rife with measurement and transcription issues in order to achieve a better understanding of intergenerational mobility. The methodology is applied to match children from the 1915 Iowa State Census to their adult-selves in the 1940 Federal Census.

Finally, Athey et al. (2021) leverage the computer science and statistics literature on *matrix completion* for imputing the missing elements in a matrix to develop a method for constructing credible counterfactuals in panel data models. In the same vein, Doudchenko and Imbens (2016) propose a more flexible version of the synthetic control method using elastic net. Both methods are described in Section 9.3.4, as their main objective is improving causal inference (see for example, Ratledge et al., 2021; Clay, Egedesø, Hansen, Jensen & Calkins, 2020; Kim & Koh, 2022).

9.2.5 Methods

The previous discussion is organized by the type of application of ML methods (source combination, granularity and dimensionality). An alternative route is to describe the contributions organized by the type of ML method they use. Table 9.3 adopts this approach and organizes the reviewed literature in Section 9.2 using in the standard classification of ML tools, in terms of 'supervised versus unsupervised learning'. To save space, the Table identifies only one paper that, in our opinion, best exemplifies the use of each technique. In the Electronic Online Supplement, Table 9.2 provides a complete list of the reviewed articles, classified by method.

9.3 Causal Inference

An empirical driving force of development studies in the last two decades is the possibility of identifying clean causal channels through which policies affect outcomes. In the last years, the flexible nature of ML methods has been exploited in many dimensions that helped improved standard causal tools. Chapter 3 in this book presents a detailed discussion of ML methods for estimating treatment effects. This section describes the implementation and contribution of such methods in PID studies, in six areas: estimation of heterogeneous effects, optimal design treatment, dealing with high-dimensional data, the construction of counterfactuals, the ability to construct otherwise unavailable outcomes and treatments, and the possibility of using ML to combine observational and experimental data.

9.3.1 Heterogeneous Treatment Effects

The 'flexible' nature of ML methods is convenient to discover and estimate heterogeneous treatment effects (HTE), i.e., treatment effects that vary among different population groups, otherwise difficult to capture in the rigid and mostly linear standard econometric specifications. Nevertheless, a crucial concern in the causal inference literature is the ability to perform valid inference. That is, the possibility of not only provide point estimations or predictions but to gauge its sampling variability, in terms of confidence intervals, standard errors or p-values that facilitate comparisons or the evaluation of relevant hypothesis. The 'multiple hypothesis testing problem' is particularly relevant when researchers search iteratively (automatically or not) for treatment effect heterogeneity over a large number of covariates. Consequently,

Table 9.3: ML methods for improving PID measurements and forecasts

Method		Papers
Supervised learning		
Trees and ensembles	Decision and regression trees	Aiken et al. (2020)
	Boosting	Chi et al. (2022)
	Random forest	Aiken et al. (2020)
Nonlinear regression methods	Generalized additive models	Burstein et al. (2018)
	Gaussian process regression	Pokhriyal and Jacques (2017)
	Lowess regression	Chetty et al. (2018)
K nearest neighbors		Yeh et al. (2020)
Naive Bayes		Venerandi et al. (2015)
Discriminant analysis		Robinson et al. (2007)
Support vector machines		Glaeser et al. (2018)
Regularization and feature selection	LASSO	Lucchetti (2018)
	Ridge regression	Jean et al. (2016)
	Elastic net	Blumenstock et al. (2015)
	Wrapper feature selector	Afzal et al. (2015)
	Correlation feature selector	Gevaert et al. (2016)
Other spatial regression methods		Burstein et al. (2018)
Deep learning	Neural networks	Jean et al. (2016)
Unsupervised learning		
Factor analysis		Gasparini et al. (2013)
PCA (including its derivations, eg., sparse PCA)		Blumenstock et al. (2015),
Clustering methods (e.g., k-means)		M. Burke et al. (2016)
Processing new data		
Natural Language Processing		Sheehan et al. (2019)
Other authomatized feature extraction (not deep learning)		Blumenstock et al. Blumenstock et al. (2015)
Network analysis		Eagle et al. (2010)

the implementation of ML methods for causal analysis must deal not only with the standard non-linearities favored by descriptive-predictive analysis, but also satisfy the inferential requirements of impact evaluation of policies.

Instead of estimating the population marginal average treatment effect (ATE),

$$ATE \equiv E[Y_i(T_i = 1) - Y_i(T_i = 0)]$$

the growing literature on HTE has proposed several parametric, semi-parametric, and non-parametric approaches to estimate the conditional average treatment effect (CATE):

$$CATE \equiv E[Y_i(T_i = 1) - Y_i(T_i = 0)|X_i = x]$$

A relevant part of the literature on HTE builds on regression tree methods, based on the seminal paper by Athey and Imbens (2016) (Chapter 3 of this book explains HTE methods in detail). Trees are a data-driven approach that finds a partition of the covariate space that groups observations with similar outcomes. When the partition groups observations with different outcomes, trees are usually called 'decision trees'. Instead, Athey and Imbens (2016) propose methods for building 'causal trees': trees that partition the data into subgroups that differ by the magnitude of their treatment effects. To determine the splits, they propose using a different objective function that rewards increases in the variance of treatment effects across leaves and penalizes splits that increase within-leave variance. Causal trees allow for valid inference for the estimated causal effects in randomized experiments and in observational studies satisfying unconfoundedness. To ensure valid estimates, the authors propose a sample-splitting approach, which they call 'honesty'. It consists of using one subset of the data to estimate the model parameters (i.e., the tree structure), and a different subset (the 'estimation sample') to estimate the average treatment effect in each leaf. Thus, the asymptotic properties of treatment effect estimates within leaves are the same as if the tree partition had been exogenously given. Finally, similar to standard regression trees, prunning proceeds by cross-validation, but in this case, the criterion for evaluating the performance of the tree in held-out data is based on treatment effect heterogeneity instead of predictive accuracy.

The causal tree method has several advantages. It is easy to explain, and, in the case of a randomized experiment, it is convenient to interpret, as the estimate in each leaf is simply the sample average treatment effect. As is the case of decision trees, an important disadvantage is their high variance. Wager and Athey (2018) extend the standard causal tree strategy to random forests, which they refer to as the *causal forests* method. Essentially, a causal forest is the average of a large number of causal trees, where trees differ from one another due to resampling. The authors establish asymptotic normality results for the estimates of treatment effects under the unconfoundedness assumption, allowing for valid statistical inference. They show that causal forest estimates are consistent for the true treatment effect, and have an asymptotically gaussian and centered sampling distribution if each individual tree in the forest is estimated using honesty as previously defined, under some more subtle assumptions regarding the size of the subsample used to grow each tree. In addition, this method outperforms other nonparametric estimation methods of HTE such as nearest neighbors or kernels as it is useful for mitigating the 'curse of dimensionality' in high-dimensional cases. Forests can be thought of as a nearest neighbors method with an adaptive neighborhood metric, where closeness between observations is defined with respect to a decision tree, and the closest points fall in the same leaf. Therefore, there is a data-driven approach to determine which dimensions of the covariate space are important to consider when selecting nearest neighbors, and which can be discarded.

Athey et al. (2019) generalize causal forests to estimate HTE in an instrumental variables framework. Let x be a specific value of the space of covariates, the goal is to estimate treatment effects τ that vary with x. To estimate the function $\tau(x)$, the authors use forest-based algorithms to define similarity weights $\alpha_i(x)$ that measure the relevance of the i-th observation for estimating the treatment effect at x. The weights are defined as the number of times that the i-th observation ended up in the same terminal leaf than x in a causal forest. Then, a local generalized method of moments is used to estimate treatment effects for a particular value of x, where observations are weighted.

In turn, Chernozhukov et al. (2018a) propose an approach that, instead of using a specific ML tool, e.g., the tree-based algorithm, applies generic techniques to explore heterogeneous effects in randomized experiments. The method focuses on providing valid inference on certain key features of CATE: linear predictors of the heterogeneous effects, average effects sorted by impact groups, and average characteristics of most and least impacted units. The empirical strategy relies on building a proxy predictor of the CATE and then developing valid inference on the key features of the CATE based on this proxy predictor. The method starts by splitting the data into an auxiliary and a main sample. Using the auxiliary sample, a proxy predictor for the CATE is constructuted with any ML method (e.g., elastic net, random forests, neural networks, etc.). Then, the main sample and the proxy predictor are used for estimating the key features of the CATE. There are other approaches to estimate HTE, and the literature is fast-growing (see Chapter 3).

Regarding applications in PID studies, Chowdhury et al. (2021) use two of the above ML approaches to investigate the heterogeneity in the effects of an RCT of an antipoverty intervention, based on the 'ultra-poor graduation model', in Bangladesh. This intervention model is composed of a sequence of supports including a grant of productive assets, hands-on coaching for 12-24 months, life-skills training, short-term consumption support, and access to financial services. The goal is that the transferred assets help to develop micro-enterprises, while all the other components are related to protecting the enterprise and/or increasing productivity. The authors explore the trade-off between immediate reduction in poverty, measured by consumption, and building assets for longer-term gains among the beneficiaries of the program. In order to handle heterogeneous impacts, they use the causal forest estimator of Athey et al. (2019) and the generic ML approach proposed by Chernozhukov et al. (2018a), and focus on two outcomes, household wealth and expenditure. They find a large degree of heterogeneity in treatment effects, especially on asset accumulation.

Chernozhukov et al. (2018a) illustrate their approach with an application to a RCT aimed at evaluating the effect of nudges on demand for immunization in India.

The intervention cross-randomized three main nudges (monetary incentives, sending SMS reminders, and seeding ambassadors). Other relevant contributions that apply Chernozhukov et al. (2018a)'s methods to PID studies include Mullally et al. (2021) who study the heterogeneous impact of a program of livestock transfers in Guatemala; and Christiansen and Weeks (2020), who analyze the heterogeneous impact of increased access to microcredit using data from three other studies in Morocco, Mongolia and Bosnia and Herzegovina. Both papers find that, for some outcomes, insignificant average effects might mask heterogeneous impacts for different groups; and they both use elastic net and random forest to create the proxy predictor. Deryugina et al. (2019) use the generic ML method to explore the HTE of air pollution on mortality, health care use, and medical costs. They find that life expectancy as well as its determinants (e.g., advanced age, presence of serious chronic conditions, and high medical spending) vary systematically with pollution vulnerability, and individuals identified as most vulnerable to pollution have significantly lower life expectancies than those identified as least vulnerable.

Other studies applied the generalized random forest method of Athey et al. (2019) on PID topics. For instance, Carter et al. (2019) employed it to understand the source of the heterogeneity in the impacts of a rural business development program in Nicaragua on three outcomes: income, investment and per-capita household consumption expenditures. First, the authors find that the marginal impact varies across the conditional distribution of the outcome variables using conditional quantile regression methods. Then, they use a generalized random forest to identify the observable characteristics that predict which households are likely to benefit from the program. Daoud and Johansson (2019) use generalized random forest to estimate the heterogeneity of the impacts of International Monetary Fund programs on child poverty. Farbmacher et al. (2021) also use this methodology to estimate how the cognitive effects of poverty vary among groups with different characteristics.

Finally, other authors have applied causal random forest to study HTE of labor market programs. For instance, Davis and Heller (2017, 2020) use this method to estimate the HTE of two randomized controlled trials of a youth summer jobs program. They find that the subgroup that improved employment after the program is younger, more engaged in school, more Hispanic, more female, and less likely to have an arrest record. Knaus et al. (2020) examine the HTE of job search programmes for unemployed workers concluding that unemployed persons with fewer employment opportunities profit more from participating in these programs (see also Bertrand, Crépon, Marguerie & Premand, 2017; Strittmatter, 2019). Finally, J. Burke et al. (2019) use causal random forests but to explore the heterogeneous effect of credit builder loan program on borrowers, providers, and credit market information, finding significant heterogeneity, most starkly with respect to baseline installment credit activity.

9.3.2 Optimal Treatment Assignment

One of the motivations for understanding treatment effect heterogeneity is informing the decision on who to treat, or optimally assigning each individual to a specific treatment. The problem of treatment assignments is present in a number of situations, for example when the government has to decide which financial-aid package (if any) should be given out to which college students, who will benefit most from receiving poverty aid or which subgroup of people enrolled in a program will benefit the most from it. Ideally, individuals should be assigned to the treatment associated with the most beneficial outcome. Identifying the needs of specific sub-groups within the target population can improve the cost effectiveness of any intervention. A growing literature has shown the power of ML methods to address this problem in both observational and randomized controlled trials. Most of these studies are still concerned with the statistical properties of their proposed methods (Kitagawa & Tetenov, 2018; Athey & Wager, 2021), hence applied studies naturally still lag behind.

When the objective is to learn a rule or policy that maps individual's observable characteristics to one of the available treatments, some authors leverage on ML methods for estimating the optimal assignment rule. Athey and Wager (2021) develop a framework to learn an optimal policy rules that not only focuses on experimental samples, but also allows for observational data. Specifically, the authors propose using a new family of algorithms for choosing who to treat by minimizing the loss from failing to use the (infeasible) ideal policy, referred to as 'the regret' of the policy. The optimization problem can be thought as a classification ML task with a different loss function to minimize. Thus, it can be solved with off-the-shelf classification tools, such as decision trees, support vector machines or recursive partitioning, among other methods. In particular, the algorithm proposed by the authors starts by computing doubly robust estimators of the ATE and then selecting the assignment rule that solves the optimization problem using decision trees, where the three depth is determined by cross-validation. The authors illustrate their method by identifying the enrollees of a program who are most likely to benefit from the intervention. The intervention is the GAIN program, a welfare-to-work program that provides participants with a mix of educational resources and job search assistance in California (USA).

Zhou et al. (2018) further study the problem of learning treatment assignment policies in the case of multiple treatment choices using observational data. They develop an algorithm based on three steps. First, they estimate the ATE via doubly robust estimators. Second, they use a K-fold algorithmic structure similar to cross-validation, where the data is divided into folds to estimate models using all data except for one fold. But instead of using the K-fold structure for selecting the hyperparameters or tuning models, they use it to estimate a score for each observation and each treatment arm. Third, they solve the policy optimization problem: selecting a policy that maximizes an objective function constructed with the outputs of the first and second steps.

The articles described above focus on a static setting where a decision-maker has just one observation for each subject and decides how to treat her. In contrast, other problems of interest may involve a dynamic component whereby the decision-maker has to decide based on time-varying covariates, for instance when to recommend mothers to stop breastfeeding to maximize infants' health or when to turn off ventilators for intensive care patients to maximize health outcomes. Nie et al. (2021) study this problem and develop a new type of doubly robust estimator for learning such dynamic treatment rules using observational data under the assumption of sequential ignorability. Sequential ignorability means that any confounders that affect making a treatment choice at certain moment have already been measured by that moment.

Finally, another application using a different approach than those described above is Björkegren et al. (2020). They develop a ML method to infer the preferences that are consistent with observed allocation decisions. They applied this method to PROGRESA, a large anti-poverty program that provides cash transfers to eligible households in Mexico. The method starts by estimating the HTE of the intervention using the causal forests ML method. Then, they consider an allocation based on a score or ranking, in terms of school attendance, child health and consumption. Next, ordinal logit is used to identify the preferences consistent with the ranking between households. Results show that the program prioritizes indigenous households, poor households and households with children. They also evaluate the counterfactual allocations that should have occurred had the policymaker placed higher priority on certain types of impacts (e.g., health vs. education) or certain types of households (e.g., lower-income or indigenous).

9.3.3 Handling High-Dimensional Data and Debiased ML

Most supervised ML methods purposedly bias estimates with the aim of improving prediction. Notably, this is the case of regularization methods. Sparse regression methods such as Ridge, LASSO or elastic net (discussed in Chapter 1) include a penalty term (a 'penalized regression adjustment') that shrinks estimations in order to avoid overfitting, resulting in a regularization bias.

The standard econometric solution to deal with biases is to include the necessary controls (Ahrens et al., 2021; Athey et al., 2018). In a high dimensional context this a delicate issue since the set of potential controls may be greater than the number of observations. The same problem can arise at the first stage of instrumental variables, as the set of instruments might be high-dimensional as well. High-dimensionality poses a problem for two reasons. First, if the number of covariates is greater than the number of observations, traditional estimation is not possible. But even if the number of potential covariates is smaller than the number of observations, including a large set of covariables may result in overfitting. ML methods for regularization help to handle high-dimensional data and are precisely designed to avoid overfitting and to allow for estimation. However, as they are designed for prediction, they bias the estimates with the aim of reducing the prediction error and they can lead to incorrect inference about model parameters (Chernozhukov et al., 2018b; Belloni, Chernozhukov & Hansen, 2014a).

Belloni et al. (2012, 2014a, 2014b) propose post double selection (PDS) methods using LASSO to address this problem. The key assumption is that the 'true model' for explaining the variation in the dependent variable is approximately sparse with respect to the available variables: either not all regressors belong in the model, or some of the correspondent coefficients are well-approximated by zero. Therefore, the effect of confounding factors can be controlled for up to a small approximation error (Belloni et al., 2014b). Under this assumption, LASSO can be used to select variables that are both relevant for the dependent variable (with the treatment variable not being subject to selection) and to the treatment variable (Ahrens et al., 2021). Then, the authors derive the conditions and procedures for valid inference after model selection using PDS. Belloni et al. (2017) extend those conditions to provide inferential procedures for parameters in program evaluation (also valid in approximately sparse models) to a general moment-condition framework and to a wide variety of traditional and ML methods. Chapter 3 of this book discusses in detail the literature on Double or Debiased ML and its extensions, such as Chernozhukov et al. (2018b) who propose an strategy applicable to many ML methods for removing the regularization bias and the risk of overfitting. Their general idea is predicting both the outcome and the value of treatment based on the other covariates with any ML model, obtaining the residuals of both predictions, and then regressing both residuals.

Apart from avoiding omitted variable bias, another reason for including control covariates is improving the efficiency of estimates, such as in randomized controlled experiments (RCT). Therefore, regularization methods may be useful in the case of RCTs with a high dimensional set of potential controls. Wager et al. (2016) show that the estimates obtained by any type of regularization method with an intercept yield unbiased estimates of the ATE in RCTs and propose a procedure for building confidence intervals for the ATE. Other strains of work tackle specific issues of high dimensionality in RCTs. The strategies proposed by Chernozhukov et al. (2018a) (see Section 9.3.1) are valid in high dimensional settings. Banerjee et al. (2021b) develop a new technique for handling high-dimensional treatment combinations. The technique allows them to determine, among a set of 75 unique treatment combinations, which of them are effective and which one is the most effective. This solves a trade-off that researchers face when implementing an RCT. On the one side, they could choose ex-ante and ad-hoc a subset of possible treatments, with the risk of letting an effective treatment combination aside. On the other side, they could implement every possible treatment combination, but treat only a few individuals with each of them, and then lack the statistical power to test whether its effects are significant. Their procedure implements many treatment combinations, but then has two steps. The first one uses the post-LASSO procedure of Belloni and Chernozhukov (2013) to pool similar treatments and prune ineffective treatments. The second step estimates the effect of the most effective one, correcting for an upward bias as suggested by Andrews et al. (2019).

Some studies apply PDS LASSO in PID studies. For instance, Banerjee et al. (2021a) use it to select control variables and to estimate the impact on poverty of the switch from in-kind food assistance to vouchers to purchase food on the market in Indonesia. Martey and Armah (2020) use it to study the effect of international migration on household expenditure, working, production hours and poverty of the left-behind household members in Ghana; and Dutt and Tsetlin (2021) to argue that poverty strongly impacts development outcomes. Others use the procedure for sensitivity analysis and robustness checks, such as Churchill and Sabia (2019), who study the effect of minimum wages on low-skilled immigrants' well-being, and Heß et al. (2021), who study of the impact of a development program in Gambia on economic interactions within rural villages. Skoufias and Vinha (2020) use the debiased ML method by Chernozhukov et al. (2018b) to study the relationships between child stature, mother's years of education, and indicators of early childhood development. To illustrate their methods, Belloni et al. (2017) estimate the average and quantile effects of eligibility and participation in the 401(k) tax-exemption Plan in the US on assets, finding that there is a larger impact at high quantiles. Finally, Banerjee et al. (2021b) apply their technique for multiple RCT treatments to a large-scale experiment in India aimed at determining which among multiple possible policies has the largest impact in increasing the number of immunizations and which is the most cost-effective, as well as at quantifying their effects. The interventions cross-randomized three main nudges (monetary incentives, sending SMS reminders, and seeding ambassadors) to promote immunization, resulting in 75 unique policy combinations. The authors found the policy with the largest impact (information hubs, SMS reminders, incentives that increase with each immunization) and the most cost-effective (information hubs, SMS reminders, no incentives).

9.3.4 Machine-Building Counterfactuals

ML methods can also be used to construct credible counterfactuals for the treated units in order to estimate causal effects. That is, to impute the potential outcomes for the treated units had not they been treated. Athey et al. (2021) and Doudchenko and Imbens (2016) propose methods for building credible counterfactuals that rely on regularization methods frequently used in ML. In general terms, these ML-based methods help to overcome multiple data and causal inference challenges, usually present in observational studies, that cannot be addressed with conventional methods. For instance, as Ratledge et al. (2021) suggest, they contribute to tackle the problem of different evolution of outcomes between targeted and untargeted populations and to deal with threats to identification such as non-parallel trends in pre-treatment outcomes, which make methods such as difference in differences unreliable.

Athey et al. (2021) leverage on the computer science and statistics literature on matrix completion. Matrix completion methods are imputation techniques for guessing at the missing values in a matrix. In the case of causal inference, the matrix is that of potential control outcomes, where values are missing for the treated unit-periods. They propose an estimator for the missing potential control outcomes based on the observed control outcomes of the untreated unit-periods. As in the matrix completion literature, missing elements are imputed assuming that the complete matrix is the sum of a low rank matrix plus noise. Briefly, they model the complete data matrix Y as $Y = L^* + \epsilon$ where $E[\epsilon|L^*] = 0$ and ϵ is a measurement error. L^* is the low-rank matrix to be estimated via regularization, by minimizing the sum of squares and adding a penalty term that consists of the nuclear norm of L multiplied by a constant λ to be chosen via cross-validation.

In turn, Doudchenko and Imbens (2016) propose a modified version of the synthetic control estimator that uses elastic net. Again, the goal is to impute the unobserved control outcomes for the treated unit, in order to estimate the causal effect. The synthetic control method constructs a set of weights such that covariates and pre-treatment outcomes of the treated unit are approximately matched by a weighted average of control units. To do so, it imposes some restrictions: the weights should be nonnegative and sum to one. That allows to define the weighs even if the number of control units exceeds the number of control outcomes. The authors propose a more general estimator that allows the weights to be nonnegative, do not restrict their sum and allow for a permanent additive difference between the treated unit and the controls. When the number of control units exceeds the number of pre-treatment periods, they propose using an elastic net penalty term for obtaining the weights.

Ratledge et al. (2021) apply both of the above methods to estimate the causal impact of electricity access on wealth asset in Uganda. The ML-based causal inference approaches help them to overcome two challenges. First, the fact that they deal with observational data, as the electrification expansion was not part of a randomized experiment. They have georreferenced data on the multi-year expansion in a grid throughout the country, so they can use the observed outcome of the control unit-periods to impute the potential outcome of the treated unit-periods, had not the electricity distribution taken place. Second, the ML-based methods help to tackle the problem of different evolution of outcomes between targeted and untargeted populations. The authors argue that they are more robust to some threats to identification such as non-parallel trends in pre-treatment outcomes. Kim and Koh (2022) apply matrix completion methods as a robustness check to study the effect of the access to health insurace coverage in subjective well-being in the United States. Finally, Clay et al. (2020) apply synthetic control with elastic net to study the immediate and long-run effects of a community-based health intervention at the beginning of the twentieth century in controlling tuberculosis and reducing mortality.

9.3.5 New Data Sources for Outcomes and Treatments

Many PID studies take advantage of the increased availability of new data sources for causal identification. New data can serve to approximate outcome variables, to identify treated units and to look for sources of exogenous variability or key control variables. In the Electronic Online Supplement, Table 9.3 lists these studies, classifying the non-traditional data source use in three non-exclusive categories: (i) outcome construction, (ii) treatment or control variable construction, and (iii) exogenous variability. It also describes the type of data source used, the evaluation method and, when applicable, the ML method employed, among other features of each paper.

Several aspects can be highlighted from Table 9.3 in the Electronic Online Supplement. First, ML methods mainly intervene in an intermediate step for data prepossessing (e.g. making use of satellite image data). However, in just a few cases ML methods also play a role in estimating the causal effect, for example, in Ratledge et al. (2021). Second, the most common contribution of new data sources to causal inference studies is to build or improve the outcome variable. In many cases, existing data provide only a partial or aggregate picture of the outcome, which makes it difficult to measure changes in the outcome over time that can be correctly attributed to the treatment. New sources of data, as well as ML methods, can contribute to solve this problem.

In the Electronic Online Supplement, Table 9.3 shows that most of the causal inference studies that rely on ML to improve the outcome measure use satellite data. A key example is the work of Huang et al. (2021), which show that relying solely on satellite data is enough to assess the impact of an anti-poverty RCT on household welfare. Specifically, they measure housing quality among treatment and control households by combining high-resolution daytime imagery and state-of-the-art deep learning models. Then, using difference-in-differences they estimate the program effects on housing quality (see also Bunte et al., 2017, and Villa, 2016, for similar studies in Liberia and Colombia respectively). In turn, Ratledge et al. (2021) stand out for using daytime satellite imagery, as well as ML-based causal inference methods, for estimating the causal impact of rural electrification expansion on welfare. In particular, they use CNNs to build a wealth index based on daytime satellite imagery. Then, to estimate the ATE, they apply matrix completion and synthetic controls with elastic net. Both methods have the advantage of being more robust against the possibility of non-parallel trends in pre-treatment outcomes (see Section 9.3.4).

Ratledge et al. (2021) note two challenges to be addressed when the outcome variable in an impact evaluation is measured using ML predictions based on new data sources. Firstly, models that predict the outcome variable of interest may include the intervention of interest as covariate itself. For instance, a satellite-based model used to predict poverty will be unreliable for estimating the causal effect of a new road construction on poverty, if the ML model considers whether a location is poor based on the road. To address this first challenge, the authors predict poverty excluding variables related to the intervention from the set of features. Secondly, for certain outcomes such as economic well-being, predictions based on ML models often have lower variance than the ground true variable to be predicted, and they over-predict for poorer individuals and under-predict for wealthier ones. Hence, if the intervention targets a specific part of the outcome distribution, as is the case of poverty programs, bias in outcome measurement could bias estimates of treatment effects. To address this second challenge, the authors predict poverty adding a term to the mean squared error loss function that penalizes bias in each quintile of the wealth distribution.

Another relevant example is Hodler and Raschky (2014), who study the phenomenon of regional favoritism by exploring whether subnational administrative regions are more developed when they are the birth region of the current political leader. They use satellite data on nighttime light intensity to approximate the outcome measure, i.e., economic activity at the subnational level. Specifically, they build a panel data set with 38,427 subnational regions in 126 countries and annual observations from 1992 to 2009, and use traditional econometric techniques to estimate the effect of birthplaces of political leaders on economic development of those areas. After including several controls, region fixed effects and country-year dummy variables, they find that being the birth region of political leaders results in a more intense nighttime light, providing evidence for regional favoritism. In turn, Alix-Garcia et al. (2013) use satellite images to study the effect of poverty-alleviation programs on environmental degradation. In particular, they assess the impact of income transfers on deforestation, taking advantage of both the threshold in the eligibility rule for Oportunidades, a conditional cash transfer program in Mexico, and random variation in the pilot phase of the program. They use satellite images to measure deforestation at the locality level, which is the outcome variable. By applying regression discontinuity, they find that additional income raises consumption of land-intensive goods and increases deforestation.

Other studies use geo-referenced data from new sources, such as innovative systems of records, to build the outcome variable. For instance, Chioda et al. (2016) take advantage of CompStat, a software used by law enforcement to map and visualize crime, to build crime rates at the school neighborhood level and explore the effects on crime of reductions in poverty and inequality associated with conditional cash transfers. Using traditional econometric techniques, such as instrumental variables and difference-in-differences approach, they find a robust and significant negative effect of cash transfers on crime resulting from lower poverty and inequality. Several studies use geo-referenced data obtained from geographic information systems (GIS), which connects data to a map, integrating location data with all types of descriptive information (see the Electronic Online Supplement, Table 9.3 for detailed examples). Articles using GIS data are part of a more extensive literature that intersects with spatial econometrics. In the Electronic Online Supplement, Table 9.3 includes some relevant causal inference studies that use this type of data in PID studies.

To a lesser extent, the growing availability of high-quality data has also been used for identifying treated units. For example, with the aim of assessing the impact of a cyclone on household expenditure, Warr and Aung (2019) use satellite images, in the weeks immediately following that event, to identify the treated and control households living in affected and not-affected areas, respectively. To build the counterfactual, they use a statistical model to predict real expenditures at the household level, based on survey data and controlling for region and year fixed effects and a large set of covariates. The estimated impact of the cyclone is computed as the difference between the observed outcome, in which the cyclone happened, and the simulated, counterfactual outcome (see the Electronic Online Supplement, Table 9.3 , for other examples).

Finally, a group of articles has used the new source of data as the source of exogenous variability. For instance, this is the case of Faber and Gaubert (2019), who assess the impact of tourism on long-term economic development. They use a novel database containing municipality-level information as well as geographic information systems (GIS) database including remote sensing satellite data. They exploit geological, oceanographic, and archaeological variation in ex ante local tourism attractiveness across the Mexican coastline as a source of exogenous variability. For that, they build variables based on the GIS and satellite data. Using a reduced-form regression approach, they find that tourism attractiveness has strong and significant positive effects on municipality total employment, population, local GDP, and wages relative to less touristic regions (see the Electronic Online Supplement, Table 9.3, for other examples).

9.3.6 Combining Observational and Experimental Data

The big data revolution facilitates researchers the access to larger, more detailed and representative data sets. The largely observational nature of the big data phenomenon clashes with RCT's, where exogenous variation is guaranteed by design. Hence, it is a considerable challenge to combine both sources to leverage the advantages of each: size, availability and, hopefully, external validity of observational data sets, and clean causal identification of RCT's.

Athey et al. (2020) focus on the particular case where both the experimental and observational data contain the same information (a secondary –e.g., short term– outcome, pre-treatment variables, and individual treatment assignments) except for the primary (often long term) outcome of interest that is only available in the observational data. The method relies on the assumptions that the RCT has both internal and external validity and that the differences in the results between the RCT and the observational study are due to endogenous selection into the treatment or lack of internal validity in the observational study. In addition, they introduce a novel assumption, named 'latent unconfoundedness', which states that the unobserved confounders that affect treatment assignment and the secondary outcome in the observational study are the same unobserved confounders that affect treatment assignment and the primary outcome. This assumption allows linking the biases in treatment-control differences in the secondary outcome (which is estimated in the experimental data) to the biases in treatment-control comparisons in the primary outcome (which the experimental data are silent about). Under these assumptions, they propose three different approaches to estimate the ATE: (1) imputing the missing primary outcome in the experimental sample, (2) weighting the units in the observational sample to remove biases, and (3) control function methods. They apply these methods to estimate the effect of class size on eight grade test scores in New York without conducting an experiment. They combine data from the New York school system (the 'observational sample') and from Project STAR (the 'experimental sample', a RCT conducted before). As described before, both samples include the same information except for test scores for the eight grade (the primary outcomes), which is only in the observational sample since the Project STAR data includes only test scores for the third grade (the secondary outcome). Both samples include pre-treatment variables (gender, whether the student gets a free lunch, and ethnicity). The authors arrive at the same results using both control function and imputation methods and conclude that the biases in the observational study are substantial, but results are more credible when they apply their approach to adjust the results based on the experimental data.

9.4 Computing Power and Tools

The increase in computational power has been essential in the contributions of new ML methods and the use of new data. To begin with, improvements in data storage and processing capabilities made it possible to work with massively large data sets. One example is Christensen et al. (2020), who needed large processing capacities to use air pollution data at high levels of disaggregation. In turn, the quality and resolution of satellite images, which are widely used in PID studies (see the Electronic Online Supplement, Table 9.1), has improved over the years thanks to the launch of new satellites (Bennett & Smith, 2017).

Also, computational power allows for more flexible models. The complexity of an algorithm measures the amount of computer time it would take to complete given an input of size n (where n is the size in units of bits needed to represent the input). It is commonly estimated by the number of elementary operations performed for each unit of input –for example, an algorithm with complexity $O(n)$ has a linear time complexity and an algorithm with complexity $O(n^2)$ has a quadratic time complexity. Improvements in computing processing capacity can reduce the time required to use an algorithm of a given complexity for processing a larger scale of data, or the time required to use algorithms of greater complexity. Thompson et al. (2020) analyze how deep learning performance depends on computational power in the domains of image classification and object detection. Deep learning is particularly dependent on computing power, due to their overparameterization and the large amount of training data used to improve performance. Circa 2010, deep learning models were ported to the GPU (Graphics Processing Unit) of the computer instead of in the CPU (Central Processing Unit), accelerating notably their processing: initially yielding a 5 to 15 times speed up, which was up to 35 times by 2012. The 2012 ImageNet competition of AlexNet, a CNN run in a GPU, was a milestone and a turning point in the widespread use of neural networks. Many of the papers reviewed in PID studies with ML use deep learning, typically, CNN; the oldest dates from 2016 (see the Electronic, Online Supplement, Table 9.1). For instance, Head et al. (2017) use CNNs to process satellite images to measure poverty in Sub-Saharan Africa. For large countries, such as Nigeria, they use GPU computing to train the CNN; this task took several days. Maiya and Babu (2018) also train a CNN using the GPU to map slums from satellite images.

In addition, a variety of new computational tools became easily accessible for PID studies. First, the assembly of *bots* to conduct experiments immediately and on a large scale via web platforms, as in Christensen et al. (2021). The authors developed a software bot at the National Center for Supercomputing Applications that sent fictitious queries to property managers on an online rental housing platform in order to test the existence of discrimination patterns. The process automation allowed them to scale data collection –they examined more than 25 thousand interactions between property managers and fictitious renters across the 50 largest cities in the United States– and to monitor discrimination at a low cost. The bot targeted listings that had been listed the day before on the platform, and sent one inquiry each of the three following day, using fictitious identities drawn in random sequence from a set of 18 names that elicited cognitive associations with one of three ethnic categories (African American, Hispanic/Latin, and White). Then, the bot followed the listing for 21 days registering whether the property was rented or not. Similarly, Carol et al. (2019) tested for the presence of discrimination in the German carpooling market by programming a bot that sent requests to drivers from fictitious profiles. Experimental designs using bots are also common in health and educational interventions (see, for example, Maeda et al., 2020; Agarwal et al., 2021).

Second, the use of interactive tools for data visualization has been increasing in PID studies, specially for mapping wealth at a granular scale (see Section 9.2.2). Table 9.1 in the Electronic Online Supplement lists many of the available links for viewing the visualization and, often, for data downloading. Many of them can be accessed for free, but not all. An illustrative example of a free data visualization and downloading tool are opportunityatlas.org from Chetty et al. (2020) (see Section 9.2.2 and a screenshot in Figure 9.1b) and tracktherecovery.org from Chetty et al. (2020a) (see Section 9.2.1).

Third, open source developments and crowd-sourced data systems make software and data available more accessible. Most of the techniques in the reviewed studies are typical ML techniques for which there are open source Python and R libraries. For example, the Inter-American Development Bank's Code for Development initiative aims at improving access to applications, algorithms and data among other tools. As for crowd-sourced data, one example is OpenStreetMap, often used for mapping wealth at a granular level such as in Soman et al. (2020) (see others in Table 9.1 of the Electronic Online Supplement).

Finally, the combination of software development and IT operations to provide continuous delivery and deployment of trained models might be a perhaps still underexploited toolkit in PID studies. That is, instead of training the model just once, re-train it on a frequent basis, automatically with new data. There are three distinct cases of model updating: periodical training and offline predictions, periodical training and online (on-demand) predictions, and online training and predictions (fine-tuning of model predictions while receiving new data as feedback). The second and third cases are more common for applications that interact with users and thus need real-time predictions. Instead, the first case could be sometimes applicable to PID studies, as PID indicators predictions might require frequent updating for two main reasons: adding more data points might improve the model's performance, and keeping the model updated with world changes, in order to deliver relevant predictions. There are different tools that facilitate model re-training commonly used in the ML industry. For example, Apache Airflow for data orchestration, that is, an automated process that can take data from multiple locations and manages the workflow of processing it to obtain the desired features, training a ML model, uploading it to some application or platform, re-running it; MLflow to keep track of experiments and model versions; Great Expectations to test the new data in order to ensure some basic data quality standards. One example of application is the Government of the City of San Diego in the United States. They started using Apache Airflow for the project StreetsSD, a map that provides an up-to-date status of paving projects. The coordination of tasks also allows the Government to send automatic notifications based on metrics and thresholds, and generate automated reports to share with different Government departments.

9.5 Concluding Remarks

This chapter provides a hopefully complete 'ecosystem' of the fast growing literature on the use of ML methods for PID studies. The suggested taxonomy classifies relevant contributions into two main categories: (1) studies aimed at providing better measurements and forecasts of PID indicators, (2) studies using ML methods and new data sources for answering causal questions. In general, ML methods have proved to be effective tools for predictions and pattern recognition; consequently, studies in the first category are more abundant. Instead, causal analysis requires to go beyond the point estimates or predictions that dominate the practice of ML, demanding inferential tools that help measure uncertainty and evaluate relevant hypothesis for policy implementation, and are, quite expectedly, of a more complex nature, which explains their relative scarcity compared to studies in the first group.

Regarding contributions in measurements and forecasts of PID indicators, ML can improve them in three ways.

First, many studies combine different data sources, especially non-traditional ones, and rely on the flexibility of ML methods to improve the prediction of a PID outcome in terms of its time frequency, granularity or coverage. Granularity has been the most extensive contribution of ML to PID studies: Figure 9.2 shows that 72 of the reviewed studies contribute in terms of granularity, whereas 29 in terms of frequency. In relation to non-traditional data, Table 9.1 in the Electronic Online Supplement shows that satellite images are the main source of data in PID research with ML. Finally, in terms of specific ML techniques, we find that tree-based methods are the most popular (Figure 9.3). Very likely, their popularity is related to their intuitive nature that facilitates communication with non-technical actors (policy makers, practitioners) and to their flexibility, that help discover non-linearities that are difficult, when not impossible, to handle ex-ante with the mostly parametric nature of standard econometric tools. Quite interestingly, OLS and its variants is the method most frequently used as a benchmark to compare results. In turn, deep learning methods dominates the use of satellite images (Figure 9.4).

Another important use of ML in PID studies is for data dimensionality, to help understand factors behind multidimensional notions like poverty or inequality to design shorter and less costly surveys, to construct indexes, and to find latent socioeconomic groups among individuals. Finally, ML is a promising alternative do deal with classic data problems, such as handling missing data or compensating for the lack of panel data.

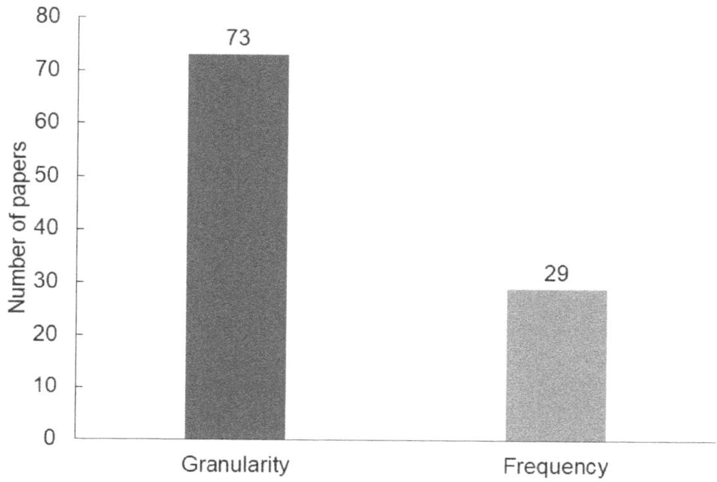

Fig. 9.2: Contributions of ML to PID studies in measurement and forecasting. Source: reviewed articles, described in Table 9.1 in the Electronic Online Supplement.

In turn, there are two main contributions of ML to causal inference in PID studies. The first one involves adapting ML techniques for causal inference, while the second one takes advantage of the increased availability of new data sources for impact evaluation analysis. The first line of work is still mainly theoretical and has contributed with many promising methods for improving existent causal inference techniques, such as methods for estimating heterogeneous treatment effects or handling high dimensional data, among others. As expected, relatively few studies have applied these methods to address causal inference questions in PID studies, although the number is growing fast. Instead, the number of studies that use ML to take advantage of new data sources for causal analysis is substantially larger, to build new outcome variables in impact evaluation studies, to look for alternative sources of exogenous variability, or to optimally combine observational and experimental data. Most commonly, the new data are used for building or improving the outcome variable, and satellite images are the most frequent source used for this purpose. For example, there is evidence that it is possible to assess the impact of anti-poverty RCTs on household welfare relying solely on satellite data, without the need of conducting baseline and follow up surveys. In addition, the reviewed literature suggests that in empirical PID studies involving causal inference, ML methods mainly intervene in an intermediate step for data processing, although in a few cases they also play a role in estimating the causal effect.

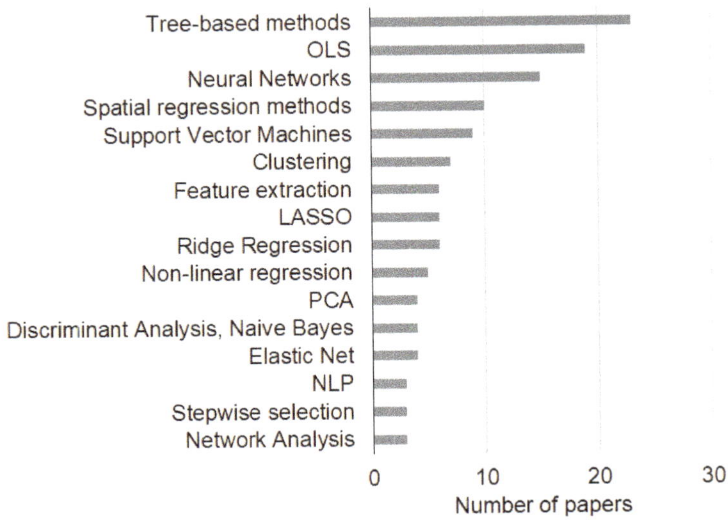

Fig. 9.3: ML methods for PID studies improving measurement and forecasting. Source: reviewed articles, described in Table 9.1 in the Electronic Online Supplement.

	Supervised					Deep	Non-supervised		Regularization				Data structuring			Traditional	
	Tree-based methods	Discriminant Analysis, Naive Bayes	Support Vector Machines	Non-linear regression	Spatial regression methods	Neural Networks	PCA	Clustering	Stepwise selection	LASSO	Ridge Regression	Elastic Net	Network Analysis	NLP	Feature extraction	OLS	Other - Econometrics
Crowdsourced maps	3	1	2		2	1							1			2	
Geolocated surveys	3		1	2	4	1		1		2						1	
Internet searches																3	1
Mobile phone	5		1	1	1		2	3		1		3			2	2	
Only traditional	3			2	1		1		1	1						1	1
Satellite imagery	11	3	6		5	13		3	2	2	6	1			5	5	2
Social media	3		2		1					1			1	2		6	
Other	1												1	1		1	

Fig. 9.4: ML methods in PID studies improving measurement and forecasting, by data source. Source: reviewed articles, described in Table 9.1 in the Electronic Online Supplement.

Several clear patterns emerge from the review. First, new data sources (such as mobile phone calls data or satellite images) do not replace traditional data (i.e., surveys and census), but rather complement them. Indeed, traditional data sets are a key part of the process. Many PID studies use traditional data as the ground truth to train a model that predicts a PID outcome based on features built from the new data, and are hence necessary to evaluate the perfomance of ML methods applied to alternative sources. They are an essential benchmark as they are usually less biased and more representative of the population. Traditional data sources are also needed in studies that use dimensionality reduction techniques to infer the structure of the underlying reality. In turn, in causal inference studies the new data sources complement traditional data to improve the source of exogeneity or to build the outcome or control variables, allowing for improved estimates.

Finally, many –if not all– of the contributions described throughout this chapter would not have been possible without the improvement in computational power. One of the most important advances is deep learning, which is extremely useful for processing satellite images, likely the main alternative data source.

The literature using ML in PID studies is growing fast, and most reviewed studies date from 2015 onwards. However, the framework we provide is hopefully useful to locate each new study in terms of its contribution relative to traditional econometric techniques, and traditional data sources.

Acknowledgements We thank the members of the Centro de Estudios para el Desarrollo Humano (CEDH) at Universidad de San Andrés for extensive discussions. Joaquín Torré provided useful insights for the section on computing power. Mariana Santi provided excellent research assistance. We specially thank the editors, Felix Chan and László Mátyás, for their comments on an earlier version, that helped improve this chapter considerably. All errors and omissions are our responsibility.

References

Afzal, M., Hersh, J. & Newhouse, D. (2015). Building a better model: Variable selection to predict poverty in Pakistan and Sri Lanka. *World Bank Research Working Paper*.

Agarwal, D., Agastya, A., Chaudhury, M., Dube, T., Jha, B., Khare, P. & Raghu, N. (2021). *Measuring effectiveness of chatbot to improve attitudes towards gender issues in underserved adolescent children in India* (Tech. Rep.). Cambridge, MA: Harvard Kennedy School.

Ahrens, A., Aitken, C. & Schaffer, M. E. (2021). Using machine learning methods to support causal inference in econometrics. In *Behavioral predictive modeling in economics* (pp. 23–52). Springer.

Aiken, E., Bedoya, G., Coville, A. & Blumenstock, J. E. (2020). Targeting development aid with machine learning and mobile phone data: Evidence from an Anti-poverty intervention in Afghanistan. In *Proceedings of the 3rd acm sigcas conference on computing and sustainable societies* (pp. 310–311).

Aiken, E., Bellue, S., Karlan, D., Udry, C. R. & Blumenstock, J. (2021). *Machine learning and mobile phone data can improve the targeting of humanitarian assistance* (Tech. Rep.). Cambridge, MA: National Bureau of Economic Research.

Alix-Garcia, J., McIntosh, C., Sims, K. R. & Welch, J. R. (2013). The ecological footprint of poverty alleviation: Evidence from Mexico's Oportunidades program. *Review of Economics and Statistics*, *95*(2), 417–435.

Andrews, I., Kitagawa, T. & McCloskey, A. (2019). *Inference on winners* (Tech. Rep.). Cambridge, MA: National Bureau of Economic Research.

Angrist, J. D. & Pischke, J.-S. (2010). The credibility revolution in empirical economics: How better research design is taking the con out of econometrics. *Journal of economic perspectives*, *24*(2), 3–30.

Antenucci, D., Cafarella, M., Levenstein, M., Ré, C. & Shapiro, M. D. (2014). *Using social media to measure labor market flows* (Tech. Rep.). Cambridge, MA: National Bureau of Economic Research.

Askitas, N. & Zimmermann, K. F. (2009). Google econometrics and unemployment forecasting. *Applied Economics Quarterly*.

Athey, S., Bayati, M., Doudchenko, N., Imbens, G. & Khosravi, K. (2021). Matrix completion methods for causal panel data models. *Journal of the American Statistical Association*, 1–15.

Athey, S., Chetty, R. & Imbens, G. (2020). Combining experimental and observational data to estimate treatment effects on long term outcomes. *arXiv preprint arXiv:2006.09676*.

Athey, S. & Imbens, G. (2016). Recursive partitioning for heterogeneous causal effects. *Proceedings of the National Academy of Sciences*, *113*(27), 7353–7360.

Athey, S., Imbens, G. W. & Wager, S. (2018). Approximate residual balancing: Debiased inference of average treatment effects in high dimensions. *Journal of the Royal Statistical Society: Series B (Statistical Methodology)*, *80*(4), 597–623.

Athey, S., Tibshirani, J. & Wager, S. (2019). Generalized random forests. *The Annals of Statistics*, *47*(2), 1148–1178.

Athey, S. & Wager, S. (2021). Policy learning with observational data. *Econometrica*, *89*(1), 133–161.

Banerjee, A., Chandrasekhar, A. G., Dalpath, S., Duflo, E., Floretta, J., Jackson, M. O., … Shrestha, M. (2021b). *Selecting the most effective nudge: Evidence from a large-scale experiment on immunization* (Tech. Rep.). Cambridge, MA: National Bureau of Economic Research.

Banerjee, A., Hanna, R., Olken, B. A., Satriawan, E. & Sumarto, S. (2021a). *Food vs. food stamps: Evidence from an at-scale experiment in Indonesia* (Tech. Rep.). Cambridge, MA: National Bureau of Economic Research.

Baylé, F. (2016). *Detección de villas y asentamientos informales en el partido de La Matanza mediante teledetección y sistemas de información geográfica* (Unpublished doctoral dissertation). Universidad de Buenos Aires. Facultad de Ciencias Exactas y Naturales.

Bedi, T., Coudouel, A. & Simler, K. (2007a). *More than a pretty picture: Using poverty maps to design better policies and interventions*. World Bank Publications.

Bedi, T., Coudouel, A. & Simler, K. (2007b). Maps for policy making: Beyond the obvious targeting applications. In *More than a pretty picture: Using poverty maps to design better policies and interventions* (pp. 3–22). World Bank Publications.

Belloni, A., Chen, D., Chernozhukov, V. & Hansen, C. (2012). Sparse models and methods for optimal instruments with an application to eminent domain. *Econometrica, 80*(6), 2369–2429.

Belloni, A. & Chernozhukov, V. (2013). Least squares after model selection in high-dimensional sparse models. *Bernoulli, 19*(2), 521–547.

Belloni, A., Chernozhukov, V., Fernández-Val, I. & Hansen, C. (2017). Program evaluation and causal inference with high-dimensional data. *Econometrica, 85*(1), 233–298.

Belloni, A., Chernozhukov, V. & Hansen, C. (2014a). High-dimensional methods and inference on structural and treatment effects. *Journal of Economic Perspectives, 28*(2), 29–50.

Belloni, A., Chernozhukov, V. & Hansen, C. (2014b). Inference on treatment effects after selection among high-dimensional controls. *The Review of Economic Studies, 81*(2), 608–650.

Bennett, M. M. & Smith, L. C. (2017). Advances in using multitemporal night-time lights satellite imagery to detect, estimate, and monitor socioeconomic dynamics. *Remote Sensing of Environment, 192*, 176–197.

Bertrand, M., Crépon, B., Marguerie, A. & Premand, P. (2017). *Contemporaneous and post-program impacts of a Public Works Program* (Tech. Rep.). Washington, DC: World Bank.

Björkegren, D., Blumenstock, J. E. & Knight, S. (2020). Manipulation-proof machine learning. *arXiv preprint arXiv:2004.03865*.

Blumenstock, J. (2018a). *Don't forget people in the use of big data for development.* Nature Publishing Group.

Blumenstock, J. (2018b). Estimating economic characteristics with phone data. In *Aea papers and proceedings* (Vol. 108, pp. 72–76).

Blumenstock, J., Cadamuro, G. & On, R. (2015). Predicting poverty and wealth from mobile phone metadata. *Science, 350*(6264), 1073–1076.

Bosco, C., Alegana, V., Bird, T., Pezzulo, C., Bengtsson, L., Sorichetta, A., ... Tatem, A. J. (2017). Exploring the high-resolution mapping of gender-disaggregated development indicators. *Journal of The Royal Society Interface, 14*(129), 20160825.

Bunte, J. B., Desai, H., Gbala, K., Parks, B. & Runfola, D. M. (2017). Natural resource sector fdi and growth in post-conflict settings: Subnational evidence from Liberia. *Williamsburg (VA): AidData.*

Burke, J., Jamison, J., Karlan, D., Mihaly, K. & Zinman, J. (2019). *Credit building or credit crumbling? A credit builder loan's effects on consumer behavior, credit scores and their predictive power* (Tech. Rep.). Cambridge, MA8: National Bureau of Economic Research.

Burke, M., Heft-Neal, S. & Bendavid, E. (2016). Sources of variation in under-5 mortality across Sub-Saharan Africa: A spatial analysis. *The Lancet Global Health*, *4*(12), e936–e945.

Burstein, R., Cameron, E., Casey, D. C., Deshpande, A., Fullman, N., Gething, P. W., ... Hay, S. I. (2018). Mapping child growth failure in Africa between 2000 and 2015. *Nature*, *555*(7694), 41–47.

Carol, S., Eich, D., Keller, M., Steiner, F. & Storz, K. (2019). Who can ride along? Discrimination in a German carpooling market. *Population, space and place*, *25*(8), e2249.

Carter, M. R., Tjernström, E. & Toledo, P. (2019). Heterogeneous impact dynamics of a rural business development program in Nicaragua. *Journal of Development Economics*, *138*, 77–98.

Caruso, G., Sosa-Escudero, W. & Svarc, M. (2015). Deprivation and the dimensionality of welfare: A variable-selection cluster-analysis approach. *Review of Income and Wealth*, *61*(4), 702–722.

Chen, X. & Nordhaus, W. (2015). A test of the new viirs lights data set: Population and economic output in Africa. *Remote Sensing*, *7*(4), 4937–4947.

Chernozhukov, V., Chetverikov, D., Demirer, M., Duflo, E., Hansen, C., Newey, W. & Robins, J. (2018b). *Double/debiased machine learning for treatment and structural parameters*. Oxford University Press Oxford, UK.

Chernozhukov, V., Demirer, M., Duflo, E. & Fernandez-Val, I. (2018a). *Generic machine learning inference on heterogeneous treatment effects in randomized experiments, with an application to immunization in India* (Tech. Rep.). Washington: National Bureau of Economic Research.

Chetty, R., Friedman, J. N., Hendren, N., Jones, M. R. & Porter, S. R. (2018). *The Opportunity Atlas: Mapping the childhood roots of social mobility* (Tech. Rep.). Cambridge, MA: National Bureau of Economic Research.

Chetty, R., Friedman, J. N., Hendren, N., Stepner, M. & Team, T. O. I. (2020a). *How did covid-19 and stabilization policies affect spending and employment? A new real-time economic tracker based on private sector data*. National Bureau of Economic Research Cambridge, MA.

Chetty, R., Hendren, N., Jones, M. R. & Porter, S. R. (2020). Race and economic opportunity in the united states: An intergenerational perspective. *The Quarterly Journal of Economics*, *135*(2), 711–783.

Chetty, R., Hendren, N., Kline, P. & Saez, E. (2014). Where is the land of opportunity? The geography of intergenerational mobility in the united states. *The Quarterly Journal of Economics*, *129*(4), 1553–1623.

Chi, G., Fang, H., Chatterjee, S. & Blumenstock, J. E. (2022). Microestimates of wealth for all low-and middle-income countries. *Proceedings of the National Academy of Sciences*, *119*(3).

Chioda, L., De Mello, J. M. & Soares, R. R. (2016). Spillovers from conditional cash transfer programs: Bolsa Família and crime in urban Brazil. *Economics of Education Review*, *54*, 306–320.

Chowdhury, R., Ceballos-Sierra, F. & Sulaiman, M. (2021). *Grow the pie or have it? using machine learning for impact heterogeneity in the Ultra-poor Graduation*

Model (Tech. Rep. No. WPS - 170). Berkeley, CA: Center for Effective Global Action, University of California, Berkeley.
Christensen, P., Sarmiento-Barbieri, I. & Timmins, C. (2020). Housing discrimination and the toxics exposure gap in the United States: Evidence from the rental market. *The Review of Economics and Statistics*, 1–37.
Christensen, P., Sarmiento-Barbieri, I. & Timmins, C. (2021). *Racial discrimination and housing outcomes in the United States rental market* (Tech. Rep.). Cambridge, MA: National Bureau of Economic Research.
Christiansen, T. & Weeks, M. (2020). *Distributional aspects of microcredit expansions* (Tech. Rep.). Cambridge, England: Faculty of Economics, University of Cambridge.
Churchill, B. F. & Sabia, J. J. (2019). The effects of minimum wages on low-skilled immigrants' wages, employment, and poverty. *Industrial Relations: A Journal of Economy and Society*, *58*(2), 275–314.
Clay, K., Egedesø, P. J., Hansen, C. W., Jensen, P. S. & Calkins, A. (2020). Controlling tuberculosis? Evidence from the first community-wide health experiment. *Journal of Development Economics*, *146*, 102510.
Dahmani, R., Fora, A. A. & Sbihi, A. (2014). Extracting slums from high-resolution satellite images. *Int. J. Eng. Res. Dev*, *10*, 1–10.
Dang, H.-A., Lanjouw, P., Luoto, J. & McKenzie, D. (2014). Using repeated cross-sections to explore movements into and out of poverty. *Journal of Development Economics*, *107*, 112–128.
Daoud, A. & Johansson, F. (2019). *Estimating treatment heterogeneity of International Monetary Fund programs on child poverty with generalized random forest.*
Davis, J. & Heller, S. (2017). Using causal forests to predict treatment heterogeneity: An application to summer jobs. *American Economic Review*, *107*(5), 546–50.
Davis, J. & Heller, S. (2020). Rethinking the benefits of youth employment programs: The heterogeneous effects of summer jobs. *Review of Economics and Statistics*, *102*(4), 664–677.
Deryugina, T., Heutel, G., Miller, N. H., Molitor, D. & Reif, J. (2019). The mortality and medical costs of air pollution: Evidence from changes in wind direction. *American Economic Review*, *109*(12), 4178–4219.
Donaldson, D. & Storeygard, A. (2016). The view from above: Applications of satellite data in economics. *Journal of Economic Perspectives*, *30*(4), 171–98.
Doudchenko, N. & Imbens, G. W. (2016). *Balancing, regression, difference-in-differences and synthetic control methods: A synthesis* (Tech. Rep.). Cambridge, MA: National Bureau of Economic Research.
Dullerud, N., Zhang, H., Seyyed-Kalantari, L., Morris, Q., Joshi, S. & Ghassemi, M. (2021). An empirical framework for domain generalization in clinical settings. In *Proceedings of the conference on health, inference, and learning* (pp. 279–290).
Dutt, P. & Tsetlin, I. (2021). Income distribution and economic development: Insights from machine learning. *Economics & Politics*, *33*(1), 1–36.
Eagle, N., Macy, M. & Claxton, R. (2010). Network diversity and economic development. *Science*, *328*(5981), 1029–1031.

Edo, M., Escudero, W. S. & Svarc, M. (2021). A multidimensional approach to measuring the middle class. *The Journal of Economic Inequality*, *19*(1), 139–162.

Elbers, C., Fujii, T., Lanjouw, P., Özler, B. & Yin, W. (2007). Poverty alleviation through geographic targeting: How much does disaggregation help? *Journal of Development Economics*, *83*(1), 198–213.

Elbers, C., Lanjouw, J. O. & Lanjouw, P. (2003). Micro-level estimation of poverty and inequality. *Econometrica*, *71*(1), 355–364.

Electronic Online Supplement. (2022). *Electronic Online Supplement of this Volume.* https://sn.pub/0ObVSo.

Elvidge, C. D., Sutton, P. C., Ghosh, T., Tuttle, B. T., Baugh, K. E., Bhaduri, B. & Bright, E. (2009). A global poverty map derived from satellite data. *Computers & Geosciences*, *35*(8), 1652–1660.

Engstrom, R., Hersh, J. S. & Newhouse, D. L. (2017). Poverty from space: Using high-resolution satellite imagery for estimating economic well-being. *World Bank Policy Research Working Paper*(8284).

Ettredge, M., Gerdes, J. & Karuga, G. (2005). Using web-based search data to predict macroeconomic statistics. *Communications of the ACM*, *48*(11), 87–92.

Faber, B. & Gaubert, C. (2019). Tourism and economic development: Evidence from Mexico's coastline. *American Economic Review*, *109*(6), 2245–93.

Farbmacher, H., Kögel, H. & Spindler, M. (2021). Heterogeneous effects of poverty on attention. *Labour Economics*, *71*, 102028.

Farrell, D., Greig, F. & Deadman, E. (2020). Estimating family income from administrative banking data: A machine learning approach. In *Aea papers and proceedings* (Vol. 110, pp. 36–41).

Feigenbaum, J. J. (2016). A machine learning approach to census record linking. *Retrieved March*, *28*, 2016.

Fraiman, R., Justel, A. & Svarc, M. (2008). Selection of variables for cluster analysis and classification rules. *Journal of the American Statistical Association*, *103*(483), 1294–1303.

Frias-Martinez, V. & Virseda, J. (2012). On the relationship between socio-economic factors and cell phone usage. In *Proceedings of the fifth international conference on information and communication technologies and development* (pp. 76–84).

Gasparini, L., Sosa-Escudero, W., Marchionni, M. & Olivieri, S. (2013). Multidimensional poverty in Latin America and the Caribbean: New evidence from the Gallup World Poll. *The Journal of Economic Inequality*, *11*(2), 195–214.

Gevaert, C., Persello, C., Sliuzas, R. & Vosselman, G. (2016). Classification of informal settlements through the integration of 2d and 3d features extracted from UAV data. *ISPRS annals of the photogrammetry, remote sensing and spatial information sciences*, *3*, 317.

Glaeser, E. L., Kominers, S. D., Luca, M. & Naik, N. (2018). Big data and big cities: The promises and limitations of improved measures of urban life. *Economic Inquiry*, *56*(1), 114–137.

González-Fernández, M. & González-Velasco, C. (2018). Can google econometrics predict unemployment? Evidence from Spain. *Economics Letters*, *170*, 42–45.

Goodfellow, I., Bengio, Y. & Courville, A. (2016). *Deep Learning*. MIT Press. (http://www.deeplearningbook.org)

Graesser, J., Cheriyadat, A., Vatsavai, R. R., Chandola, V., Long, J. & Bright, E. (2012). Image based characterization of formal and informal neighborhoods in an urban landscape. *IEEE Journal of Selected Topics in Applied Earth Observations and Remote Sensing*, *5*(4), 1164–1176.

Graetz, N., Friedman, J., Friedman, J., Osgood-Zimmerman, A., Burstein, R., Biehl, C., Molly H.and Shields, ... Hay, S. I. (2018). Mapping local variation in educational attainment across Africa. *Nature*, *555*(7694), 48–53.

Gulrajani, I. & Lopez-Paz, D. (2020). In search of lost domain generalization. *arXiv preprint arXiv:2007.01434*.

Hastie, T., Tibshirani, R., Friedman, J. H. & Friedman, J. H. (2009). *The elements of statistical learning: data mining, inference, and prediction* (Vol. 2). Springer.

Head, A., Manguin, M., Tran, N. & Blumenstock, J. E. (2017). Can human development be measured with satellite imagery? In *Ictd* (pp. 8–1).

Henderson, J. V., Storeygard, A. & Weil, D. N. (2012). Measuring economic growth from outer space. *American economic review*, *102*(2), 994–1028.

Heß, S., Jaimovich, D. & Schündeln, M. (2021). Development projects and economic networks: Lessons from rural Gambia. *The Review of Economic Studies*, *88*(3), 1347–1384.

Hodler, R. & Raschky, P. A. (2014). Regional favoritism. *The Quarterly Journal of Economics*, *129*(2), 995–1033.

Hristova, D., Rutherford, A., Anson, J., Luengo-Oroz, M. & Mascolo, C. (2016). The international postal network and other global flows as proxies for national wellbeing. *PloS one*, *11*(6), e0155976.

Huang, L. Y., Hsiang, S. & Gonzalez-Navarro, M. (2021). Using satellite imagery and deep learning to evaluate the impact of anti-poverty programs. *arXiv preprint arXiv:2104.11772*.

Huang, X., Liu, H. & Zhang, L. (2015). Spatiotemporal detection and analysis of urban villages in mega city regions of China using high-resolution remotely sensed imagery. *IEEE Transactions on Geoscience and Remote Sensing*, *53*(7), 3639–3657.

ILO-ECLAC. (2018). *Child labour risk identification model. Methodology to design preventive strategies at local level.* https://dds.cepal.org/redesoc/publicacion?id=4886.

Jean, N., Burke, M., Xie, M., Davis, W. M., Lobell, D. B. & Ermon, S. (2016). Combining satellite imagery and machine learning to predict poverty. *Science*, *353*(6301), 790–794.

Jurafsky, D. & Martin, J. H. (2014). Speech and language processing. *US: Prentice Hall*.

Kavanagh, L., Lee, D. & Pryce, G. (2016). Is poverty decentralizing? quantifying uncertainty in the decentralization of urban poverty. *Annals of the American Association of Geographers*, *106*(6), 1286–1298.

Khelifa, D. & Mimoun, M. (2012). Object-based image analysis and data mining for building ontology of informal urban settlements. In *Image and signal processing for remote sensing xviii* (Vol. 8537, p. 85371I).

Kim, S., Kim, H., Kim, B., Kim, K. & Kim, J. (2019). Learning not to learn: Training deep neural networks with biased data. In *Proceedings of the ieee/cvf conference on computer vision and pattern recognition* (pp. 9012–9020).

Kim, S. & Koh, K. (2022). Health insurance and subjective well-being: Evidence from two healthcare reforms in the United States. *Health Economics*, *31*(1), 233–249.

Kitagawa, T. & Tetenov, A. (2018). Who should be treated? empirical welfare maximization methods for treatment choice. *Econometrica*, *86*(2), 591–616.

Knaus, M. C., Lechner, M. & Strittmatter, A. (2020). Heterogeneous employment effects of job search programmes: A machine learning approach. *Journal of Human Resources*, 0718–9615R1.

Kohli, D., Sliuzas, R., Kerle, N. & Stein, A. (2012). An ontology of slums for image-based classification. *Computers, Environment and Urban Systems*, *36*(2), 154–163.

Kohli, D., Sliuzas, R. & Stein, A. (2016). Urban slum detection using texture and spatial metrics derived from satellite imagery. *Journal of spatial science*, *61*(2), 405–426.

Kuffer, M., Pfeffer, K. & Sliuzas, R. (2016). Slums from space—15 years of slum mapping using remote sensing. *Remote Sensing*, *8*(6), 455.

Lansley, G. & Longley, P. A. (2016). The geography of Twitter topics in London. *Computers, Environment and Urban Systems*, *58*, 85–96.

Lazer, D. M., Pentland, A., Watts, D. J., Aral, S., Athey, S., Contractor, N., ... Wagner, C. (2020). Computational social science: Obstacles and opportunities. *Science*, *369*(6507), 1060–1062.

Liu, J.-H., Wang, J., Shao, J. & Zhou, T. (2016). Online social activity reflects economic status. *Physica A: Statistical Mechanics and its Applications*, *457*, 581–589.

Llorente, A., Garcia-Herranz, M., Cebrian, M. & Moro, E. (2015). Social media fingerprints of unemployment. *PloS one*, *10*(5), e0128692.

Lucchetti, L. (2018). What can we (machine) learn about welfare dynamics from cross-sectional data? *World Bank Policy Research Working Paper*(8545).

Lucchetti, L., Corral, P., Ham, A. & Garriga, S. (2018). *Lassoing welfare dynamics with cross-sectional data* (Tech. Rep. No. 8545). Washington, DC: World Bank.

Luzzi, G. F., Flückiger, Y. & Weber, S. (2008). A cluster analysis of multidimensional poverty in Switzerland. In *Quantitative approaches to multidimensional poverty measurement* (pp. 63–79). Springer.

Maeda, E., Miyata, A., Boivin, J., Nomura, K., Kumazawa, Y., Shirasawa, H., ... Terada, Y. (2020). Promoting fertility awareness and preconception health using a chatbot: A randomized controlled trial. *Reproductive BioMedicine Online*, *41*(6), 1133–1143.

Mahabir, R., Croitoru, A., Crooks, A. T., Agouris, P. & Stefanidis, A. (2018). A critical review of high and very high-resolution remote sensing approaches for detecting and mapping slums: Trends, challenges and emerging opportunities.

Urban Science, *2*(1), 8.
Maiya, S. R. & Babu, S. C. (2018). Slum segmentation and change detection: A deep learning approach. *arXiv preprint arXiv:1811.07896*.
Martey, E. & Armah, R. (2020). Welfare effect of international migration on the left-behind in Ghana: Evidence from machine learning. *Migration Studies*.
McBride, L. & Nichols, A. (2018). Retooling poverty targeting using out-of-sample validation and machine learning. *The World Bank Economic Review*, *32*(3), 531–550.
Merola, G. M. & Baulch, B. (2019). Using sparse categorical principal components to estimate asset indices: New methods with an application to rural Southeast Asia. *Review of Development Economics*, *23*(2), 640–662.
Michalopoulos, S. & Papaioannou, E. (2014). National institutions and subnational development in Africa. *The Quarterly journal of economics*, *129*(1), 151–213.
Mohamud, J. H. & Gerek, O. N. (2019). Poverty level characterization via feature selection and machine learning. In *2019 27th signal processing and communications applications conference (siu)* (pp. 1–4).
Mullally, C., Rivas, M. & McArthur, T. (2021). Using machine learning to estimate the heterogeneous effects of livestock transfers. *American Journal of Agricultural Economics*, *103*(3), 1058–1081.
Nie, X., Brunskill, E. & Wager, S. (2021). Learning when-to-treat policies. *Journal of the American Statistical Association*, *116*(533), 392–409.
Njuguna, C. & McSharry, P. (2017). Constructing spatiotemporal poverty indices from big data. *Journal of Business Research*, *70*, 318–327.
Okiabera, J. O. (2020). *Using random forest to identify key determinants of poverty in kenya.* (Unpublished doctoral dissertation). University of Nairobi.
Owen, K. K. & Wong, D. W. (2013). An approach to differentiate informal settlements using spectral, texture, geomorphology and road accessibility metrics. *Applied Geography*, *38*, 107–118.
Pokhriyal, N. & Jacques, D. C. (2017). Combining disparate data sources for improved poverty prediction and mapping. *Proceedings of the National Academy of Sciences*, *114*(46), E9783–E9792.
Quercia, D., Ellis, J., Capra, L. & Crowcroft, J. (2012). Tracking "gross community happiness" from tweets. In *Proceedings of the acm 2012 conference on computer supported cooperative work* (pp. 965–968).
Ratledge, N., Cadamuro, G., De la Cuesta, B., Stigler, M. & Burke, M. (2021). *Using satellite imagery and machine learning to estimate the livelihood impact of electricity access* (Tech. Rep.). Cambridge, MA: National Bureau of Economic Research.
Robinson, T., Emwanu, T. & Rogers, D. (2007). Environmental approaches to poverty mapping: An example from Uganda. *Information development*, *23*(2-3), 205–215.
Rosati, G. (2017). Construcción de un modelo de imputación para variables de ingreso con valores perdidos a partir de ensamble learning: Aplicación en la encuesta permanente de hogares (EPH). *SaberEs*, *9*(1), 91–111.
Rosati, G., Olego, T. A. & Vazquez Brust, H. A. (2020). Building a sanitary vulner-

ability map from open source data in Argentina (2010-2018). *International Journal for Equity in Health*, *19*(1), 1–16.
Schmitt, A., Sieg, T., Wurm, M. & Taubenböck, H. (2018). Investigation on the separability of slums by multi-aspect Terrasar-x dual-co-polarized high resolution spotlight images based on the multi-scale evaluation of local distributions. *International journal of applied earth observation and geoinformation*, *64*, 181–198.
Sen, A. (1985). *Commodities and Capabilities*. Oxford University Press.
Sheehan, E., Meng, C., Tan, M., Uzkent, B., Jean, N., Burke, M., ... Ermon, S. (2019). Predicting economic development using geolocated wikipedia articles. In *Proceedings of the 25th acm sigkdd international conference on knowledge discovery & data mining* (pp. 2698–2706).
Skoufias, E. & Vinha, K. (2020). *Child stature, maternal education, and early childhood development* (Tech. Rep. Nos. Policy Research Working Paper, No. 9396). Washington, DC: World Bank.
Sohnesen, T. P. & Stender, N. (2017). Is random forest a superior methodology for predicting poverty? an empirical assessment. *Poverty & Public Policy*, *9*(1), 118–133.
Soman, S., Beukes, A., Nederhood, C., Marchio, N. & Bettencourt, L. (2020). Worldwide detection of informal settlements via topological analysis of crowdsourced digital maps. *ISPRS International Journal of Geo-Information*, *9*(11), 685.
Soto, V., Frias-Martinez, V., Virseda, J. & Frias-Martinez, E. (2011). Prediction of socioeconomic levels using cell phone records. In *International conference on user modeling, adaptation, and personalization* (pp. 377–388).
Steele, J. E., Sundsøy, P. R., Pezzulo, C., Alegana, V. A., Bird, T. J., Blumenstock, J., ... Bengtsson, L. (2017). Mapping poverty using mobile phone and satellite data. *Journal of The Royal Society Interface*, *14*(127), 20160690.
Strittmatter, A. (2019). Heterogeneous earnings effects of the job corps by gender: A translated quantile approach. *Labour Economics*, *61*, 101760.
Sutton, P. C., Elvidge, C. D. & Ghosh, T. (2007). Estimation of gross domestic product at sub-national scales using nighttime satellite imagery. *International Journal of Ecological Economics & Statistics*, *8*(S07), 5–21.
Taubenböck, H., Kraff, N. J. & Wurm, M. (2018). The morphology of the Arrival City, A global categorization based on literature surveys and remotely sensed data. *Applied Geography*, *92*, 150–167.
Thompson, N. C., Greenewald, K., Lee, K. & Manso, G. F. (2020). The computational limits of deep learning. *arXiv preprint arXiv:2007.05558*.
Thoplan, R. (2014). Random forests for poverty classification. *International Journal of Sciences: Basic and Applied Research (IJSBAR), North America*, *17*.
UN Global. (2016). *Building proxy indicators of national wellbeing with postal data.* Project Series, no. 22, https://www.unglobalpulse.org/document/building-proxy-indicators-of-national-wellbeing-with-postal-data/.
Venerandi, A., Quattrone, G., Capra, L., Quercia, D. & Saez-Trumper, D. (2015). Measuring urban deprivation from user generated content. In *Proceedings of the 18th acm conference on computer supported cooperative work & social*

computing (pp. 254–264).
Villa, J. M. (2016). Social transfers and growth: Evidence from luminosity data. *Economic Development and Cultural Change*, *65*(1), 39–61.
Wager, S. & Athey, S. (2018). Estimation and inference of heterogeneous treatment effects using random forests. *Journal of the American Statistical Association*, *113*(523), 1228–1242.
Wager, S., Du, W., Taylor, J. & Tibshirani, R. J. (2016). High-dimensional regression adjustments in randomized experiments. *Proceedings of the National Academy of Sciences*, *113*(45), 12673–12678.
Wald, Y., Feder, A., Greenfeld, D. & Shalit, U. (2021). On calibration and out-of-domain generalization. *arXiv preprint arXiv:2102.10395*.
Warr, P. & Aung, L. L. (2019). Poverty and inequality impact of a natural disaster: Myanmar's 2008 cyclone Nargis. *World Development*, *122*, 446–461.
Watmough, G. R., Marcinko, C. L., Sullivan, C., Tschirhart, K., Mutuo, P. K., Palm, C. A. & Svenning, J.-C. (2019). Socioecologically informed use of remote sensing data to predict rural household poverty. *Proceedings of the National Academy of Sciences*, *116*(4), 1213–1218.
Wurm, M., Taubenböck, H., Weigand, M. & Schmitt, A. (2017). Slum mapping in polarimetric SAR data using spatial features. *Remote sensing of environment*, *194*, 190–204.
Yeh, C., Perez, A., Driscoll, A., Azzari, G., Tang, Z., Lobell, D., . . . Burke, M. (2020). Using publicly available satellite imagery and deep learning to understand economic well-being in Africa. *Nature communications*, *11*(1), 1–11.
Zhou, Z., Athey, S. & Wager, S. (2018). Offline multi-action policy learning: Generalization and optimization. *arXiv preprint arXiv:1810.04778*.
Zhuo, J.-Y. & Tan, Z.-M. (2021). Physics-augmented deep learning to improve tropical cyclone intensity and size estimation from satellite imagery. *Monthly Weather Review*, *149*(7), 2097–2113.

Chapter 10
Machine Learning for Asset Pricing

Jantje Sönksen

Abstract This chapter reviews the growing literature that describes machine learning applications in the field of asset pricing. In doing so, it focuses on the additional benefits that machine learning – in addition to, or in combination with, standard econometric approaches – can bring to the table. This issue is of particular importance because in recent years, improved data availability and increased computational facilities have had huge effects on finance literature. For example, machine learning techniques inform analyses of conditional factor models; they have been applied to identify the stochastic discount factor and purposefully to test and evaluate existing asset pricing models. Beyond those pertinent applications, machine learning techniques also lend themselves to prediction problems in the domain of empirical asset pricing.

10.1 Introduction

Research in the domain of empirical asset pricing traditionally has embraced the fruitful interaction of finance and the development of econometric/statistical models, pushing the envelope of both economic theory and empirical methodology. This effort is personified by Lars P. Hansen, who received the Nobel Prize in Economics for his contributions to asset pricing, and who also developed empirical methods that form the pillars of modern econometric analysis: the generalized method of moments (GMM, Hansen, 1982), along with its variant, the simulated method of moments (SMM, Duffie & Singleton, 1993). Both GMM and SMM are particularly useful for empirical asset pricing and closely connected to it, but these methods also have many more applications (Hall, 2005).

Asset pricing is concerned with answering the question of why some assets pay higher average returns than others. Two influential monographs, by Cochrane (2005)

Jantje Sönksen ✉
Eberhard Karls University, Tübingen, Germany, e-mail: jantje.soenksen@uni-tuebingen.de

F. Chan, L. Mátyás (eds.), *Econometrics with Machine Learning*, Advanced Studies in Theoretical and Applied Econometrics 53, https://doi.org/10.1007/978-3-031-15149-1_10

and Singleton (2006), provide comprehensive synopses of the state of empirical asset pricing research in the mid-2000s, in which they emphasize the close interactions of asset pricing theory/financial economics, econometric modeling, and method development. Notably, neither book offers a discussion of machine learning methods. Evidently, even though machine learning models have been employed in finance research since the 1990s – and particularly in efforts to forecast financial time series – their connection with theory-based asset pricing had not yet been adequately worked out at that time. Applications of machine learning methods appeared more like transfers, such that researchers leveraged models that had proven successful for forecasting exercises in environments with more favorable signal-to-noise ratios (e.g., artificial neural networks) to financial time series data. Because they did not seek to establish clear connections with financial economic theory, early adoptions of machine learning in finance may well be characterized as measurement without theory.

In turn, machine learning methods did not become part of the toolkit used in theory-based empirical finance/asset pricing. However, in more recent years, a literature sparked that consistently connects theory-based empirical asset pricing with machine learning techniques, resulting in contributions, published in leading finance journals, that substantially augment the standard econometric toolbox available for empirical finance.

Superficially, this surge of machine learning applications for asset pricing might seem due mostly to increased data availability (i.e., big data in finance), together with increasingly powerful computational resources. But with this chapter, I identify and argue for deeper reasons. The recent success of machine learning stems from its unique ability to help address and alleviate some long-standing challenges associated with empirical asset pricing, which standard econometric methods have had problems dealing with. Therefore, with this chapter, I outline which aspects of empirical asset pricing benefit particularly from machine learning, then detail how machine learning can complement standard econometric approaches.[1]

Identification of the Stochastic Discount Factor and Machine Learning

I adopt an approach that mimics the one pursued by Cochrane (2005) and Singleton (2006). Both authors take the basic asset pricing equation of financial economics as a theoretical starting point for their discussion of empirical methodologies. Accordingly, I consider the following version of the basic asset pricing equation:

$$P_t^i = \mathbb{E}_t[X_{t+1}^i m_{t+1}(\delta)], \tag{10.1}$$

where $\mathbb{E}_t$ denotes the expected value conditional on time t information, and P_t^i is the price of a future payoff X_{t+1}^i offered by asset i, which is the sum of the price of the asset and cash payments in $t+1$.[2] Next, m_{t+1} is the stochastic discount factor (SDF)

[1] Surveys on the question how machine learning techniques can be applied in the field of empirical asset pricing are also provided by Weigand (2019), Giglio, Kelly and Xiu (2022), and Nagel (2021).

[2] For stocks, X_{t+1}^i amounts to $P_{t+1}^i + D_{t+1}^i$, where D_{t+1}^i denotes the dividend in $t+1$. Chapter 1 in Cochrane (2005) provides a useful introduction to standard asset pricing notation and concepts.

also referred to as the pricing kernel. The SDF is *the* central element in empirical asset pricing, because the same discount factor must be able to price *all* assets of a certain class (e.g., there is an m_{t+1} that fulfills Equation (10.1) for *all* stocks and their individual payoffs X^i_{t+1}). In preference-based asset pricing, the SDF represents the marginal rate of substitution between consumption in different periods, and the vector δ contains parameters associated with investors' risk aversion or time preference. Imposing less economic structure, the fundamental theorem of financial economics states that in the absence of arbitrage, a positive SDF exists.

In traditional empirical asset pricing, one finds stylized structural models that imply a specific SDF, possibly nonlinear, the parameters of which can be estimated with either GMM or SMM. The power utility SDF used by Hansen and Singleton (1982) is a canonical example. Ad hoc specifications of the SDF also appear in the empirical asset pricing literature, for which Cochrane (2005) emphasizes the need to remain mindful of the economic meaning of the SDF associated with investor utility.

Because they can provide a negotiating balance between these opposites – theory-based versus ad hoc specifications – machine learning methods have proven useful in identifying and recovering the SDF. I review and explain these types of contributions in Section 10.2.

Selection of Test Assets and Machine Learning

Two types of payoffs are most important for empirical asset pricing. The first is the gross return of asset i, $R^i_{t+1} = X^i_{t+1}/P^i_t$, for which Equation 10.1 becomes:

$$\mathbb{E}_t\left[R^i_{t+1} m_{t+1}(\delta)\right] = 1. \tag{10.2}$$

The price of any gross return thus is 1. Considering a riskless payoff, Equation (10.2) yields an expression for the risk-free rate, given by[3]

$$R^f_{t+1} = \frac{1}{\mathbb{E}_t\left[m_{t+1}(\delta)\right]}.$$

The second type of payoff that is important for empirical asset pricing is the excess return of asset i, defined as the return of asset i in excess of a reference return, $R^{e,i}_{t+1} = R^i_{t+1} - R^b_{t+1}$. The chosen reference return is often the risk-free rate R^f_{t+1}. It follows that the price of any excess return is 0:

$$\mathbb{E}_t\left[R^{e,i}_{t+1} m_{t+1}(\delta)\right] = 0. \tag{10.3}$$

Equation (10.3) also can be conveniently rewritten such that

$$\mathbb{E}_t(R^{e,i}_{t+1}) = \mathbb{E}_t(R^i_{t+1}) - R^f_{t+1} = -R^f_{t+1} \cdot \text{cov}_t(m_{t+1}, R^i_{t+1}), \tag{10.4}$$

[3] Note that a riskless payoff implies that X^f_{t+1} is known with certainty at some point in time t. Thus, the return R^f_{t+1} is part of the information set in t and can be drawn out of the $\mathbb{E}_t\left[\cdot\right]$ operator.

which yields an expression for the risk-premium associated with an asset i. The sign and size of the premium, reflected in the conditional expected excess return on asset i, are determined by the conditional covariance of the asset return and the SDF on the right-hand side of Equation (10.4). By inserting R^m_{t+1}, the return of a market index, for R^i_{t+1}, Equation (10.4) gives an expression for the market equity premium.

Equation (10.3) and its reformulation in Equation (10.4) thus represent a cornerstone of empirical asset pricing. Because Equation (10.3) is a conditional moment constraint, it provides the natural starting point for the application of moment-based estimation techniques. In particular, using instrumental variables z_t (part of the econometrician's and investor's information set, observed at time t), it is possible to generate unconditional moment conditions,

$$\mathbb{E}_t\,[R^{e,i}_{t+1} z_t m_{t+1}(\delta)] = 0. \tag{10.5}$$

Cochrane (2005) notes that Equation (10.5) can be conceived of as the conditioned down version (using the law of total expectation) of the basic asset pricing equation for a payoff $R^{e,i}_{t+1} z_t$ of an asset class, which Cochrane refers to as *managed portfolios*. By choosing a set of test assets and their excess returns, as well as a set of instruments, the modeler can create a set of moment conditions that form the basis for GMM or SMM estimations.

In this second application, standard econometric analysis for empirical asset pricing again benefits from machine learning. It is not obvious which test assets and which instruments to select to generate managed portfolios. Among the plethora of test assets to choose from, and myriad potential instrumental variables, Cochrane (2005) simply recommends choosing meaningful test assets and instruments, which begs the underlying question. The selection of test assets and instruments serves two connected purposes: to ensure that the parameters of the SDF can be efficiently estimated, such that the moment conditions are informative about those parameters, and to challenge the asset pricing model (i.e., the SDF model) with meaningful test assets. In Section 10.3, I explain how machine learning can provide useful assistance for these tasks.

Conditional Linear Factor Models and Machine Learning

Linear factor models have traditionally been of paramount importance in empirical asset pricing; they imply an SDF that can be written as a linear function of K risk factors:

$$m_{t+1} = \delta_0 + \delta_1 f_{1,t+1} + \delta_2 f_{2,t+1} + \cdots + \delta_K f_{K,t+1}. \tag{10.6}$$

The GMM approach works for asset pricing models with linear and nonlinear SDF alike. However, linear factor models also can lend themselves to linear regression-based analysis (see Chapter 12, Cochrane, 2005), provided that the test asset payoffs are excess returns. The risk factors f_k can be excess returns themselves, and in many empirical asset pricing models, they are. Empirical implementations of the capital asset pricing model (CAPM) use the excess return of a wealth portfolio proxy as a single factor. Another well-known example is the Fama-French three-factor model, which in addition uses the risk factors HML (value; high-minus-low book-to-market) and SMB (size; small-minus-big), both of which are constructed as excess returns of long-short portfolios. Using the basic asset pricing equation as it applies to an excess return and a linear factor model in which the factors are excess returns provides an alternative way to write Equation (10.3), using $z_t = 1$ in an expected return-beta-lambda representation:

$$\mathbb{E}[R_t^{e,i}] = \beta_1\lambda_1 + \beta_2\lambda_2 + \cdots + \beta_K\lambda_K, \tag{10.7}$$

such that the β_k for $k = 1, \ldots, K$ are linear projection coefficients that result from a population regression of $R^{e,i}$ on the K factors, and λ_k denotes the expected excess return of the k'th risk factor.

Although they are empirically popular, linear factor models induce both theoretical and methodological problems, and again, machine learning has proven useful for addressing them. In particular, a notable methodological issue arises because the search for linear factor model specifications has created a veritable *factor zoo*, according to Harvey, Liu and Zhu (2015) and Feng, Giglio and Xiu (2020). Cochrane (2005) calls for discipline in selecting factors and emphasizes the need to specify their connections to investor preferences or the predictability of components of the SDF, yet their choice often is ad hoc. Harvey et al. (2015) identify 316 risk factors introduced in asset pricing literature since the mid-1960s, when the CAPM was first proposed (Sharpe, 1964, Lintner, 1965, Mossin, 1966), many of which are (strongly) correlated, such that it rarely is clear whether a candidate factor contains genuinely new information within the vast factor zoo.[4] Another methodological concern involves time-varying parameters in the SDF. In many situations, it may be argued that the parameters in Equation (10.6) should be time-dependent, and therefore,

$$m_{t+1} = \delta_{0,t} + \delta_{1,t} f_{1,t+1} + \delta_{2,t} f_{2,t+1} + \cdots + \delta_{K,t} f_{K,t+1}. \tag{10.8}$$

The CAPM is a prominent example in which the market excess return is the only factor, such that $K = 1$ in Equation (10.8). However, a theory-consistent derivation of the CAPM's SDF implies that indeed the parameters in the linear SDF must be time-dependent. Regardless of the rationale for including time-varying parameters of the SDF, conditioning down the conditional moment restriction in Equation (10.3) by using the law of total expectations is not possible, so an expected return β-λ representation equivalent to Equation (10.7) must have time-varying β and λ. Time-varying β also emerge if the test assets are stocks for firms with changing business models. Jegadeesh, Noh, Pukthuanthong, Roll and Wang (2019) and Ang, Liu and Schwarz (2020) argue that an aggregation of stocks into portfolios according to firm characteristics – which arguably alleviates the non-stationarity of the return series – may not be innocuous.

[4] Harvey et al. (2015) restrict their analysis to factors published in a leading finance journal or proposed in outstanding working papers. There are many other factors, not accounted for by Harvey et al., that have been suggested.

As a result, neither GMM nor regression-based analysis are directly applicable. In the particular case of the CAPM, the Hansen-Richard critique states that the CAPM is not testable in the first place (Hansen & Richard, 1987). To provide a solution to the problem, Cochrane (1996) proposes scaling factors by using affine functions of time t variables for the parameters in Equation (10.6); this approach has been applied successfully by Lettau and Ludvigson (2001). Yet to some extent, these notions represent ad hoc and partial solutions; prior theory does not establish clearly which functional forms should be used.

Dealing with the challenges posed by linear factor models thus constitutes the third area in which machine learning proves useful for empirical asset pricing. In Section 10.4, I outline how machine learning can 'tame the factor zoo' and address the problem of time-varying parameters in conditional factor asset pricing models.

Predictability of Asset Returns and Machine Learning

Cochrane (2005) emphasizes that the basic asset pricing Equation (10.1) does not rule out the predictability of asset returns. In fact, the reformulated basic asset pricing equation for an excess return in (10.4) not only provides an expression for the risk premium associated with asset i but also establishes the optimal prediction of the excess return $R^{e,i}_{t+1}$, provided the loss function is the mean squared error (MSE) of the forecast. The conditional expected value is the MSE-optimal forecast. Accordingly, to derive theory-consistent predictions, Equation (10.4) represents a natural starting point. Conceiving the conditional covariance $\text{cov}_t(m_{t+1}, R^i_{t+1})$ as a function of time t variables, one could consider flexible functional forms and models to provide MSE-optimal excess return predictions. This domain represents a natural setting for the application of machine learning methods.

Exploiting the return predictability that is theoretically possible has always been an active research area, with obvious practical interest; machine learning offers intriguing possibilities along these lines. In the present context with its low signal to noise ratio, diligence is required in creating training and validation schemes when those highly flexible multidimensional statistical methods are employed. Moreover, limitations implied by theory-based empirical assessments of risk premia using information contained in option data should be considered. I present this fourth area of asset pricing using machine learning by giving best practice examples in Section 10.5 of this chapter.

In summary, and in line with the outline provided in the preceding sections, Section 10.2 contains an explanation of how machine learning techniques can help identify the SDF. Section 10.3 elaborates on the use of machine learning techniques to test and evaluate asset pricing models. Then Section 10.4 reports on applications of machine learning to estimate linear factor models. In Section 10.5, I explain how machine learning techniques can be applied to prediction problems pertaining to empirical asset pricing. Finally, I offer some concluding remarks in Section 10.6.

10.2 How Machine Learning Techniques Can Help Identify Stochastic Discount Factors

Chen, Pelger and Zhu (2021) aim to identify an SDF that is able to approximate risk premia at the stock level. Their study is particularly interesting, in that they (i) propose a loss function based on financial economic theory; (ii) use three distinct types of neural networks, each of which is tailored to a particular aspect of the identification strategy; and (iii) relate that strategy to the GMM framework. Furthermore, Chen et al. argue that whilst machine learning methods are suitable for dealing with the vast set of conditioning information an SDF might depend on (e.g., firm characteristics, the business cycle) and generally can model complex interactions and dependencies, their application to empirical asset pricing may be hampered by the very low signal-to-noise ratio of individual stock returns. To counteract this problem and to provide additional guidance in the training of machine learning models, Chen et al. propose imposing economic structure on them through the no-arbitrage condition. Instead of relying on the loss functions conventionally used in machine learning applications (e.g., MSE minimization), they revert to the basic asset pricing Equation (10.3). This theory-guided choice of the loss function represents a key novelty and contribution of their study. Two further contributions refer to the way in which Chen et al. extract hidden states of the economy from a panel of macroeconomic variables and to the data-driven technique that is used to construct managed portfolios (as I describe in more detail subsequently).

Regarding the SDF, Chen et al. (2021) assume a linear functional form, using excess returns as factors and accounting for time-varying factor weights:

$$m_{t+1} = 1 - \sum_{i=1}^{N} \omega_{t,i} R_{t+1}^{e,i} = 1 - \omega_t' \mathbf{R}_{t+1}^{e}, \tag{10.9}$$

where N denotes the number of excess returns. Note that the SDF depicted in Equation (10.9) is a special case of the SDF representation in Equation (10.8), in which the factors are excess returns. When using solely excess returns as test assets, the mean of the SDF is not identified and can be set to any value. Chen et al. (2021) choose $\delta_{0,t} = 1$, which has the interesting analytical implication that $\omega_t = \mathbb{E}_t[\mathbf{R}_{t+1}^{e} \mathbf{R}_{t+1}^{e}{}']^{-1} \mathbb{E}_t[\mathbf{R}_{t+1}^{e}]$.

The weights in Equation (10.9) are both time-varying and asset-specific. They are functions of firm-level characteristics ($\mathbf{I}_{t,i}$) and information contained in macroeconomic time series ($\mathbf{I}_t$). By including not only stock-level information but also economic time series in the information set, the SDF can capture the state of the economy (e.g., business cycle, crises periods). Chen et al. (2021) rely on more than 170 individual macroeconomic time series, many of which are strongly correlated. From this macroeconomic panel, they back-out a small number of hidden variables that capture the fundamental dynamics of the economy using a recurrent long-short-term-memory network (LSTM), such that $\mathbf{h}_t = h(\mathbf{I}_t)$ and $\omega_{t,i} = \omega(\mathbf{h}_t, \mathbf{I}_{t,i})$. As suggested by its name, this type of neural network is particularly well suited to detecting both short- and long-term dependencies in the macroeconomic panel and thus for capturing business cycle dynamics. To account for nonlinearities and possibly complex interactions between $\mathbf{I}_t$ and $\mathbf{h}_t$, $\omega_{t,i}$ can be modeled using a feedforward neural network. For an introduction to neural networks, see Chapter 4.

To identify the SDF weights ω, Chen et al. (2021) rely on managed portfolios. The idea behind using managed portfolios is that linear combinations of excess returns are excess returns themselves. Thus, instead of exclusively relying on the excess returns of individual stocks $R^{e,i}_{t+1}$ for estimation purposes, it is possible to use information available at time t to come up with portfolio weights $\mathbf{g}_t$ and new test assets $\mathbf{R}^e_{t+1}\mathbf{g}_t$. Historically, managed portfolios have been built in an ad hoc fashion, such as by using time t information on the price-dividend ratio. Choosing a data-driven alternative, Chen et al. (2021) construct each element of $\mathbf{g}_{t,i}$ as a nonlinear function of $\mathbf{I}_{t,i}$ and a set of hidden macroeconomic state variables $\mathbf{h}^g_t = h^g(\mathbf{I}_t)$, where the hidden states in $\mathbf{h}^g_t$ and $\mathbf{h}_t$ may differ. However, just as with the SDF specification, Chen et al. infer $h^g(\cdot)$ with an LSTM, and they model $\mathbf{g}_t$ using a feedforward neural network.

Putting all of these elements together – no-arbitrage condition, managed portfolios, SDF and portfolio weights derived from firm-level and macroeconomic information – leads to the moment constraints:

$$\mathbb{E}\left[\left(1-\sum_{i=1}^{N}\omega_{t,i}R^{e,i}_{t+1}\right)\mathbf{R}^e_{t+1}\mathbf{g}_t\right]=\mathbf{0},$$

which serve as a basis for identifying the SDF. However, $g(\mathbf{h}^g_t, \mathbf{I}_{t,i})$ would allow for constructing infinitely many different managed portfolios. To decide which portfolios are most helpful for pinning down the SDF, Chen et al. build on Hansen and Jagannathan's (1997) finding that the SDF proxy that minimizes the largest possible pricing error is the one that is closest to an admissible true SDF in least square distance. Therefore, Chen et al. set up the empirical loss function:

$$L(\omega|\hat{\mathbf{g}},\mathbf{h}^g_t,\mathbf{h}_t,\mathbf{I}_{t,i})=\frac{1}{N}\sum_{i=1}^{N}\frac{T_i}{T}\left\|\frac{1}{T_i}\sum_{t\in T_i}\left(1-\sum_{i=1}^{N}\omega_{t,i}R^{e,i}_{t+1}\right)R^{e,i}_{t+1}\hat{\mathbf{g}}_{t,i}\right\|^2$$

and formulate a minimax optimization problem in the spirit of a generative adversarial network (GAN):

$$\{\hat{\omega},\hat{\mathbf{h}}_t,\hat{\mathbf{g}},\hat{\mathbf{h}}^g_t\}=\min_{\omega,\mathbf{h}_t}\max_{\mathbf{g},\mathbf{h}^g_t}L(\omega|\hat{\mathbf{g}},\mathbf{h}^g_t,\mathbf{h}_t,\mathbf{I}_{t,i}).$$

Within this framework, Chen et al. (2021) pit two pairs of neural networks against each other: The LSTM that extracts $\mathbf{h}_t$ from macroeconomic time series and the feedforward network that generates ω from $\mathbf{h}_t$ together with the asset-specific characteristics $\mathbf{I}_{t,i}$ are trained to minimize the loss function, whereas the other LSTM and feedforward network serve as adversaries, aiming to construct managed portfolios that are particularly hard to price, such that they maximize pricing errors.

For their empirical analysis, Chen et al. (2021) use monthly excess returns on $N = 10,000$ stocks for the period from January 1967 to December 2016 and thereby establish five main findings. First, performance of highly flexible machine learning models in empirical asset pricing can be improved by imposing economic structure, such as in the form of a no-arbitrage constraint. Second, interactions between stock-level characteristics matter. Third, the choice of test assets is important. This finding is in line with the literature presented in Section 10.3. Fourth, macroeconomic states matter and it is important to consider the full time series of macroeconomic variables instead of reduced-form information, such as first differences or the last observation. Fifth, the performance of the flexible neural network approach proposed by Chen et al. (2021) might be improved, if it were combined with multifactor models of the IPCA-type, as discussed in Section 10.4.2.

An alternative approach to identifying SDFs is brought forward by Korsaye, Quaini and Trojani (2019) who enforce financial economic theory on the SDF by minimizing various concepts of SDF dispersion whilst imposing constraints on pricing errors. The approach by Korsaye et al. is inspired by the works of Hansen and Jagannathan (1991) on identifying the admissible minimum variance SDF. This SDF constitutes a lower bound on the variance of any admissible SDF and thus upper-bounds the Sharpe ratio that can be attained using linear portfolios of traded securities. The authors argue that the constraints which they impose on the pricing errors can be justified, for example, by market frictions and they name the thus resulting model-free SDFs *minimum dispersion smart SDFs (S-SDFs)*. Korsaye et al. consider different measures of dispersion, as well as multiple economically motivated penalties on the pricing errors. Furthermore, they develop the econometric theory required for estimation and inference of S-SDFs and propose a data-driven method for constructing minimum variance SDFs.

10.3 How Machine Learning Techniques Can Test/Evaluate Asset Pricing Models

The idea that the choice of test assets is important when evaluating and testing asset pricing models – either by standard econometric techniques or incorporating machine learning methods – is not novel. For example, Lewellen, Nagel and Shanken (2010) criticize a common test of asset pricing models that uses book-to-market sorted portfolios as test assets. In a similar vein, Ahn, Conrad and Dittmar (2009) warn that constructing portfolios based on characteristics that are known to be correlated with returns might introduce a data-snooping bias. They suggest an alternative strategy, which forms base assets by grouping securities that are more strongly correlated, and compare the inferences drawn from this set of basis assets with those drawn from other benchmark portfolios.

In more recent, machine learning-inspired literature on empirical asset pricing, Chen et al. (2021) reiterate the importance of selecting test assets carefully. In another approach, with a stronger focus on model evaluation, Bryzgalova, Pelger and Zhu (2021a) argue that test assets conventionally used in empirical asset pricing studies do not only fail to provide enough of a challenge to the models under consideration but also contribute to the growing factor zoo. To circumvent this issue, they propose so-called asset pricing trees (AP-Trees) that can construct easy-to-interpret and hard-to-price test assets. An overview of other tree-based approaches is available in Chapter 2 of this book.

Bryzgalova, Pelger and Zhu (2021a) consider a linear SDF, spanned by J managed portfolios, constructed from N excess returns:

$$m_{t+1} = 1 - \sum_{j=1}^{J} \omega_{t,j} \mathbf{R}^{e,\text{man}}_{t+1,j} \qquad \text{with} \qquad \mathbf{R}^{e,\text{man}}_{t+1,j} = \sum_{i=1}^{N} f(C_{t,i}) R^{e,i}_{t+1}, \tag{10.10}$$

where $f(C_{t,i})$ denotes a nonlinear function of the time t stock characteristics $C_{t,i}$.

The representation in Equation (10.10) resembles the linear SDF used by Chen et al. (2021), as described in Section 10.2 of this chapter. However, the SDF weights they applied were also functions of macroeconomic time series. Additionally, Bryzgalova, Pelger and Zhu approach Equation (10.10) from the perspective of trying to find the tangency portfolio, that is, the portfolio on the mean-variance frontier that exhibits the highest Sharpe ratio:

$$SR = \frac{\mathbb{E}[R^e]}{\sigma(R^e)} \qquad \text{with} \qquad |SR| \leq \frac{\sigma(m)}{\mathbb{E}(m)},$$

where $\sigma(m)$ denotes the unconditional standard deviation of the SDF. The mean-variance frontier is the boundary of the mean-variance region, which contains all accessible combinations of means and variances of the assets' excess returns. Put differently, for any given variance of R^e, the mean-variance frontier answers the question of which mean excess returns are accessible. Importantly, all excess returns that lie exactly on the mean-variance frontier are perfectly correlated with all other excess returns on the boundary and also perfectly correlated with the SDF. Furthermore, $\frac{\sigma(m)}{\mathbb{E}(m)}$ is the slope of the mean-variance frontier at the tangency portfolio and upper-bounds the Sharpe ratio of any asset.[5] In this sense, Bryzgalova, Pelger and Zhu (2021a) attempt to span the SDF by forming managed portfolios through $f(C_{t,i})$ and selecting weights ω, such that $\sum_{j=1}^{J} \omega_{t,i} \mathbf{R}^{e,\text{man}}_{t+1,j}$ approximates the tangency portfolio. This theoretical concept serves as a foundation for their pruning strategy. Bryzgalova, Pelger and Zhu (2021a) construct the managed portfolios in Equation (10.10) using trees, which group stocks into portfolios on the basis of their characteristics. Trees lend themselves readily to this purpose, because they are reminiscent of the stock characteristic-based double- and triple-sorts that have served portfolio construction purposes for decades. Applying the tree-based approach, a large number of portfolios could be constructed.

To arrive at a sensible number of interpretable portfolios that help recover the SDF, Bryzgalova, Pelger and Zhu (2021a) also introduce a novel pruning strategy that aims at maximizing the Sharpe ratio. To apply this technique, the authors compute the variance-covariance matrix of the cross-section of the candidate managed portfolios

[5] The upper bound is derived in Appendix 1.

constructed with AP-Trees, $\hat{\boldsymbol{\Sigma}}$, then assemble the corresponding portfolio means in $\hat{\boldsymbol{\mu}}$. Furthermore, they rephrase the problem of maximizing the Sharpe ratio of the tangency portfolio proxy as a variance minimization problem, subject to a minimum mean excess return of the portfolio proxy. An L_1-penalty imposed on portfolio weights helps select those portfolios relevant to spanning the SDF. Additionally, the sample average portfolio returns are shrunken to their cross-sectional average value by an L_2-penalty, accounting for the fact that extreme values are likely due to over- or underfitting (see Chapter 1 of this book for an introduction to L_1 and L_2 penalties). The optimization problem thus results in

$$\begin{aligned} \min_{\omega} \quad & \frac{1}{2}\omega'\hat{\boldsymbol{\Sigma}}\omega + \lambda_1||\omega||_1 + \frac{1}{2}\lambda_2||\omega||_2^2, \\ s.t. \quad & \omega'\mathbf{1} = 1 \\ & \omega'\hat{\mu} \geq \mu_0, \end{aligned} \tag{10.11}$$

where $\mathbf{1}$ denotes a vector of ones, and λ_1, λ_2, and μ_0 are hyperparameters. Apart from the inclusion of financial economic theory, a key difference between the pruning technique outlined in Equation (10.11) and standard techniques used in related literature (see Chapter 2 for more details on tree-based methods and pruning) is that in the former case, the question of whether to collapse children nodes cannot be answered solely on the basis of these nodes (e.g., by comparing the Sharpe ratios of the children nodes to that of the parent node). Instead, the decision depends on the Sharpe ratio of $\omega\mathbf{R}_{t+1}^{e,\mathrm{man}}$, so *all* other nodes must be taken into account too.[6]

Noting some parallels between the studies by Chen et al. (2021) and Bryzgalova, Pelger and Zhu (2021a) – both adapt highly nonlinear machine learning approaches with financial economic theory to identify the SDF, both highlight the importance of selecting test assets that help span the SDF, and both propose strategies for the construction of such managed portfolios – the relevant differences in their proposed strategies and the focus of their studies are insightful as well. In particular, in the neural network-driven approach by Chen et al. (2021), constructing hard-to-price test assets is critical, because they adopt Hansen and Jagannathan's (1997) assertion that minimizing pricing errors for these assets corresponds to generating an SDF proxy that is close to a true admissible SDF. The managed portfolios constructed by the GAN are complex, nonlinear functions of the underlying macroeconomic time series and stock-specific characteristics, such that they are not interpretable in a meaningful way. In contrast, the portfolios obtained from the AP-Trees proposed by Bryzgalova, Pelger and Zhu (2021a) can be straightforwardly interpreted as a grouping of stocks that share certain characteristics. The thus identified portfolios have value of their own accord and can be used as test assets in other studies.

Bryzgalova, Pelger and Zhu (2021a) evaluate the ability of their framework to recover the SDF and construct hard-to-price test assets from monthly data between January 1964 and December 2016. They consider 10 firm-specific characteristics and arrive at three pertinent conclusions. First, a comparison with conventionally

[6] Further methodological details are available from Bryzgalova, Pelger and Zhu (2021b).

sorted portfolios shows that the managed portfolios constructed from AP-Trees yield substantially higher Sharpe ratios and thus recover the SDF more successfully. Second, the way AP-Trees capture interactions between characteristics is particularly important. Third, according to robustness checks, imposing economic structure on the trees by maximizing the Sharpe ratio is crucial to out-of-sample performance.

Giglio, Xiu and Zhang (2021) are also concerned with the selection of test assets for the purpose of estimating and testing asset pricing models. They point out that the identification of factor risk premia hinges critically on the test assets under consideration and that a factor may be labeled *weak* because only a few of the test assets are exposed to it, making standard estimation and inference incorrect. Therefore, their novel proposal for selecting assets from a wider universe of test assets and estimating the risk premium of a factor of interest, as well as the entire SDF, explicitly accounts for weak factors and assets with highly correlated risk exposures. The procedure consists of two steps that are conducted iteratively: Given a particular factor of interest, the first step selects those test assets that exhibit the largest (absolute) correlation with that factor. These test assets are then used in principal component analysis (PCA) for the construction of a latent factor. Then, a linear projection is used to make both the test asset returns and the factor orthogonal to the latent factor before returning to the selection of the most correlated test assets. Giglio, Xiu and Zhang refer to their proposed methodology as supervised principal component analysis (SPCA) and argue that the iterative focus on those test assets that exhibit the highest (absolute) correlation with the (residual) factor of interest ensures that weak factors are also captured. In establishing the asymptotic properties of the SPCA estimator, and comparing its limiting behavior with that of other recently proposed estimators (e.g., Ridge, LASSO, and partial least squares), Giglio, Xiu and Zhang report that the SPCA outperforms its competitors in the presence of weak factors not only in theory, but also in finite samples.

10.4 How Machine Learning Techniques Can Estimate Linear Factor Models

This section illustrates how machine learning techniques can alleviate the issues faced in the context of linear factor models outlined in the introduction: dealing with time-varying parameters and the selection of factors. In Section 10.4.1, I present a two-step estimation approach proposed by Gagliardini, Ossola and Scaillet (2016). Their initial methodology was not strictly machine learning (the boundaries with 'standard' econometric method development are blurred anyway), but in later variants, the identification strategy included more elements of machine learning. In Section 10.4.2, I present the instrumented principal components analysis (IPCA) proposed by Kelly, Pruitt and Su (2019), which modifies standard PCA, a basic machine learning method. A workhorse method, IPCA can effectively analyze conditional linear factor models. Section 10.4.3 presents an extension of IPCA that adopts more machine learning aspects, namely, Gu, Kelly and Xiu's (2021) autoencoder approach. In

Section 10.4.4, I outline the regularized Bayesian approach introduced by Kozak, Nagel and Santosh (2018). Finally, Section 10.4.5 lists some recent contributions that address the selection of factors, including the problem of weak factors. Appendix 2 contains an overview of different PCA-related approaches.

10.4.1 Gagliardini, Ossola, and Scaillet's (2016) Econometric Two-Pass Approach for Assessing Linear Factor Models

Gagliardini et al. (2016) propose a novel econometric methodology to infer time-varying equity risk premia from a large unbalanced panel of individual stock returns under conditional linear asset pricing models. Their weighted two-pass cross-sectional estimator incorporates conditioning information through instruments – some of which are common to all assets and others are asset-specific – and its consistency and asymptotic normality are derived under simultaneously increasing cross-sectional and time-series dimensions. In their empirical analysis, Gagliardini et al. (2016) consider monthly return data between July 1964 and December 2009 for about 10,000 stocks. They find that risk premia are large and volatile during economic crises and appear to follow the macroeconomic cycle.

A competing approach comes from Raponi, Robotti and Zaffaroni (2019), who consider a large cross-section, but – in contrast with Gagliardini et al. (2016) – use a small and fixed number of time-series observations. Gagliardini, Ossola and Scaillet (2019) build on these developments, focusing on omitted factors. They propose a diagnostic criterion for approximate factor structure in large panel data sets. A misspecified set of observable factors will turn risk premia estimates obtained by means of 2-pass regression worthless. Under correct specification, however, errors will be weakly cross-sectionally correlated and this serves as the foundation of the newly proposed criterion, which checks for observable factors whether the errors are weakly cross-sectionally correlated or share one or more unobservable common factor(s). This approach to determine the number of omitted common factors can also be applied in a time-varying context.

Bakalli, Guerrier and Scaillet (2021) still work with a two-pass approach and aim at time-varying factor loadings, but this time, they make use of machine learning methods (L_1-penalties) to ensure sparsity in the first step. In doing so, they address a potential weakness of the method put forward by Gagliardini et al. (2016), which refers to the large number of parameters required to model time-varying factor exposures and risk premia. Bakalli et al. thus develop a penalized two-pass regression with time-varying factor loadings. In the first pass, a group LASSO is applied to target the time-invariant counterpart of the time-varying models thereby maintaining compatibility with the no arbitrage restrictions. The second pass delivers risk premia estimates to predict equity excess returns. Bakalli et al. derive the consistency for their estimator and exhibit its good out-of-sample performance using a simulation study and monthly return data on about 7,000 stocks in the period from July 1963 to December 2019.

10.4.2 Kelly, Pruitt, and Su's (2019) Instrumented Principal Components Analysis

Kelly et al. (2019) propose a modeling approach for the cross-section of returns. Their IPCA method is motivated by the idea that stock characteristics might line up with average returns, because they serve as proxies for loadings on common (but latent) risk factors. The methodology allows for latent factors and time-varying loadings by introducing observable characteristics that instrument for the unobservable dynamic loadings. To account for this idea, the authors model excess returns as:

$$\begin{aligned} R^{e,i}_{t+1} &= \alpha_{i,t} + \boldsymbol{\beta}_{i,t}\mathbf{f}_{t+1} + \varepsilon_{i,t+1} \\ \text{where} \quad \alpha_{i,t} &= \mathbf{z}'_{i,t}\boldsymbol{\Gamma}_{\alpha} + \nu_{\alpha,i,t} \\ \text{and} \quad \beta_{i,t} &= \mathbf{z}'_{i,t}\boldsymbol{\Gamma}_{\beta} + \nu_{\beta,i,t}, \end{aligned} \tag{10.12}$$

where $\mathbf{f}_{t+1}$ denotes a vector of latent risk factors, and $\mathbf{z}_{i,t}$ is a vector of characteristics that is specific to stock i and time t. $\boldsymbol{\Gamma}_{\alpha}$ and $\boldsymbol{\Gamma}_{\beta}$ are time-independent matrices. The time dependency of $\alpha_{i,t}$ and $\boldsymbol{\beta}_{i,t}$ reflects the ways that the stock characteristics themselves may change over time. The key idea of Kelly et al. (2019) is to estimate $\mathbf{f}_{t+1}$, $\boldsymbol{\Gamma}_{\alpha}$, and $\boldsymbol{\Gamma}_{\beta}$ jointly via:

$$\begin{aligned} \text{vec}\left(\hat{\boldsymbol{\Gamma}}'\right) &= \left(\sum_{t=1}^{T-1} \mathbf{Z}'_t\mathbf{Z}_t \otimes \hat{\tilde{\mathbf{f}}}_{t+1}\hat{\tilde{\mathbf{f}}}'_{t+1}\right)^{-1} \left(\sum_{t=1}^{T-1} \left[\mathbf{Z}_t \otimes \hat{\tilde{\mathbf{f}}}'_{t+1}\right]' \mathbf{R}^e_{t+1}\right) \\ \text{and} \quad \hat{\mathbf{f}}_{t+1} &= \left(\hat{\boldsymbol{\Gamma}}'_{\beta}\mathbf{Z}'_t\mathbf{Z}_t\hat{\boldsymbol{\Gamma}}_{\beta}\right)^{-1} \hat{\boldsymbol{\Gamma}}'_{\beta}\mathbf{Z}'_t(\mathbf{R}^e_{t+1} - \mathbf{Z}_t\hat{\boldsymbol{\Gamma}}_{\alpha}), \\ \text{where} \quad \hat{\boldsymbol{\Gamma}} &= \left[\hat{\boldsymbol{\Gamma}}_{\alpha}, \hat{\boldsymbol{\Gamma}}_{\beta}\right] \quad \text{and} \quad \hat{\tilde{\mathbf{f}}}_{t+1} = [1, \hat{\mathbf{f}}'_{t+1}]'. \end{aligned}$$

This specification allows for variation in returns to be either attributed to factor exposure (by means of $\boldsymbol{\Gamma}_{\beta}$) or to an anomaly intercept (through $\boldsymbol{\Gamma}_{\alpha}$).[7] The IPCA framework also supports tests regarding the characteristics under consideration, such as whether a particular characteristic captures differences in average returns that are not associated with factor exposure and thus with compensation for systematic risk. With IPCA, it is possible to assess the statistical significance of one characteristic while controlling for all others. In addressing the problem of the growing factor zoo, Kelly et al. (2019) argue that such tests can reveal the genuine informational content of a newly discovered characteristic, given the plethora of existing competitors. For both types of tests, the authors rely on bootstrap inference. They outline the methodological theory behind IPCA in Kelly, Pruitt and Su (2020).

In their empirical analysis, Kelly et al. (2019) use a data set provided by Freyberger, Neuhierl and Weber (2020) that contains 36 characteristics of more than 12,000 stocks in the period between July 1962 and May 2014. Allowing for five latent factors, the authors find that IPCA outperforms the Fama-French five-factor model. They also

[7] Note that a unique identification requires additional restrictions, which Kelly et al. (2019) impose through the orthogonality constraint $\boldsymbol{\Gamma}'_{\alpha}\boldsymbol{\Gamma}_{\beta} = \mathbf{0}$.

find that only 10 of the 36 characteristics are statistically significant at the 1% level and that applying IPCA to just these 10 characteristics barely affects model fit.

10.4.3 Gu, Kelly, and Xiu's (2021) Autoencoder

Kelly et al. (2019) make use of stock covariates to allow for time-varying factor exposures. However, they impose that covariates affect factor exposures in a linear fashion. Accounting for interactions between the covariates or higher orders is generally possible, but it would have to be implemented manually (i.e., by computing the product of two covariates and adding it as an additional instrument).

Gu et al. (2021) offer a generalization of IPCA that still is applicable at the stock level and accounts for time-varying parameters but that does not restrict the relationship between covariates and factor exposures to be linear, as is implied by Equation (10.12). According to their problem formulation, identifying latent factors and their time-varying loadings can be conceived of as an autoencoder that consists of two feedforward neural networks. One network models factor loadings from a large set of firm characteristics; the other uses excess returns to flexibly construct latent factors. Alternatively, a set of characteristic-managed portfolios can be used to model the latent factors. At the output layer of the autoencoder, the $\beta_{t,i,k}$ estimates (specific to time t, asset i, and factor k) of the first network combine with the K factors constructed by the second network, thereby providing excess return estimates. This model architecture is based on the no-arbitrage condition and thus parallels the SDF identification strategy by Chen et al. (2021) outlined in Section 10.2 of this chapter.

The term *autoencoder* signifies that the model's target variables, the excess returns of different stocks, are also input variables. Hence, the model first *encodes* information contained in the excess returns (by extracting latent factors from them), before *decoding* them again to arrive at excess return predictions. Gu et al. (2021) then confirm that IPCA results as a special case of their autoencoder specification.

The autoencoder is trained to minimize the MSE between realized excess returns and their predictions, and an L_1-penalty together with *early stopping* helps avoid overfitting. Using monthly excess returns for all NYSE-, NASDAQ-, and AMEX-traded stocks from March 1957 to December 2016, as well as 94 different firm-level characteristics, Gu et al. (2021) retrain the autoencoder on an annual basis to provide out-of-sample predictions. This retraining helps ensure the model can adapt to changes in the factors or their loadings, but it constitutes a computationally expensive step, especially compared with alternative identification strategies, such as IPCA.

Contrasting the out-of-sample performance of their network-based approach with that of competing strategies, including IPCA, Gu et al. (2021) establish that the autoencoder outperforms IPCA in terms of both the out-of-sample R^2 and the out-of-sample Sharpe ratio. Thus, the increased flexibility with which factors and their loadings can be described by the autoencoder translates into improved predictive abilities.

10.4.4 Kozak, Nagel, and Santosh's (2020) Regularized Bayesian Approach

Kozak, Nagel and Santosh (2020) present a Bayesian alternative to Kelly et al.'s (2019) IPCA approach. They argue against characteristic-sparse SDF representations and suggest constructing factors by first computing principal components (PCs) from the vast set of cross-sectional stock return predictors, then invoking regularization later, to select a small number of PCs to span the SDF. Building on their findings in Kozak et al. (2018), the authors argue that a factor earning high excess returns should also have a high variance. This theoretical consideration is incorporated in a Bayesian prior on the means of the factor portfolios, which are notoriously hard to estimate. As a consequence, more shrinkage gets applied to the weights of PC-factors associated with low eigenvalues.

Kozak et al. (2020) identify the weights ω, by pursuing an elastic net mean-variance optimization:

$$\min_{\omega} \quad \frac{1}{2}(\hat{\mu} - \hat{\Sigma}\omega)'\hat{\Sigma}^{-1}(\hat{\mu} - \hat{\Sigma}\omega) + \lambda_1||\omega||_1 + \frac{1}{2}\lambda_2||\omega||_2^2, \tag{10.13}$$

where $\hat{\Sigma}$ is the variance-covariance matrix of PC-factors, and $\hat{\mu}$ refers to its mean. The L_2-penalty is inherited from the Bayesian prior; including the L_1-penalty in Equation (10.13) supports factor selection.

For their empirical analyses, Kozak et al. (2020) use different sets of stock characteristics and transform them to account for interactions or higher powers before extracting principal components. Comparing the performance of their PC-sparse SDFs to that of models that rely on only a few characteristics, they find that PC-sparse models perform better.

The pruning function used by Bryzgalova, Pelger and Zhu (2021a) in Equation (10.11) is similar to Equation (10.13); and indeed, Bryzgalova, Pelger and Zhu argue that their weights $\hat{\omega}$ and those proposed by Kozak et al. (2020) coincide in the case of uncorrelated assets. Linking the Bayesian setup of Kozak et al. (2020) to the IPCA approach of Kelly et al. (2019), Nagel (2021, p. 89) points out that the IPCA methodology (see Section 10.4.2) requires a prespecification of the number of latent factors, which could be understood within the framework of Kozak et al. (2020) "as a crude way of imposing the prior beliefs that high Sharpe ratios are more likely to come from major sources of covariances than from low eigenvalue PCs."

10.4.5 Which Factors to Choose and How to Deal with Weak Factors?

As mentioned in the introduction, two important issues associated with working with factor models are the choice of factors under consideration and how to deal with weak factors. These problems are by no means new, and plenty of studies attempt to tackle them from traditional econometric perspectives. However, the ever-increasing *factor zoo* and number of candidate variables aggravate the issues and make applications of machine learning methodologies, e.g., as they relate to regularization techniques, highly attractive. One method frequently used in analyses of (latent) factors is PCA. Interestingly, this technique is an inherent part of both classic econometrics and also the toolbox of machine learning methods. For this reason, I highlight in this subsection those studies as invoking machine learning methods that not only apply PCA (and analyze its respective strengths and weaknesses), but that extend PCA by some other component that is used in the context of machine learning, such as a regularization or penalty term.

Amongst the econometric approaches, Bai and Ng (2002) propose panel criteria to determine the right number of factors in a model in which both the time-series and cross-sectional dimension are very large. Onatski (2012) analyzes the finite sample distribution of principal components when factors are weak and proposes an approximation of the finite sample biases of PC estimators in such settings. Based on these findings, he develops an estimator for the number of factors for which the PCA delivers reliable results and applies this methodology to U.S. stock return data, leading him to reject the hypothesis that the Fama and French (1993) factors span the entire space of factor returns. Instead of focusing on the number of factors, Bailey, Kapetanios and Pesaran (2021) aim at estimating the strength of individual (observed and unobserved) factors. For this purpose, they propose a measure of factor strength that builds on the number of statistically significant factor loadings and for which consistency and asymptotics can be established if the factors in question are at least moderately strong. Monte Carlo experiments serve to study the small sample properties of the proposed estimator. In an empirical analysis, Bailey et al. evaluate the strength of the 146 factors assembled by Feng et al. (2020) and find that factor strength exhibits a high degree of time variation and that only the market factor qualifies as *strong*. On a related issue, Pukthuanthong, Roll and Subrahmanyam (2018) propose a protocol for identifying genuine risk factors and find that many characteristics-based factors do not pass their test.[8] Notably, the market factor is amongst those candidate factors that comply with the protocol.

Gospodinov, Kan and Robotti (2014) note that inference fails in linear factor models containing irrelevant factors, in the sense that the irrelevant factors have a high probability of being mistaken for being priced. Proposing a method that establishes proper inference in such misspecified models, Gospodinov et al. (2014) find little evidence that macro factors are important – a finding that conflicts with Chen et al.'s (2021) application of LSTM models to macro factors, which reveals that these factors are important but that it is crucial to consider them as time series. In line with some of the studies previously mentioned, Gospodinov et al. identify the market factor as one of few factors that appear to be priced.

Anatolyev and Mikusheva (2022) describe three major challenges in dealing with factor models. First, factors might be weak, but still priced, such that the resulting betas and estimation errors are of the same order of magnitude. Second, the error terms might be strongly correlated in the cross-section (e.g., due to mismeasurement

[8] The protocol comprises the correlation between factors and returns, the factor being priced in the cross-section of returns, and also a sensible reward-to-risk ratio.

of the true factors), thereby interfering with both estimation and inference. Third, in empirical applications, the number of assets or portfolios considered is often of a similar size as the time-series dimension. Anatolyev and Mikusheva show that the two-pass estimation procedure conventionally applied in the context of linear factor models results in inconsistent estimates when confronted with the aforementioned challenges. Relying on sample-splitting and instrumental variables regression, they come up with a new estimator that is consistent and can be easily implemented. However, the model under consideration by Anatolyev and Mikusheva (2022) does not account for time-varying factor loadings.[9]

Also on the topic of weak factors, but more closely related to machine learning methods, Lettau and Pelger (2020b) extend PCA by imposing economic structure through a penalty on the pricing error, thereby extracting factors not solely based on variation, but also on the mean of the data. They name this approach to estimating latent factors *risk-premium PCA* (RP-PCA) and argue that it allows to fit both the cross-section and time series of expected returns. Furthermore, they point out that RP-PCA – in contrast to conventional PCA – allows to identify weak factors with high Sharpe ratios. RP-PCA is described in more detail in Appendix 10.6; the statistical properties of the estimator are derived in Lettau and Pelger (2020a). In an alternative approach of combining financial economic structure and the flexibility of machine learning techniques, Feng, Polson and Xu (2021) train a feedforward neural network to study a characteristics-sorted factor model. Their objective function focuses on the sum of squared pricing errors and thus resembles an equally weighted version of the GRS test statistic by Gibbons, Ross and Shanken (1989). Importantly, Feng et al. pay particular attention to the hidden layers of the neural network, arguing that these can be understood to generate deep-learning risk factors from the firm characteristics that serve as inputs. Furthermore, they propose an activation function for the estimation of long-short portfolio weights. Whilst similar to IPCA and RP-PCA, an important difference between PCA-based approaches and that by Feng et al. is that the application of a neural network allows for nonlinearities on the firm characteristics instead of extracting linear components.

Whilst Lettau and Pelger (2020b) impose sparsity regarding the number of factors (but notably not regarding the number of characteristics from which these factors might be constructed), there are other studies which assume a sparse representation in terms of characteristics. For example, DeMiguel, Martín-Utrera, Uppal and Nogales (2020) are concerned with the issue of factor selection for portfolio optimization. They incorporate transaction costs into their consideration and – using L_1-norm penalties – find that the number of relevant characteristics increases compared to a setting without transaction costs. Also focusing on individual characteristics, Freyberger et al. (2020) address the need to differentiate between factor candidates that contain incremental information about average returns and others, with no such independent informational content. They propose to use a group LASSO for the purpose of selecting characteristics and use these characteristics in a nonparametric setup to model the cross-section of expected returns, thereby avoiding strong functional form

[9] An overview of recent econometric developments in factor model literature is provided by Fan, Li and Liao (2021).

assumptions. Freyberger et al. assemble a data set of 62 characteristics and find that only 9 to 16 of these provide incremental information in the presence of the other factors and that the predictive power of characteristics is strongly time-varying.

Giglio and Xiu (2021) propose a three-step approach that invokes PCA to deal with the issue of omitted factors in asset pricing models. They argue that standard estimators of risk premia in linear asset pricing models are biased if some priced factors are omitted, and that their method can correctly recover the risk premium of any observable factor in such a setting. The approach augments the standard two-pass regression method by PCA, such that the first step of the procedure is the construction of principal components of test asset returns to recover the factor space. In a related study, Giglio, Liao and Xiu (2021) address the challenges of multiple testing of many alphas in linear factor models. With multiple testing, there is a danger of producing a high number of false positive results just by chance. Additionally, omitted factors – a frequent concern with linear asset pricing models – hamper existing false discovery control approaches (e.g., Benjamini & Hochberg, 1995). Giglio, Liao and Xiu develop a framework that exploits various machine learning methods to deal with omitted factors and missing data (using matrix completion) and provide asymptotic theory required for their proposed estimation and testing strategy.

Using Bayesian techniques that shrink weak factors according to their correlation with asset returns, Bryzgalova, Huang and Julliard (2021) put forward a unified framework for analyzing linear asset pricing models, which allows for traded and non-traded factors and possible model misspecification. Their approach can be applied to the entire factor zoo and is able to identify the dominant model specification – or, in the absence of such – will resort to Bayesian model averaging. Considering about 2.25 quadrillion models, Bryzgalova, Huang and Julliard find that there appears to be no unique best model specification. Instead, hundreds of different linear factor models exhibit almost equivalent performance. They also find that only a small number of factors robustly describe the cross-section of asset returns. Pelger and Xiong (2020) combine nonparametric kernel projection with PCA and develop an inferential theory for state-varying factor models of a large cross-sectional and time-series dimension. In their study, factor loadings are functions of the state-process, thus increasing the model's flexibility and, compared with constant factor models, making it more parsimonious regarding the number of factors required to explain the same variation in the data. They derive asymptotic results and develop a statistical test for changes in the factor loadings in different states. Applying their method to U.S. data, Pelger and Xiong (2020) find that the factor structures of U.S. Treasury yields and S&P 500 stock returns exhibit a strong time variation.

In work related to Kelly et al. (2019), Kim, Korajczyk and Neuhierl (2021) separate firm characteristics' ability to explain the cross-section of asset returns into a risk component (factor loadings) and a mispricing component, whilst allowing for a time-varying functional relationship – a notable difference to IPCA, where the time variation of factor loadings results from changing characteristics. To do so, they extend the projected principal components approach proposed by Fan, Liao and Wang (2016). Applying their technique to U.S. equity data, they find that firm characteristics are informative regarding potential mispricing of stocks.

10.5 How Machine Learning Can Predict in Empirical Asset Pricing

Assessing return predictability has always been an active pursuit in finance. A comprehensive survey of studies that provide return predictions using – to a large extent – more traditional econometric approaches is provided by Rapach and Zhou (2013). Lewellen (2015) assesses return predictability using Fama-MacBeth regressions. With the advent of the second generation of machine learning methods in empirical finance, predictability literature experienced another boost.

Because this chapter centers on machine learning in the context of asset pricing, it makes sense to start the discussion of return predictability with the reformulated basic asset pricing Equation (10.4). As mentioned in the introduction, the conditional covariance $\text{cov}_t(m_{t+1}, R^i_{t+1})$ could be conceived of as a function of time t variables, and one could consider flexible functional forms and models to provide MSE-optimal excess return forecasts. This notion provides the starting point for Gu, Kelly and Xiu (2020), who examine a variety of machine learning methods, including artificial neural networks, random forests, gradient-boosted regression trees, and elastic nets – to provide those flexible functional forms. They are prudent to avoid overfitting by working out a dynamic training and validation scheme. Considering a vast stock universe that also includes penny stocks, the authors find that feedforward networks are particularly well suited for excess return prediction at the one-month investment horizon. Grammig, Hanenberg, Schlag and Sönksen (2021) compare the data-driven techniques considered by Gu et al. (2020) with option-based approaches for approximating stock risk premia (and thus MSE-optimal forecasts from a theoretical point of view). The latter are based on financial economic theory and advocated by Martin and Wagner (2019). Grammig et al. (2021) also assess the potential of hybrid strategies that employ machine learning algorithms to pin down the approximation error inherent in the theory-based approach. Their results indicate that random forests in particular offer great potential for improving the performance of the pure theory-based approach.

The IPCA approach outlined in Section 10.4 also can serve prediction purposes. Kelly, Moskowitz and Pruitt (2021) use IPCA focusing on momentum and address the question to what extent the momentum premium can be explained by time-varying risk exposure. They show that stock momentum (and other past return characteristics that predict future returns) help predict future realized betas, but that momentum no longer significantly contributes to understanding conditional expected stock returns once the conditional factor risk channel has been accounted for. Furthermore, Büchner and Kelly (2022) adopt IPCA to predict option returns. Compared with other asset classes, options pose particular challenges for return prediction, because their short lifespans and rapidly changing characteristics, such as moneyness (i.e., the intrinsic value of an option in its current state), make them ill-suited for off-the-shelf methods. Using IPCA in this context allows to view all attributes of an option contract as pricing-relevant characteristics that translate into time-varying latent factor loadings. Moreover, Kelly, Palhares and Pruitt (2021) exploit IPCA to model corporate bond returns using a five-factor model and time-varying factor loadings. They find that this approach outperforms competing empirical strategies for bond return prediction. With another assessment of bond return predictability, Bianchi, Büchner and Tamoni (2021) report that tree-based approaches and neural networks, two highly nonlinear approaches, prove particularly useful for this purpose. Studying the neural network forecasts in more detail, Bianchi et al. further find these forecasts to exhibit countercyclicality and to be correlated with variables that proxy for macroeconomic uncertainty and (time-varying) risk aversion.

Wu, Chen, Yang and Tindall (2021) use different machine learning approaches for cross-sectional return predictions pertaining to hedge fund selections, using features that contain hedge fund-specific information. Like Gu et al. (2020), they find that the flexibility of feedforward neural networks is particularly well suited for return predictions. Wu et al. (2021) note that their empirical strategy outperforms prominent hedge fund research indices almost constantly. Guijarro-Ordonez, Pelger and Zanotti (2021) seek an optimal trading policy by exploiting temporal price differences among similar assets. In addition to proposing a framework for such statistical arbitrage, they use a convolutional neural network combined with a transformer to detect commonalities and time-series patterns from large panels. The optimal trading strategy emerging from these results outperforms competing benchmark approaches and delivers high out-of-sample Sharpe ratios; the profitability of arbitrage trading appears to be non-declining over time. Cong, Tang, Wang and Zhang (2021) apply deep reinforcement learning directly to optimize the objectives of portfolio management, instead of pursuing a traditional two-step approach (i.e., training machine learning techniques for return prediction first, then translating the results into portfolio management considerations), such that they employ multisequence, attention-based neural networks. Studying return predictability at a high frequency, Chinco, Clark-Joseph and Ye (2019) consider the entire cross-section of lagged returns as potential predictors. They find that making rolling one-minute-ahead return forecasts using LASSO to identify a sparse set of short-lived predictors increases both out-of-sample fit and Sharpe ratios. Further analysis reveals that the selected predictors tend to be related to stocks with news about fundamentals.

The application of machine learning techniques to empirical asset pricing also encompasses unstructured or semi-structured data. For example, Obaid and Pukthuanthong (2022) predict market returns using the visual content of news. For this purpose, they develop a daily market-level investor sentiment index that is defined as the fraction of negative news photos and computed from a large sample of news media images provided by the Wall Street Journal. The index negatively predicts next day's market returns and captures a reversal in subsequent days. Obaid and Pukthuanthong compare their index with an alternative derived from text data and find that both options appear to contain the same information and serve as substitutes. Relying on news articles distributed via Dow Jones Newswires, Ke, Kelly and Xiu (2021) also use textual data, but their focus is on the prediction of individual stock returns. To this end, they propose a three-step approach to assign sentiment scores on an article basis without requiring the use of pre-existing sentiment dictionaries. Ke et al. find that their proposed text-mining approach detects a reasonable return predictive signal in the data and outperforms competing commercial sentiment indices. Jiang, Kelly and Xiu (2021) apply convolutional neural networks (CNN) to images of stock-level price charts to extract price patterns that provide better return predictions than competing trend signals. They argue that the detected trend-predictive signals generalize well to other time-scales and also to other financial markets. In particular, Jiang et al. transfer a CNN trained using U.S. data to 26 different international markets; in 19 of these 26 cases, the transferred model produced higher Sharpe ratios than models trained on local data. This result implies that the richness of U.S. stock market data could be meaningfully exploited to assist the analysis of other financial markets with fewer stocks and shorter time series.

Avramov, Cheng and Metzker (2021) offer a skeptical view of predictability. Their objective is not about finding the best method for trading purposes or about conducting a comprehensive comparison between different machine learning approaches. Rather, the authors aim at assessing whether machine learning approaches can be successfully applied for the purpose of a profitable investment strategy under realistic economic restrictions, e.g., as they relate to transaction costs. In doing so, Avramov et al. build on prior work by Gu et al. (2020), Chen et al. (2021), Kelly et al. (2019), and Gu et al. (2021) and compare the performance of these approaches on the full universe of stocks considered in each of the studies, respectively, to that obtained after reducing the stock universe by microcaps, firms without credit rating coverage, and financially distressed firms. They find that the predictability of deep learning methods deteriorates strongly whilst that of the IPCA approach, which assumes a linear relationship between firm characteristics and stock returns and is outperformed by the other techniques when evaluated on the full universe of stocks, is somewhat less affected. This leads the authors to conclude that the increased flexibility of the deep learning algorithms proves especially useful for the prediction of difficult-to-value and difficult-to-arbitrage stocks. Further analysis reveals that the positive predictive performance hinges strongly on periods of low market liquidity. Additionally, all machine learning techniques under consideration imply a high portfolio turnover, which would be rather expensive given realistic trading costs. However, Avramov et al. (2021) also find that machine learning-based trading strategies display less downside risk and yield considerable profit in long positions. Machine learning methods successfully identify mispriced stocks consistent with most anomalies and remain viable in the post-2001 period, during which traditional anomaly-based trading strategies weaken.

Brogaard and Zareei (2022) are also concerned with evaluating whether machine learning algorithms' predictive strengths translate into a realistic setting that includes transaction costs. In particular, they use machine learning techniques to contribute to the discussion whether technical trading rules can be profitably applied by practitioners; specifically, whether a practitioner could have identified a profitable trading rule ex ante. This notion is contradicted by the efficient market hypothesis according to which stock prices contain all publicly available information. One key difference between Brogaard and Zareei's (2022) study and that by Avramov et al. (2021) is that the former focus on the identification of profitable trading rules. Such trading rules are manifold – trading based on momentum constitutes one of them – and they are easy to implement. In contrast, the approaches considered by Avramov et al. use vast sets of firm characteristics that are in some cases related to trading rules, but not necessarily so. Exploiting a broad set of different machine learning techniques that include both evolutionary genetic algorithms and standard loss-minimizing approaches, Brogaard and Zareei search for profitable trading rules. Controlling for data-snooping and transaction costs, they indeed identify such trading rules, but find their out-of-sample profitability to be decreasing over time.

10.6 Concluding Remarks

The Nobel Prize in Economics shared by E. Fama, L.P. Hansen, and R. Shiller in 2013 made it clear that empirical asset pricing represents a highly developed field within the economics discipline, with contributions that help define the forefront of empirical method development. A rich data environment – both quantitatively and qualitatively – and the interest from both academia and practice provides a fertile ground.

The notable spread of data-intensive machine learning techniques in recent years shows that research in empirical asset pricing remains as innovative and vibrant as ever. It is noteworthy that the adoption of the methods did not entail just improved measurement without theory. Instead, the integration of machine learning techniques has proceeded in such a way that unresolved methodological issues have been addressed creatively. This chapter provides a review of prominent, recent contributions, thus revealing how machine learning techniques can help identify the stochastic discount factor, the elusive object of core interest in asset pricing, as well as how pertinent methods can be employed to improve efforts to test and evaluate asset pricing approaches, and how the development of conditional factor models benefits from the inclusion of machine learning techniques. It also showed how machine learning can be applied for asset return prediction, keeping in mind the limitations implied by theory and the challenges of a low signal-to-noise environment.

Appendix 1: An Upper Bound for the Sharpe Ratio

The upper bound of the Sharpe ratio can be derived directly from the basic asset pricing equation. Assume that a law of total expectation has been applied to Equation (10.3), such that the expectation is conditioned down to:

$$\begin{aligned}
\mathbb{E}[m_{t+1}R^e_{t+1}] &= 0 \\
\operatorname{cov}[m_{t+1}, R^e_{t+1}] + \mathbb{E}[m_{t+1}]\mathbb{E}[R^e_{t+1}] &= 0 \\
\mathbb{E}[m_{t+1}]\mathbb{E}[R^e_{t+1}] &= -\operatorname{cov}(m_{t+1}, R^e_{t+1}) \\
\mathbb{E}[m_{t+1}]\mathbb{E}[R^e_{t+1}] &= -\rho(m_{t+1}, R^e_{t+1})\sigma(m_{t+1})\sigma(R^e_{t+1}),
\end{aligned}$$

where $\sigma(\cdot)$ refers to the standard deviation, and $\rho(\cdot,\cdot)$ denotes a correlation. Further reformulation yields:

$$\frac{\mathbb{E}[R^e_{t+1}]}{\sigma(R^e_{t+1})} = -\frac{\sigma(m_{t+1})}{\mathbb{E}[m_{t+1}]}\rho(m_{t+1}, R^e_{t+1}).$$

Because $|\rho(\cdot,\cdot)| \leq 1$, and $\mathbb{E}[m_{t+1}] > 0$, we obtain:

$$\frac{|\mathbb{E}[R^e_{t+1}]|}{\sigma(R^e_{t+1})} \leq \frac{\sigma(m_{t+1})}{\mathbb{E}[m_{t+1}]}.$$

Appendix 2: A Comparison of Different PCA Approaches

For this overview of key differences among the various PCA-related techniques mentioned in the chapter, denote as $\mathbf{X}$ an $(N \times T)$ matrix of excess returns, $\mathbf{F}$ is an $(T \times K)$ matrix of (latent) factors, $\tilde{\mathbf{X}}$ and $\tilde{\mathbf{F}}$ are the demeaned counterparts, and $\mathbf{Z}$ is a $(T \times P)$ matrix of (observed) characteristics. I assume the general factor structure:

$$\mathbf{X} = \mathbf{F}\boldsymbol{\Lambda}' + \mathbf{e}$$

and try to identify $\mathbf{F}$ and $\boldsymbol{\Lambda}$ from it.

Principal Component Analysis (PCA)

With conventional PCA, factor loadings $\boldsymbol{\Lambda}$ are obtained by using PCA directly on the sample variance-covariance matrix of $\mathbf{X}$, which is $\frac{1}{T}\mathbf{X}'\mathbf{X} - \bar{\mathbf{X}}'\bar{\mathbf{X}}$, where $\bar{\mathbf{X}}$ denotes sample means. Regressing $\mathbf{X}$ on $\hat{\boldsymbol{\Lambda}}$ results in estimates of the factors. Alternatively, estimates of $\boldsymbol{\Lambda}$ and $\tilde{\mathbf{F}}$ might derive from

$$\min_{\boldsymbol{\Lambda},\tilde{\mathbf{F}}} \frac{1}{NT}\sum_{i=1}^{N}\sum_{t=1}^{T}\left(\tilde{X}_{t,i} - \tilde{F}_t'\Lambda_i\right)^2.$$

Risk-Premium Principal Component Analysis (RP-PCA)

Lettau and Pelger (2020a) extend the conventional PCA by overweighting the mean, such that they apply PCA not to $\frac{1}{T}\mathbf{X}'\mathbf{X} - \bar{\mathbf{X}}'\bar{\mathbf{X}}$ but instead to $\frac{1}{T}\mathbf{X}'\mathbf{X} + \gamma\bar{\mathbf{X}}'\bar{\mathbf{X}}$, where γ can be interpreted as a penalty parameter. To see this, note that we could alternatively consider the problem:

$$\min_{\boldsymbol{\Lambda},\tilde{\mathbf{F}}} \frac{1}{NT}\sum_{i=1}^{N}\sum_{t=1}^{T}\left(\tilde{X}_{t,i} - \tilde{F}_t'\Lambda_i\right)^2 + (1+\gamma)\frac{1}{N}\sum_{i=1}^{N}\left(\bar{X}_i - \bar{F}'\Lambda_i\right)^2.$$

The first part of the objective function deals with minimizing unexplained variation, and the second part imposes a penalty on pricing errors.

Projected Principal Component Analysis
Fan et al. (2016) consider daily data and propose, instead of applying PCA to the variance-covariance matrix of $\mathbf{X}$, projecting $\mathbf{X}$ on a set of time-invariant asset-specific characteristics $\mathbf{Z}$. Then the PCA would be applied to the variance-covariance matrix of the thus smoothed $\hat{\mathbf{X}}$.

Instrumented Principal Component Analysis (IPCA)
The IPCA approach introduced by Kelly et al. (2019) is described in detail in Section 10.4.2 of this Chapter. It is related to projected principal component analysis, but Kelly et al. (2019) consider time-varying instruments – an important difference, because it translates into time-varying factor loadings.

Supervised Principal Component Analysis (SPCA)
Giglio, Xiu and Zhang (2021) introduce supervised principal component analysis to counteract issues associated with weak factors. The term *supervised* comes from supervised machine learning. The authors propose that, rather than using PCA on the variance-covariance matrix of the entire $\mathbf{X}$, they should select stocks that exhibit a strong correlation (in absolute value) with the instruments in $\mathbf{Z}$. So, instead of using all N stocks, Giglio, Xiu and Zhang (2021) limit themselves to a fraction q of the sample and select those stocks with the strongest correlation:

$$\hat{I} = \left\{ i \,\middle|\, \frac{1}{T} \left| \tilde{\mathbf{X}}_{[i]} \tilde{\mathbf{Z}}' \right| \geq c_q \right\},$$

where c_q denotes the $(1-q)$ correlation quantile. Then the PCA can be conducted on $\tilde{\mathbf{X}}_{[\hat{I}]}$.

References

Ahn, D.-H., Conrad, J. & Dittmar, R. F. (2009). Basis Assets. *The Review of Financial Studies*, *22*(12), 5133–5174. Retrieved from http://www.jstor.org/stable/40468340

Anatolyev, S. & Mikusheva, A. (2022). Factor Models with Many Assets: Strong Factors, Weak Factors, and the Two-Pass Procedure. *Journal of Econometrics*, *229*(1), 103–126. Retrieved from https://www.sciencedirect.com/science/article/pii/S0304407621000130 doi: 10.1016/j.jeconom.2021.01.002

Ang, A., Liu, J. & Schwarz, K. (2020). Using Stocks or Portfolios in Tests of Factor Models. *Journal of Financial and Quantitative Analysis*, *55*(3), 709–750. doi: 10.1017/S0022109019000255

Avramov, D., Cheng, S. & Metzker, L. (2021). *Machine Learning versus Economic Restrictions: Evidence from Stock Return Predictability.* Retrieved from http://dx.doi.org/10.2139/ssrn.3450322 (forthcoming: Management Science)

Bai, J. & Ng, S. (2002). Determining the Number of Factors in Approximate Factor Models. *Econometrica, 70*(1), 191–221. doi: 10.1111/1468-0262.00273

Bailey, N., Kapetanios, G. & Pesaran, M. H. (2021). Measurement of Factor Strength: Theory and Practice. *Journal of Applied Econometrics, 36*(5), 587–613. doi: 10.1002/jae.2830

Bakalli, G., Guerrier, S. & Scaillet, O. (2021). *A Penalized Two-Pass Regression to Predict Stock Returns with Time-Varying Risk Premia.* Retrieved from http://dx.doi.org/10.2139/ssrn.3777215 (Working Paper, accessed February 23, 2022)

Benjamini, Y. & Hochberg, Y. (1995). Controlling the False Discovery Rate: A Practical and Powerful Approach to Multiple Testing. *Journal of the Royal Statistical Society: Series B (Methodological), 57*(1), 289–300. Retrieved from https://rss.onlinelibrary.wiley.com/doi/abs/10.1111/j.2517-6161.1995.tb02031.x doi: https://doi.org/10.1111/j.2517-6161.1995.tb02031.x

Bianchi, D., Büchner, M. & Tamoni, A. (2021). Bond Risk Premiums with Machine Learning. *The Review of Financial Studies, 34*(2), 1046–1089. doi: 10.1093/rfs/hhaa062

Brogaard, J. & Zareei, A. (2022). *Machine Learning and the Stock Market.* Retrieved from http://dx.doi.org/10.2139/ssrn.3233119 (forthcoming: Journal of Financial and Quantitative Analysis)

Bryzgalova, S., Huang, J. & Julliard, C. (2021). *Bayesian Solutions for the Factor Zoo: We Just Ran Two Quadrillion Models.* Retrieved from http://dx.doi.org/10.2139/ssrn.3481736 (Working Paper, accessed February 22, 2022)

Bryzgalova, S., Pelger, M. & Zhu, J. (2021a). *Forest through the Trees: Building Cross-Sections of Stock Returns.* Retrieved from http://dx.doi.org/10.2139/ssrn.3493458 (Working Paper, accessed February 23, 2022)

Bryzgalova, S., Pelger, M. & Zhu, J. (2021b). *Internet Appendix for Forest through the Trees: Building Cross-Sections of Stock Returns.* Retrieved from https://ssrn.com/abstract=3569264

Büchner, M. & Kelly, B. (2022). A Factor Model for Option Returns. *Journal of Financial Economics, 143*(3), 1140–1161. Retrieved from https://www.sciencedirect.com/science/article/pii/S0304405X21005249 doi: 10.1016/j.jfineco.2021.12.007

Chen, L., Pelger, M. & Zhu, J. (2021). *Deep Learning in Asset Pricing.* Retrieved from http://dx.doi.org/10.2139/ssrn.3350138 (Working Paper, accessed July 27, 2021)

Chinco, A., Clark-Joseph, A. D. & Ye, M. (2019). Sparse Signals in the Cross-Section of Returns. *The Journal of Finance, 74*(1), 449–492. doi: 10.1111/jofi.12733

Cochrane, J. (1996). A Cross-Sectional Test of an Investment-Based Asset Pricing Model. *Journal of Political Economy, 104*(3), 572–621. Retrieved from http://www.jstor.org/stable/2138864

Cochrane, J. (2005). *Asset Pricing.* Princeton University Press, New Jersey.

Cong, L. W., Tang, K., Wang, J. & Zhang, Y. (2021). *AlphaPortfolio: Direct Construction Through Deep Reinforcement Learning and Interpretable AI.* Retrieved from http://dx.doi.org/10.2139/ssrn.3554486 (Working Paper, accessed February 23, 2022)

DeMiguel, V., Martín-Utrera, A., Uppal, R. & Nogales, F. (2020, 1st). A Transaction-Cost Perspective on the Multitude of Firm Characteristics. *Review of Financial Studies*, *33*(5), 2180–2222. Retrieved from https://academic.oup.com/rfs/article-abstract/33/5/2180/5821387 doi: 10.1093/rfs/hhz085

Duffie, D. & Singleton, K. J. (1993). Simulated Moments Estimation of Markov Models of Asset Prices. *Econometrica*, *61*(4), 929–952. doi: 10.2307/2951768

Fama, E. F. & French, K. R. (1993). Common Risk Factors in the Returns on Stocks and Bonds. *Journal of Financial Economics*, *33*(1), 3–56. Retrieved from https://www.sciencedirect.com/science/article/pii/0304405X93900235 doi: 10.1016/0304-405X(93)90023-5

Fan, J., Li, K. & Liao, Y. (2021). Recent Developments in Factor Models and Applications in Econometric Learning. *Annual Review of Financial Economics*, *13*(1), 401–430. doi: 10.1146/annurev-financial-091420-011735

Fan, J., Liao, Y. & Wang, W. (2016). Projected Principal Component Analysis in Factor Models. *The Annals of Statistics*, *44*(1), 219 – 254. doi: 10.1214/15-AOS1364

Feng, G., Giglio, S. & Xiu, D. (2020). Taming the Factor Zoo: A Test of New Factors. *The Journal of Finance*, *75*(3), 1327–1370. doi: 10.1111/jofi.12883

Feng, G., Polson, N. & Xu, J. (2021). *Deep Learning in Characteristics-Sorted Factor Models.* Retrieved from http://dx.doi.org/10.2139/ssrn.3243683 (Working Paper, accessed February 22, 2022)

Freyberger, J., Neuhierl, A. & Weber, M. (2020). Dissecting Characteristics Nonparametrically. *The Review of Financial Studies*, *33*(5), 2326–2377. doi: 10.1093/rfs/hhz123

Gagliardini, P., Ossola, E. & Scaillet, O. (2016). Time-Varying Risk Premium in Large Cross-Sectional Equity Data Sets. *Econometrica*, *84*(3), 985–1046. doi: 10.3982/ECTA11069

Gagliardini, P., Ossola, E. & Scaillet, O. (2019). A Diagnostic Criterion for Approximate Factor Structure. *Journal of Econometrics*, *212*(2), 503–521. doi: 10.1016/j.jeconom.2019.06.001

Gibbons, M. R., Ross, S. A. & Shanken, J. (1989). A Test of the Efficiency of a Given Portfolio. *Econometrica*, *57*(5), 1121–1152. Retrieved from http://www.jstor.org/stable/1913625

Giglio, S., Kelly, B. T. & Xiu, D. (2022). *Factor Models, Machine Learning, and Asset Pricing.* Retrieved from http://dx.doi.org/10.2139/ssrn.3943284 (forthcoming: Annual Review of Financial Economics)

Giglio, S., Liao, Y. & Xiu, D. (2021). Thousands of Alpha Tests. *The Review of Financial Studies*, *34*(7), 3456–3496. doi: 10.1093/rfs/hhaa111

Giglio, S. & Xiu, D. (2021). Asset Pricing with Omitted Factors. *Journal of Political Economy*, *129*(7), 1947–1990. doi: 10.1086/714090

Giglio, S., Xiu, D. & Zhang, D. (2021). *Test Assets and Weak Factors.* Retrieved from http://dx.doi.org/10.2139/ssrn.3768081 (Working Paper, accessed February 23, 2022)

Gospodinov, N., Kan, R. & Robotti, C. (2014). Misspecification-Robust Inference in Linear Asset-Pricing Models with Irrelevant Risk Factors. *The Review of*

Financial Studies, *27*(7), 2139–2170. doi: 10.1093/rfs/hht135

Grammig, J., Hanenberg, C., Schlag, C. & Sönksen, J. (2021). *Diverging Roads: Theory-Based vs. Machine Learning-Implied Stock Risk Premia.* Retrieved from http://dx.doi.org/10.2139/ssrn.3536835 (Working Paper, accessed February 22, 2022)

Gu, S., Kelly, B. & Xiu, D. (2020). Empirical Asset Pricing via Machine Learning. *The Review of Financial Studies*, *33*(5), 2223–2273. doi: 10.1093/rfs/hhaa009

Gu, S., Kelly, B. & Xiu, D. (2021). Autoencoder Asset Pricing Models. *Journal of Econometrics*, *222*(1), 429–450. doi: 10.1016/j.jeconom.2020.07

Guijarro-Ordonez, J., Pelger, M. & Zanotti, G. (2021). *Deep Learning Statistical Arbitrage.* Retrieved from http://dx.doi.org/10.2139/ssrn.3862004 (Working Paper, accessed July 27, 2021)

Hall, A. (2005). *Generalized Method of Moments.* Oxford University Press.

Hansen, L. P. (1982). Large Sample Properties of Generalized Method of Moments Estimators. *Econometrica*, *50*(4), 1029–1054. Retrieved from http://www.jstor.org/stable/1912775

Hansen, L. P. & Jagannathan, R. (1991). Implications of Security Market Data for Models of Dynamic Economies. *Journal of Political Economy*, *99*(2), 225–262. Retrieved from http://www.jstor.org/stable/2937680

Hansen, L. P. & Jagannathan, R. (1997). Assessing Specification Errors in Stochastic Discount Factor Models. *The Journal of Finance*, *52*(2), 557–590. doi: 10.1111/j.1540-6261.1997.tb04813.x

Hansen, L. P. & Richard, S. F. (1987). The Role of Conditioning Information in Deducing Testable Restrictions Implied by Dynamic Asset Pricing Models. *Econometrica*, *55*(3), 587–613. Retrieved from http://www.jstor.org/stable/1913601

Hansen, L. P. & Singleton, K. J. (1982). Generalized Instrumental Variables Estimation of Nonlinear Rational Expectations Models. *Econometrica*, *50*(2), 1296–1286.

Harvey, C. R., Liu, Y. & Zhu, H. (2015). ...and the Cross-Section of Expected Returns. *The Review of Financial Studies*, *29*(1), 5–68. doi: 10.1093/rfs/hhv059

Jegadeesh, N., Noh, J., Pukthuanthong, K., Roll, R. & Wang, J. (2019). Empirical Tests of Asset Pricing Models with Individual Assets: Resolving the Errors-in-Variables Bias in Risk Premium Estimation. *Journal of Financial Economics*, *133*(2), 273–298. Retrieved from https://www.sciencedirect.com/science/article/pii/S0304405X19300431 doi: 10.1016/j.jfineco.2019.02.010

Jiang, J., Kelly, B. T. & Xiu, D. (2021). *(Re-)Imag(in)ing Price Trends.* Retrieved from http://dx.doi.org/10.2139/ssrn.3756587 (Working Paper, accessed February 22, 2022)

Ke, Z., Kelly, B. T. & Xiu, D. (2021). *Predicting Returns with Text Data.* Retrieved from http://dx.doi.org/10.2139/ssrn.3389884 (Working Paper, accessed February 22, 2022)

Kelly, B. T., Moskowitz, T. J. & Pruitt, S. (2021). Understanding Momentum and Reversal. *Journal of Financial Economics*, *140*(3), 726–743. Retrieved from https://www.sciencedirect.com/science/article/pii/S0304405X21000878 doi:

10.1016/j.jfineco.2020.06.024

Kelly, B. T., Palhares, D. & Pruitt, S. (2021). *Modeling Corporate Bond Returns.* Retrieved from http://dx.doi.org/10.2139/ssrn.3720789 (Working Paper, accessed February 22, 2022)

Kelly, B. T., Pruitt, S. & Su, Y. (2019). Characteristics are Covariances: A Unified Model of Risk and Return. *Journal of Financial Economics*, *134*(3), 501–524. Retrieved from https://www.sciencedirect.com/science/article/pii/S0304405X19301151 doi: https://doi.org/10.1016/j.jfineco.2019.05.001

Kelly, B. T., Pruitt, S. & Su, Y. (2020). *Instrumented Principal Component Analysis.* Retrieved from http://dx.doi.org/10.2139/ssrn.2983919 (Working Paper, accessed February 23, 2022)

Kim, S., Korajczyk, R. A. & Neuhierl, A. (2021). Arbitrage Portfolios. *The Review of Financial Studies*, *34*(6), 2813–2856. doi: 10.1093/rfs/hhaa102

Korsaye, S. A., Quaini, A. & Trojani, F. (2019). *Smart SDFs.* Retrieved from http://dx.doi.org/10.2139/ssrn.3475451 (Working Paper, accessed February 22, 2022)

Kozak, S., Nagel, S. & Santosh, S. (2018). Interpreting Factor Models. *The Journal of Finance*, *73*(3), 1183–1223. doi: 10.1111/jofi.12612

Kozak, S., Nagel, S. & Santosh, S. (2020). Shrinking the Cross-Section. *Journal of Financial Economics*, *135*(2), 271–292. Retrieved from https://www.sciencedirect.com/science/article/pii/S0304405X19301655 doi: 10.1016/j.jfineco.2019.06.008

Lettau, M. & Ludvigson, S. (2001). Resurrecting the (C)CAPM: A Cross-Sectional Test When Risk Premia Are Time-Varying. *Journal of Political Economy*, *109*(6), 1238–1287. Retrieved from http://www.jstor.org/stable/10.1086/323282

Lettau, M. & Pelger, M. (2020a). Estimating Latent Asset-Pricing Factors. *Journal of Econometrics*, *218*(1), 1–31. Retrieved from https://www.sciencedirect.com/science/article/pii/S0304407620300051 doi: https://doi.org/10.1016/j.jeconom.2019.08.012

Lettau, M. & Pelger, M. (2020b). Factors That Fit the Time Series and Cross-Section of Stock Returns. *The Review of Financial Studies*, *33*(5), 2274–2325. doi: 10.1093/rfs/hhaa020

Lewellen, J. (2015). The Cross-Section of Expected Stock Returns. *Critical Finance Review*, *4*(1), 1–44. doi: 10.1561/104.00000024

Lewellen, J., Nagel, S. & Shanken, J. (2010). A Skeptical Appraisal of Asset Pricing Tests. *Journal of Financial Economics*, *96*(2), 175–194. Retrieved from https://www.sciencedirect.com/science/article/pii/S0304405X09001950 doi: https://doi.org/10.1016/j.jfineco.2009.09.001

Lintner, J. (1965). The Valuation of Risk Assets and the Selection of Risky Investments in Stock Portfolios and Capital Budgets. *The Review of Economics and Statistics*, *47*, 13–37.

Martin, i. W. R. & Wagner, C. (2019). What Is the Expected Return on a Stock? *The Journal of Finance*, *74*(4), 1887–1929. doi: https://doi.org/10.1111/jofi.12778

Mossin, J. (1966). Equilibrium in a Capital Asset Market. *Econometrica, 34*, 768–783.

Nagel, S. (2021). *Machine Learning in Asset Pricing*. Princeton University Press, New Jersey.

Obaid, K. & Pukthuanthong, K. (2022). A Picture is Worth a Thousand Words: Measuring Investor Sentiment by Combining Machine Learning and Photos from News. *Journal of Financial Economics, 144*(1), 273–297. Retrieved from https://www.sciencedirect.com/science/article/pii/S0304405X21002683 doi: https://doi.org/10.1016/j.jfineco.2021.06.002

Onatski, A. (2012). Asymptotics of the Principal Components Estimator of Large Factor Models with Weakly Influential Factors. *Journal of Econometrics, 168*(2), 244–258. Retrieved from https://www.sciencedirect.com/science/article/pii/S0304407612000449 doi: https://doi.org/10.1016/j.jeconom.2012.01.034

Pelger, M. & Xiong, R. (2020). *State-Varying Factor Models of Large Dimensions*. Retrieved from http://dx.doi.org/10.2139/ssrn.3109314 (Working Paper, accessed July 27, 2021)

Pukthuanthong, K., Roll, R. & Subrahmanyam, A. (2018). A Protocol for Factor Identification. *The Review of Financial Studies, 32*(4), 1573–1607. doi: 10.1093/rfs/hhy093

Rapach, D. & Zhou, G. (2013). Chapter 6 - Forecasting Stock Returns. In G. Elliott & A. Timmermann (Eds.), *Handbook of Economic Forecasting* (Vol. 2, pp. 328–383). Elsevier. Retrieved from https://www.sciencedirect.com/science/article/pii/B9780444536839000062 doi: https://doi.org/10.1016/B978-0-444-53683-9.00006-2

Raponi, V., Robotti, C. & Zaffaroni, P. (2019). Testing Beta-Pricing Models Using Large Cross-Sections. *The Review of Financial Studies, 33*(6), 2796–2842. doi: 10.1093/rfs/hhz064

Sharpe, W. F. (1964). Capital Asset Prices: A Theory of Market Equilibrium under Conditions of Risk. *The Journal of Finance, 19*(3), 425-–442. doi: 10.1111/j.1540-6261.1964.tb02865.x

Singleton, K. J. (2006). *Empirical Dynamic Asset Pricing*. Princeton University Press, New Jersey.

Weigand, A. (2019). Machine Learning in Empirical Asset Pricing. *Financial Markets and Portfolio Management, 33*(1), 93–104. doi: 10.1007/s11408-019-00326-3

Wu, W., Chen, J., Yang, Z. B. & Tindall, M. L. (2021). A Cross-Sectional Machine Learning Approach for Hedge Fund Return Prediction and Selection. *Management Science, 67*(7), 4577–4601. doi: 10.1287/mnsc.2020.3696

Appendix A
Terminology

A.1 Introduction

As seen in this volume, terminologies used in the Machine Learning literature often have equivalent counterparts in statistics and/or econometrics. These differences in terminologies create potential hurdles in making machine learning methods more readily applicable in econometrics. The purpose of this Appendix is to provide a list of terminologies typically used in machine learning and explain them in the language of econometrics or, at the very least, explain in a language that is more familiar to economists or econometricians.

A.2 Terms

Adjacency matrix. *A square matrix which can represent nodes and edges in a network.* See Chapter 6.

Agents. *Entities in an econometric analysis (e.g., individuals, households, firms).*

Bagging. *Using Bootstrapping for selecting (multiple) data and AGGregating the results.*

Boosting. *Model averaging: sequentially learn from models: use residuals from each models and fit data to them in a systematic way.*

Bots. *A program that interacts with systems or users.*

Causal forest. *A causal forest estimator averages several individual causal trees grown on different samples (just as a standard random forest averages several individual regression trees).* See Chapter 3.

Causal tree. *A regression tree used to estimate heterogeneous treatment effects.* See Chapter 3.

F. Chan, L. Mátyás (eds.), *Econometrics with Machine Learning*, Advanced Studies in Theoretical and Applied Econometrics 53, https://doi.org/10.1007/978-3-031-15149-1

Classification problem. *Modelling/Predicting discrete nominal outcomes, sometimes the random variables are ordinal but not cardinal.* See Chapter 2.

Clustering. *Algorithms for grouping data points.*

Confusion matrix. *A* 2×2 *matrix measuring the prediction performance of a binary choice model.* See Chapter 2.

Convolutional Neural Network. *A class of neural networks with an architecture that has at least one convolution layer, which consists of performing matrix products between the matrix in the input or previous layer and other matrices called convolution filters.*

Covariates. *Explanatory Variables.* See Chapter 1.

Cross-fitting. *A variant of the DML procedure in which the sample is used in a similar way as in a cross-validation exercise.* See Chapter 3.

Cross-validation. *Method used to estimated the tuning parameter (based on dividing the sample randomly into partitions and randomly select from these to estimate the tuning parameter).* See Chapter 1.

Crowd-sourced data. *Data collected with a participatory method, with the help of a large group of people.*

Debiased Machine Learning (DML). *A way to employ ML methods to estimate average treatment effect parameters without regularization bias.* See Chapter 3.

Deep learning. *Subfield of ML grouping algorithms organized in layers (an input layer, an output layer, and hidden layers connecting the input and output layers).*

Dimensionality reduction/embedding. *Algorithms which aim at representing the data in a lower dimensional space while retaining useful information.*

Domain generalization. *Also known as out-of-distribution generalization. A set of techniques aiming to create the best prediction models on random variables which joint distributions are different, but related, to those in the training set.*

Double Machine Learning (DML). *Alternative terminology for debiased ML. The adjective 'double' is due to the fact that in some simple settings, such as the partially linear regression model, debiased parameter estimation requires the implementation of two ML procedures.* See Chapter 3.

Edge. *Connection between nodes in a network.* See Chapter 6.

Elastic Net. *A Regularised Regression Method which regularizer is an affine combination of LASSO and Ridge.* See Chapter 1.

Ensemble method. *Combine predictions from different models. A form of models averaging.*

Feature. *Covariate or explanatory variable.*

Feature engineering. *Transforming variables and manipulating (imputing) observations.*

Feature extraction. *Obtaining features from the raw data that will be used for preditction.*

Forest. *A collection of trees.* See Chapter 2.

Graph Laplacian eigenmaps. *A matrix factorization technique primarily used to learn lower dimensional representations of graphs.* See Chapter 6.

Graph representation learning. *Assigning representation vectors to graph elements (e.g.,: nodes, edges) containing accurate information on the structure of a large graph.* See Chapter 6.

Hierarchical bayesian geostatistical models.

Honest approach. *A sample splitting approach to estimating a causal tree. One part of the sample is used to find the model specification and the other part to estimate the parameter(s) of interest. This method ensures unbiased parameter estimates and standard inference.* See Chapter 3.

Instance. *A data point (or sometimes a small subset of the database). Can also be called Records.*

K **Nearest Neighbors**. *Algorithms to predict a group or label based on the distance of a data point to the other data points.*

Labelled data. *Each observation is linked with a particular value of the response variable (opposite of unlabelled data).*

LASSO. *The Least Absolute Shrinkage and Selection Operator, or LASSO imposes a penalty function that limits the size of its estimators by absolute value.* See Chapter 1.

Learner. *Model.*

Loss function. *A function mapping the observed data and model estimates to loss. Learning is usually equivalent to minimizing a loss function.*

Machine. *Algorithm (usually used for learning).*

Network/graph. *A set of agents (nodes) connected by edges.* See Chapter 6.

Neural network. *Highly nonlinear model.*

Node. *Agents in a network.* See Chapter 6.

Open-source data. *Data publicly available for analysis and redistribution.*

Overfitting. *When predictions are fitted too closely to a certain set of data, resulting in potentially failure to predict with other sets of data reliably.*

Post-selection inference. *Computing standard errors, confidence intervals, p-values, etc., after using an ML method to find a model.* See Chapter 3.

Principal component analysis. *An algorithm which uses a linear mapping to maximize variance in the lower dimensional space.*

Predictors. *Features, covariates or explanatory variables.* See Chapter 1.

Prune. *Methods to control the size of a tree to avoid overfitting.* See Chapter 2.

Random Forest. *A forest with each tree consists of random selection of regressors (covariates).* See Chapter 2.

Regularization. *Estimator with regularizer.* See Chapter 1.

Regularizer. *Penalty function.* See Chapter 1.

Scoring. *Re-estimation of a model on some new/additional data. Not to be confused with the score (vector).*

Shrinkage. *Restrict the magnitude of the parameter vector for purpose of identifying the important features/covariates.*

Spatiotemporal deep learning. *Deep learning using spatial panel data.*

Stepwise selection. *Fitting regression models by adding or eliminating variables through automatic procedure based on their significance, typically determined via F-tests.*

Supervised learning. *Estimation with/of a model.*

Target. *Dependent variable.*

Test set. *Out-of-Sample (subset of data used for prediction evaluation).*

Training. *Estimation of a (predictive) model.*

Training set. *With-in-sample (subset of data used for estimation).*

Transfer learning. *Algorithms to store results from one ML problem and apply them to solve other related or similar ML problems.*

Tree. *Can be viewed as a threshold regression model.* See Chapter 2.

Tuning parameter. *Parameter defining the constrain in a shrinkage estimator.* See Chapter 1.

Unlabelled data. *Each observation is not linked with a particular value of the response variable (opposite of labelled data).*

Unstructured data. *Data that are not organized in a predefined data format such as tables or graphs.*

Unsupervised learning. *Estimation without a model. Often related to non-parametric techniques (No designated dependent variable).*

Variable importance. *A visualisation showing the impact of excluding a particular variable on the prediction performance, typically measured in MSE or Gini index.* See Chapter 2.

Wrapper feature selector. *A method to identify relevant variables in a regression setting using subset selection.*

www.ingramcontent.com/pod-product-compliance
Ingram Content Group UK Ltd.
Pitfield, Milton Keynes, MK11 3LW, UK
UKHW020712071026
15590UKWH00010B/474

* 9 7 8 3 0 3 1 1 5 1 5 1 4 *